JOURNAL OF POLYMER SCIENCE: Polymer Symposia No. 64

Unsolved Problems of Co- and Graft Polymerization

Proceedings of the
First S.R. Romanian–U.S. Seminar on Polymer Science
Held under the sponsorship of
the Romanian National Council for Science and Technology (NCST) and
the U.S. National Science Foundation (NSF)
21–25 September 1976

Editors:
Otto Vogl
Professor of Polymer Science and Engineering
University of Massachusetts
Amherst, Massachusetts
Cristofor I. Simionescu
Professor of Organic and Macromolecular Chemistry
Institute of Macromolecular Chemistry
Jassy, Romania

an Interscience® Publication
published by JOHN WILEY & SONS

JOURNAL OF POLYMER SCIENCE: Polymer Symposia

Editors:

H. Mark · C. G. Overberger

Editorial Board:

J. J. Hermans · H. W. Melville · G. Smets

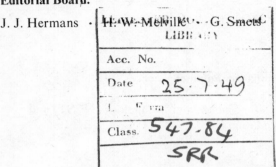
O. Vogl and Cr. Simionescu have been appointed Editors for this Symposium by the Editorial Board of the *Journal of Polymer Science*.

Published by John Wiley & Sons, Inc., this book constitutes a part of the annual subscription of the *Journal of Polymer Science* and as such is supplied without additional charge to the subscribers. Single copies can be purchased from the Subscription Department, John Wiley & Sons.

Subscription price, *Journal of Polymer Science,* Vol. 16, 1978: $485.00. Postage and handling outside U.S.A.: $37.00. Please allow four to six weeks for processing a change of address. Back volumes, microfilm, and microfiche are available for previous years. Request price list from publisher.

Printed in the United States of America.

Contents

Preface v

Developments in Radical Polymerization
    O. Vogl 1

Radical Ring-Opening Polymerization and Copolymerization with Expansion in Volume
    W. J. Bailey and T. Endo 17

Some Rheological Properties of Polycarbonate Blends
    V. Dobrescu and V. Cobzaru 27

Inorganic Fragments in Graft and Block Copolymers: A Review of Solved and Unsolved Problems
    I. Haiduc 43

Some Unsolved Problems Concerned with Radiation Grafting to Natural and Synthetic Polymers
    V. Stannett and T. Memetea 57

Radical Cyclo- and Cyclocopolymerization
    G. B. Butler 71

Some Problems Concerning the Degradation of Linear Vinyl Polymers and Copolymers in Nonisothermal Conditions
    I. A. Schneider 95

Free Radical and Ionic Grafting
    J. P. Kennedy 117

Grafting Process in Vinyl Acetate Polymerization in the Presence of Nonionic Emulsifiers
    I. Gavăt, V. Dimonie, D. Donescu, C. Hagiopol, M. Munteanu, K. Goşa, and Th. Deleanu 125

Topological Model of the Crystalline Morphology in Polyethylene with Grafted Defects
    Gh. Drăgan 141

Mechanochemical Syntheses
    C. Simionescu, C. Vasiliu Oprea, and C. Negulianu 149

Some Relationships between Physical Properties and Morphology of Heterogeneous Polymeric Systems
    P. H. Lindenmeyer 181

Some Aspects Regarding the Structure and the Mechanism of the PVC Synthesis
    A. Caraculacu, E. Buruiană, and G. Robilă 189

Isomeric Copolymers of Acetylenes
    C. Simionescu, Sv. Dumitrescu, and V. Percec 209

Polymerization of Vinylpyridinium Salts. X. Copolymerization Studies of Cationic–Anionic Monomer Pairs
    J. C. Salamone, A. C. Watterson, T. D. Hsu, C. C. Tsai, M. U. Mahmud, A. W. Wisniewski, and S. C. Israel 229

Some Physicochemical Aspects Regarding the Synthesis of Polyvinyl Alcohol
    M. Dimonie, C. Cincu, C. Oprescu, and Gh. Hubcă 245

Solute Binding and Catalysis by Poly(Vinylbenzo-18-Crown-6)
    J. Smid, S. C. Shah, A. J. Varma, and L. Wong 267
Abiotic Synthesis and the Properties of Some Protobiocopolymers
    C. I. Simionescu, F. Dénes, and I. Negulescu 281
Considerations of Some Copolymers Obtained by Radical Mecha-
    nisms
    C. I. Simionescu 305
Copolymerization of Vinyl Chloride with Bis(Beta-Chloroethyl) Vinyl-
    phosphonate
    R. Gallagher and J. C. H. Hwa 329
Contributions to Vinyl Chloride Suspension Polymerization with Constant
    Rate
    D. Feldman and A. Macoveanu 339
Thermoplastic Urethane Chemical Crosslinking Effects
    C. S. Schollenberger and K. Dinbergs 351
Author Index 369

Preface

The first S.R. Romanian–U.S. Seminar on Polymer Science was held in Iasi, Romania, from September 21, 1976 to September 25, 1976, under the title, "Unsolved Problems of Co- and Graft Polymerization." Two opening lectures and twenty-five discussion papers were given by ten U.S. and nineteen Romanian participants.

The Seminar was jointly sponsored by the Romanian National Council for Science and Technology (NCST) and the U.S. National Science Foundation (NSF).

The Seminar had many novel and unique features from its conception through the preparation and actual proceedings. Long and careful negotiations and preparation resulted in a meeting that was an unqualified success. We were very fortunate in having the full support of a number of organizations and the enthusiastic cooperation of many individuals.

The Seminar was under the primary guidance of the Central Institute of Chemical Research and its Director, Mme. Elena Ceausescu, and the Academy of S.R. Romania.

We would also like to acknowledge the help of Academician Professor I. Ursu, President, Mr. O. Groza, Prime Vice-President, and Mr. A. Magureanu, Councilor for Chemistry of NCST; the Ministry of Chemical Industry, especially Mr. M. Florescu, Minister, Mr. Gh. Manolescu, Deputy Minister, and Mr. C. Craiu, Director; members of the Administration of ICECHIM, the Institute of Chemical Research, and the Institute of Macromolecular Chemistry, "Petru Poni"; and the Polytechnic Institute of Iasi.

The organization of the Seminar would not have been possible without the generous support of the Province of Iasi, in particular, Mr. I. Iliescu and Mr. G. Brehauscu; the Province of Suceava, especially Mr. M. Dobrescu and Mr. V. Gabor; and Mr. I. Manciuc of the City of Iasi. The cooperation of the Romanian Embassy in Washington, D.C., particularly that of Mr. G. Ionita, and the Romanian Library in New York, especially Mrs. E. Gheorghe, is greatly appreciated.

Many scientists at the Institute of Macromolecular Chemistry, "P. Poni," the Polytechnic Institute of the Academy of S.R. Romania in Iasi, gave countless hours to the preparation and running of the program of the Seminar to ensure its success.

The talks consisted of two-thirds presentation and one-third discussion. The discussions were brief and effective. Wherever possible, the emphasis of the presentations and discussions was on "unsolved" problems and future aspects of the subjects discussed rather than on already established facts.

We were also very fortunate in having the strong support of the U.S. Embassy in Bucharest, especially of the Ambassador, Mr. H. Barnes, and the Scientific Attaché, Mr. S. Smith. Mr. R. Hull, Program Director at the National Science Foundation, gave us continued and unwavering support during the period of preparation for the S.R. Romanian–U.S. Seminar.

O. Vogl
Co-Chairman (U.S.A.)
Cr. Simionescu
Co-Chairman (S.R. Romania)

LEGEND: 1. Adrian Caraculacu; 2, William J. Bailey; 3, Jesse C. H. Hwa; 4, Dorel Feldman; 5, George B. Butler; 6, Cristofor I. Simionescu; 7, Otto Vogl; 8, Viorica Dobrescu; 9, Paul H. Lindenmeyer; 10, Ionel Haiduc; 11, Adrian Carpov; 12, Ioan I. Negulescu; 13, Mihai Dimonie; 14, Ferencz S. Dénes; 15, Joseph P. Kennedy; 16, Ioan A. Schneider; 17, Nicolae Asandei; 18, Virgil Percec; 19, Vivian Stannett; 20, Ioan Gavat; 21, Joseph C. Salamone; 22, Gheorghe Drăgan; 23, Johannes Smid; 24, Charles S. Schollenberger; 25, Adelaida Liga; 26, Cleopatra Vasiliu-Oprea.

DEVELOPMENTS IN RADICAL POLYMERIZATION

O. VOGL

Polymer Science and Engineering, University of Massachusetts, Amherst, Massachusetts 01003

SYNOPSIS

Developments in the field of radical polymerization including co- and graft polymerization are evaluated from a personal viewpoint. A number of important advances have been made in the last few years which present the opportunity of opening new ways for further scientific and technological progress. Also included in this critical evaluation of developments in co- and graft polymerization primarily by radical initiators is work that is peripheral to the subject but constitutes possible important new directions of development.

INTRODUCTION

Radical polymerization has traditionally played a central role in the development of polymer chemistry. Many concepts of polymerization have initially been developed for radical polymerization [1,2] and methods using radical polymerization have lent themselves very conveniently to the development of technical processes; they had flexibility and could be used for various processes such as bulk polymerization, suspension polymerization, and emulsion polymerization, most of which are polymerization techniques based on radical processess.

Many technically important polymers are copolymers and a number of the more recently developed polymers are graft copolymers. In Table I we see the tonnage production of polymers in the United States in 1974 and the breakdown in the important individual polymer categories. As can be seen, 10.5 million tons or more than 60% of all polymers produced in this year were produced by radical processes.

The thermoplastic resins, the main portion of polymer production, are prepared by radical processes to an even greater degree. Thermoplastic polymers produced commercially on a large scale by radical processes include high pressure polyethylene, poly(vinyl chloride), poly(vinyl acetate) and its derivative poly(vinyl alcohol), polystyrene, poly(methyl methacrylate), ABS, butadiene/styrene copolymers, and fluorocarbon resins. The individual polymerization processes employed for the production of these resins vary slightly from producer to producer because of different demands for polymer particle size and shape and increasingly rigid specifications for product color and minimum monomer

Journal of Polymer Science: Polymer Symposium 64, 1–16 (1978)
0360-8905/78/0064-0001$01.00

TABLE I

Plastics Production, 1974, in the United States (in 10^6 tons)

Thermoplastics	9.7 (76% by radical processes)
Thermosets	1.6
Fibers	3.7
Rubbers	2.5
Total	17.5

Polymers produced by radical processes: 10.5×10^6 tons (60%)

content; probably most importantly, producers use various ways to remove the heat of polymerization. As a consequence, different polymerization processes such as bulk polymerization, emulsion and suspension polymerization, and subtle variations of these are used for polymer production. Most fibers are condensation polymers and also the thermosetting resins are primarily made by non-radical processes.

More and more demands are now being made to develop polymerization processes which have no detrimental effects on the environment. Rising costs of common organic solvents make the development of processes based on bulk polymerization particularly attractive. The bulk polymerization of vinyl chloride, a process developed by St. Gobain, has received considerable attention. In this polymerization process, the heat of polymerization is removed by evaporation of the monomer. The polymer precipitates during the polymerization, and the polymerization reactor contains a refrigeration unit to recondense the monomeric vinyl chloride [3]. The process does, however, not always produce polymer particles of desired shape and bulk density.

A substantial portion of styrene produced is used in preparing "high impact" polystyrene by suspension polymerization. It has become a very attractive goal to carry out the polymerization of styrene in bulk in the presence of the toughening agent, for example, polybutadiene. This problem is not very simple as the 10% content of polybutadiene needed as the toughening agent tends to undergo cross-linking early during the styrene bulk polymerization, giving products without the desired toughness [4]; in contrast the desired limited amount of cross-linking can be relatively easily achieved in suspension polymerization. Many efforts are still underway to develop polystyrene compositions using varying amounts of different rubbers with different particle sizes and shapes to obtain maximum toughening. Another desirable goal in this area is to prepare toughened, transparent polystyrene.

Important work has also been done in the development of some effective acrylonitrile/styrene co- and graft polymers for use in improved ABS resins, and for "barrier" resins for use in bottles for carbonated beverages [5].

We still have many unsolved problems in polymerization chemistry: needs for new initiator systems, systems that can be used in a flexible manner under varying economic conditions to allow maximum utilization of reactors and equipment, and the changing of individual product properties by simple modification. These extensive demands require our continued efforts for a better understanding of individual initiator systems.

It would furthermore be desirable to control reaction conditions in order to prepare easily and predictably linear polymers, polymers with controlled number and length of branches, or copolymers of controlled and predetermined randomness or alternation of comonomer units in an effort to prepare polymeric products with new combinations of properties.

When entirely new properties or combinations of properties are desired, new monomers will still have to be prepared and polymerized and the properties of their polymers studied. New principles for the development of specific polymer properties and ultimately the tailoring of properties to obtain the ultimate properties of our materials must be achieved.

Numerous advances have been made recently in radical polymerization or are now under way. In the area of initiation, control of initiation has allowed the preparation of block copolymers by preparing bifunctional peroxide initiators which could be decomposed in two stages by thermal or selective electron transfer mechanisms. New block copolymers have also been prepared by the sequential use of two or more monomers in radical emulsion polymerization processes. It has also recently been found that homo- and copolymerizations of some phenol group containing monomers are not inhibited by these groups and high molecular weight polymers have been prepared. Charge transfer polymerization is also an area of great interest and the possibility of initiation of *radical* polymerization by charge transfer is one of the unsolved problems in radical polymerization. Much work has recently been done on the development and understanding of polymerization and copolymerization systems initiated by complexed radicals. The possibility of varying the relative rate in copolymerizations of individual monomers is very important because for many years radical polymerizations of individual monomers were believed to be independent of the solvent used, and by implication their reactivity could not be significantly influenced.

The existence of living polymer radicals has now been clearly demonstrated and living radicals have been used for the preparation of block copolymers by radical polymerization. Emulsion polymerization, under some conditions that are very similar to living systems, has been used for the preparation of block copolymers. Macroradicals, polymer radicals of limited solubility and swellability, are a most important new development.

Graft and comb polymers in which polymer backbone and side chains are of drastically different polarity have become important as compatibilization agents. It has been found that macromers, monomers with an active polymerizable group and long side chains, can be readily prepared and conveniently homo- and copolymerized to polymers with new and unusual properties. At the same time it has been shown that similar monomers with long side chains actually have increased polymerization rates because of the special nature of the monomer associates prior to polymerization.

For the preparation of polymer blends where at least one of the polymers is insoluble and infusible, a chemical technique of preparing polymer blends by sequential polymerization has been developed. One polymer which desirably precipitates as a gel is prepared by ionic, usually anionic polymerization, and the second polymer is then prepared by radical polymerization of an olefinic

monomer within the solid gel matrix of the first polymer. If the radical polymerization is carried out in the presence of a cross-linking agent, interpenetrating
networks can be prepared.

A few other specific problems in radical polymerization and in the preparation
of co- and graft polymers are also awaiting solutions. They include various aspects of alternating copolymerization, specific problems of the achievement of
polymer regularity based on the stereochemistry and the bulkiness of side groups
and their interrelationships with polar factors and inductive effects in radical
polymerization and copolymerizations.

SPECIFIC RESULTS, UNSOLVED PROBLEMS, AND INTERESTING POSSIBILITIES

It has been known for a long time [6, 7] that radicals from peroxides, hydroperoxides, and peroxy acid derivatives can be generated at various temperatures
depending upon the individual substituents of the —O—O— bond. Suggestions
existed for the preparation of block copolymers using a bifunctional radical
source, for example, an unsymmetric bisperoxide whose peroxide groups have
substantially different thermal stability. At lower temperature only the less stable
peroxide group of the bisperoxide will decompose and radical polymerization
can be carried out with one monomer; the other radical may be generated at a
different, usually higher temperature to give block polymers. It seems that this
principle has now been demonstrated and peroxides of suitable structures have
been prepared. Peroxy acid derivatives with two different substituents have been
found to be suitable for the preparation of these A–B block copolymers [eq. (1)]
by radical polymerization.

$$\overline{R}\text{—O—O}\text{\large\sim}\text{O—OH} + \ m \ CH_2{=}CH \cdot R \ \longrightarrow \ AAAAAAAA_n\text{—O}\text{\large\sim}\text{O—O—H}$$
$$(A)$$

$$\xrightarrow[+ \ nB]{\Delta \ or \ e^{\ominus}} \ AAAAAAAAA_n\text{—O}\text{\large\sim}\text{O—BBBBBBBBB}_n \qquad (1)$$

After the radical is generated by decomposing the first peroxy group at lower
temperatures, polymerization of monomer (A) is allowed to proceed. When the
first polymerization is complete, the second monomer is added and the peroxy
group which is now attached to the end of the homopolymer chain is decomposed
to the radical, from which the polymerization of the second monomer can be
accomplished.

It is evident that the generation of the two radicals, the first and particularly
the second radical, can be done not only by thermal means but also with electron
donors such as tertiary aromatic amines, allowing an additional dimension to
utilize the individual stability of such bisperoxides; the tertiary perester of a
peroxy acid on one end and a peroxy acid on the other end of the initiator molecular have been used. It is also necessary that the two different peroxide groups
which are the actual initiators be separated by a carbon chain of sufficient chain
length. The only possible drawback in this particular technique to prepare A–B

block copolymers is when transfer to initiator occurs; in this case initiator is used up [8].

This concept of preparing block copolymers by radical polymerization is similar to recent work on cationic block copolymer formation where a chain of eight carbon atoms is necessary between the individual initiating chlorine or bromine end groups of the compound which is used as the cationic initiator.

In radical polymerization there is another problem: A–B block copolymers can only be made when the polymerizations of (A) and (B) are terminated by disproportionation. When (A) terminates by recombination and (B) terminates by disproportionation, B–A–B block copolymers are formed. When (A) terminates by disproportionation and (B) by recombination, A–B–A block copolymers can be formed. Should (A) and (B) terminate by recombination only, –A–B–A–B– block copolymers could be formed. None of these examples have been demonstrated but such are within the present possibilities of radical polymerization; especially when resonance stabilized monomers such as styrene or methyl methacrylate are used as monomers (A) and (B).

Another possible technique of initiation of radical polymerization is by charge transfer complexes [9]. In general, charge transfer complexes initiate the cationic polymerization of donor monomers as has been demonstrated in many cases in N-vinyl carbazole polymerization. In some cases, particularly when the acceptor was vinylidene cyanide, anionic polymerization was also observed. Another reaction that occurs sometimes between donor and acceptor olefins is the formation of cyclobutane derivatives, which was studied most extensively in the reaction of vinyl carbazole and tetracyanoethylene. It has been shown that radical recombination of the initially formed cation radicals and anion radicals occurs first followed by the recombination of cations and anions. The reverse, the formation of biradical intermediates, has not been demonstrated; such would permit an effective charge transfer initiation of radical polymerization by biradicals as in eq. (2):

$$A + D \longrightarrow [AD] \xrightarrow[\substack{\Delta \\ \text{temp}}]{h\nu} [AD]^* \longrightarrow [A^{\ominus} \cdot D^{\oplus} \cdot] \longrightarrow A^{\ominus} \cdot + D^{\oplus} \cdot$$

$$A \text{ or } D \text{ as } A^{\ominus} \cdot \text{ or } D^{\oplus} \cdot \text{ may initiate polymerization} \qquad (2)$$
$$A \text{ or } D \text{ may be monomers}$$

$$A^{\ominus} \cdot + B^{\oplus} \cdot \longrightarrow \cdot A\text{–}B \cdot$$

It has been believed for many years that free phenolic OH groups act as radical polymerization inhibitors but actually they interfere only with oxygen radicals. Recently it has been shown that p-vinylphenol could be polymerized with AIBN as the initiator [10]. o-Vinylphenol gave polymers of very low molecular weight; even the homopolymers of m- and p-vinylphenol were not of very high molecular weight. Vinylphenols could, however, be copolymerized with various acrylates and methacrylates and copolymerization reactivity ratios determined. Viscosity measurements indicated that the molecular weight of these copolymers was adequately high [11]. Some interesting polymerizations were carried out with *N*-(3,5-di*tert*-butyl-4-hydroxy benzyl) acrylamide and *N*-(4-diphenylamino) acrylamide [12]. These two monomers which contained groups which are potentially useful antioxidants have been polymerized with a number of different initiators and the results, if confirmed, seem to be very interesting. With AIBN or 4,4′-azobis(4-cyanopentanoic acid) as the initiators, either monomer polymerized normally and gave homo- and copolymers of good molecular weight. However, when benzoyl peroxide or cumene hydroperoxide were used as the initiators, no polymerization occurred or oligomers of low molecular weights were obtained [eq. (3)]:

$$CH_2=CH-CO$$
$$|$$
$$NH \xrightarrow{\text{AIBN}} \text{High polymer (a)}$$
$$|$$
$$CH_2 \quad \text{Bz}_2\text{O}_2 \tag{3}$$

$$\text{or PhC(Me)}_2\text{OOH} \longrightarrow \text{No polymer or (b)}$$
$$\text{low MW polymer}$$

$$OH$$

We have investigated [13] the polymerization of the previously unknown 5-vinyl derivatives of methyl salicylate, methyl acetylsalicylate, salicyclic acid, and acetyl salicyclic acid and have found that these monomers are very reactive and polymerize even in the solid state below their melting point or during the course of purification by sublimation or recrystallization. It could be argued that the polymerization of salicyclic acid derivatives (free OH group) with AIBN is not surprising because the hydroxy group is hydrogen bonded to the carbonyl group of the ester or acid or that AIBN has been shown to polymerize also other phenolic vinyl compounds, as in eq. (4):

$$CH=CH_2 \qquad \qquad \left(CH-CH_2\right)_n$$

$$\xrightarrow{\text{AIBN}} \tag{4}$$

$$COOR' \qquad \qquad COOR'$$
$$OR \qquad \qquad OR$$

$$R = H \qquad R' = CH_3$$
$$H \qquad \qquad H$$
$$CH_3CO \qquad H$$
$$CH_3CO \qquad CH_3$$

We have also prepared 2,4-dihydroxy-4′-vinylbenzophenone [14] and found that this compound also polymerizes and copolymerizes readily. In fact, it is very difficult to isolate pure monomer from the synthesis mixture and if not carefully done, only oligomers of this compound are obtained. The homopolymer and the copolymers with styrene and methacrylic acid are of high molecular weight and of molecular weight distribution identical to that of the homopolymers of styrene or methacrylic acid prepared under the same conditions. It can be argued that the 2-hydroxy group may be hydrogen bonded to the carbonyl group of benzo-phenone. The 4-hydroxy group, however, is a true phenol group and if free phenol groups would hinder the polymerization of vinyl compounds, this monomer would certainly inhibit the polymerization. We have also checked the copolymerization of the 2,4-dihydroxy-4'-vinylbenzophenone with styrene by different initiators and have found that polymers of high molecular weight can be obtained not only by initiation with AIBN but also by initiation with peroxides such as benzoyl peroxide [eq. (5)]. As a consequence, the earlier work which suggested that there

$$(5)$$

is a difference in the effectiveness of individual initiators in the case of acrylates which have aromatic amine and antioxidant type phenolic groups in the molecule does not coincide with the results of our work. Our results seem to indicate that there is a possibility that radical polymerizations can be carried out in the presence of phenolic groups of the type that might be present in commercial antioxidants. This result should encourage renewed efforts in the area of radical polymerization in the presence of stabilizers which might simplify the purifi-cation procedures of monomers and obviate the need for the protective gas.

The preparation of telechelic polymers [15], oligomers or polymers of mod-erate molecular weights which have a functional group at each end of the polymer chain, has recently been actively investigated. For practical purposes, it would be desirable to have polymers or oligomers with carboxyl or OH groups as the terminal groups [16]. Such polymers, particularly those which contain polymer chains of low glass transition temperature and can provide a rubbery link in condensation polymers, have been investigated as potential candidates for their use in polyurethanes and other applications [16]. They can be made by control of the radical polymerization in direction of termination completely by recombination [17], or very effective transfer to initiator [16] [eq. (6)]:

$$
\underset{\underset{CN}{|}}{HOOC\!-\!(CH_2)_n\!-\!\overset{\overset{CH_3}{|}}{C}\cdot} \; + \; A \quad \xrightarrow[\substack{DMSO \\ recombin- \\ ation \; only}]{Solvent} \quad
\begin{array}{l} HOOCAA\text{----}AACOOH \\ HOAA\text{----}AAOH \end{array}
$$

$$
\xrightarrow[\substack{\text{transfer to} \\ \text{excess} \\ \text{initiator}}]{}
\begin{array}{l}
HOOCAA\text{----}AACOOH \\[6pt]
HOOC\text{----}\underset{\underset{CN}{|}}{\overset{\overset{CH_3}{|}}{C}}\cdot
\end{array} \qquad (6)
$$

For many years it has been believed that the kinetics of radical polymerization are relatively independent of the solvent used for the polymerization. It has also been found that complexation of one monomer, particularly the acceptor monomer, can dramatically change the copolymerization behavior to the point where the copolymerization becomes alternating. These copolymerizations have usually required the use of relatively strong donor monomers and resonance stabilized monomers such as vinyl ethers, styrene and dienes as the donor monomers.

$$
\begin{array}{c}
CH_2\!=\!CH_2 \\
+ \\
CH_2\!=\!CH\cdot OCOCH_3
\end{array}
\quad \xrightarrow[\substack{Et_3Al/butyrolactone \\ 50 \; atm}]{RCO_3H} \quad -\!(CH_2\!-\!CH_2)_n\!- \qquad (7)
$$

$$
\downarrow BF_3, RCO_3H
$$

Alternating
copolymer

It has recently been found that ethylene can be homopolymerized at 50 atmospheres pressure and room temperature in the presence of triethylaluminum/butyrolactone as the complexing agents when the polymerization is initiated with aliphatic acyl peroxides or hydroperoxides [20] [eq. (7)]. The coordinative radical polymerization of a complexed activated ethylene is very similar to the method which was used by Bier [21] for the polymerization of ethylene at or near room temperature and at 20–40 atmospheres pressure in water with silver perchlorate or other silver salts as complexing agent for the ethylene.

It was also found that copolymerization of ethylene and vinyl acetate [22] could be accomplished when equimolar amounts of BF_3 were used and an alternating ethylene/vinyl acetate copolymer could be prepared. It has furthermore been found that vinyl acetate copolymerizes with methyl acrylate [23] with a peroxy acid as the initiator and triethylaluminum/butyrolactone as the complexing agent to give an alternating copolymer. With aluminum alkyl chlorides alone as the complexing agent, random copolymers were obtained. Two important points seem to be significant for these polymerizations. The first one is the importance of the use of specific aluminum compounds to direct the overall copolymerization behavior from alternating to random copolymerization; but another more important point is that vinyl acetate can be made to copolymerize readily with monomers which form radicals that are very much more resonance stabilized than the vinyl acetate radical. Under normal copolymerization conditions of vinyl acetate with styrene or methyl methacrylate, essentially pure homopolymer of styrene or methyl methacrylate is produced. This principle of complexed radical polymerization, particularly the possibility of modifying monomers and/or growing radicals to cause a smooth copolymerization of monomers which give resonance stabilized and unstabilized radicals, permits a new aspect of control coordinative radical copolymerizations [eq. (8)]:

$$CH_2{=\!\!=}CH \cdot OCOCH_3 \ + \ CH_2{=\!\!=}CH{-\!\!-}COOCH_3 \ \xrightarrow[\text{Et}_3\text{Al/butyrolactone}]{\text{RCO}_3\text{H}} \ \begin{array}{c}\text{Alternating}\\ \text{copolymer}\end{array}$$

$$(8)$$

with $R_x AlCl_y$: Random copolymers

Another newer development in the initiation of polymerization is the use of polymeric initiators for the preparation of A–B copolymers, although the principle of using polymeric initiators is not new and has been used for the preparation of block copolymers by anionic initiation. For example, living polystyrene was used as the initiator for the polymerization of methyl methacrylate or ethylene oxide [18, 19, 24]. More recently, cationic polymerization has also provided examples of living polymerizations and such have been used for the preparation of block- and graft polymers. In the case of the living cationic polymerization of THF [25, 26] it was recognized that monomers which produce relatively nonreactive growing cations are needed to produce a living cationic system. Should the growing ion be too reactive, side reactions such as rearrangements and terminations are observed to occur and the polymerizations are not truly living.

Radical polymerization is terminated by basically two reactions: recombination and disproportionation. In principle, living radical polymerization could be obtained when growing radicals are prevented from undergoing these reactions. The problem of preventing termination by a bimolecular reaction of two polymer radicals is different from that of preventing termination in ionic polymerization because in ionic polymerizations, both anionic and cationic, the growing polymer chains cannot terminate by recombination of one chain with another. In radical polymerization, the termination by bimolecular recombination or disproportionation, can in principle be avoided by eliminating the diffusion of two growing radicals to meet each other and effect the bimolecular termination reactions [27]. This has been seen in the case of popcorn polymerization, the Trommsdorff effect (at high viscosity of the polymerizing mixture), emulsion polymerization and polymerizations where particles precipitate during the polymerization in a swollen state.

It has now been shown that macroradicals [28] can be produced which can be isolated and even dried when certain physical conditions are met. The macroradical may be a homo- or a copolymer radical and the polymerization can be continued with a different monomer to produce block copolymers [eq. (9)]:

~~~P· + nM ⟶ ~~~PM$_n$· Macroradicals may be homo- or copolymer radicals

2~~~P· ⟶ ~~~PPM~~~ Radical polymerization may be continued by different monomers

Prevent termination by controlling mobility of macroradical   (9)
  • Growing macroradical must precipitate from solution
  • Monomer must still be able to diffuse to growing macroradical
  • Solubility parameter of polymer and solvent/monomer 2–3 units different

In order for termination of the growing macroradicals to be prevented in this manner, the mobility of the macroradical must be controlled by precipitation of the polymer radical from solution. The second monomer must, however, swell the polymer radical and be able to diffuse to the growing radical. Under the actual conditions of the reaction, the solubility parameter of polymer and solvent/monomer must be different by two to three solubility parameter units. If the solubility parameter difference is less than two, the growing radicals are too mobile and termination by recombination or disproportionation occurs. If the difference of the solubility parameters is larger than three, the growing polymer radical will be too insoluble in the monomer or solvent/monomer mixture for the monomer to diffuse to the site and polymerization will stop because of lack of monomer. Specific cases of A–B block copolymers obtained by this route include the preparation of a copolymer of methyl acrylate/styrene in benzene followed by the polymerization of styrene, methyl methacrylate, or acrylonitrile by the resultant macroradical. Other examples are the preparation of a living methyl methacrylate radical in cyclohexane followed by block copolymerization of acrylonitrile/sytrene or methyl acrylate/styrene onto the methyl methacrylate macroradical and the preparation of an acrylonitrile macroradical in benzene followed by a block copolymerization of a mixture of methyl acrylate and styrene [28] [eq. (10)]:

$$(MA/St)_n \text{ in } C_6H_6 \longrightarrow St, MMA, AN$$
$$(MMA)_n \text{ in } C_6H_{14} \longrightarrow AN, MA/St, St \qquad (10)$$
$$(AN)_n \quad \text{ in } C_6H_6 \longrightarrow MA/St$$

Another possible way to produce macroradicals for the preparation of block copolymers is by properly performed emulsion polymerization. Polymers of very large molecular weights can be obtained because the whole micelle in the emulsion polymerization becomes one single macromolecule. When either a chain transfer agent is added or initiator fragments diffuse into the micelle termination occurs.

Practical ways have now been found to prepare block copolymers by carrying out a sequence of two emulsion homopolymerizations [29]. After the first monomer is polymerized, a second monomer is added and the polymerization is continued. This method of polymerization could be practical for the preparation of block copolymers which have blocks with different solubility characteristics, for example, hydrophilic and hydrophobic components.

In both cases of block copolymer preparation, the initial formation of living macroradicals by precipitation or by emulsion polymerization causes the resulting polymers to have broad molecular weight distributions, substantially different from the molecular weight distributions of classic living anionic and cationic polymerization. Living ionic polymers are prepared in solution under equilibrium conditions and are nearly monodisperse. Copolymers made by "living" radical polymerization utilize immobilized macroradicals of the initial homopolymers. They are made specifically under nonequilibrium conditions and consequently the molecular weight distribution both of the individual blocks and of the whole A–B copolymer is relatively broad.

Block, graft, and comb polymers consist of two or three types of monomer units which are of different polarity and consequently provide aspects which contribute to the compatibility of such macromolecules. In some block copolymers it has been demonstrated that the individual components aggregate in individual phases and cause the formation of microphase separated systems. On the other hand, blends of homopolymers of different solubility parameters may be very incompatible but may be compatibilized by addition of co-, graft, and comb polymers of sufficiently long sequence length of individual comonomers. Such combinations improve mechanical properties, toughness, and impact strength of the polymer composites. It has now been found that specific comb polymers can be readily prepared from an anionic living polymer which is reacted with a specific reactive molecule which provides the polymerizable group for the specific molecule to form a "Macromer" [30]. Only about 20% of the macromer in copolymers is needed or even desired for the preparation of copolymers with unusual and novel properties. A variety of macromers based on styrene and diene living polymers were prepared which could be used for the preparation of macromer monomers for ionic or radical polymerization. One example for the preparation of a macromer which can be used for radical polymerization is here described. Living polystyrene of appropriate molecular weight and narrow molecular weight distribution was prepared by traditional means and reacted with one molecule of ethylene oxide to provide an alkoxide anion. The polymeric alkoxide was then allowed to react with acryloyl chloride or methacryloyl chloride which gave a macromeric acrylate or methacrylate which could be polymerized or copolymerized with other acrylates or methacrylates, for example, butyl methacrylate, to prepare copolymers with properties which are a combination and an extension of the properties of the polyacrylate main chain and the polystyrene side chain [eq. (11)]:

(1)   Preparation of macromer

$$R'^{\ominus} \; + \; n\,CH_2\!=\!CH\cdot R \; \longrightarrow \; R'\!-\!(\!CH_2\!-\!CH\cdot R\,)_{\overline{n-1}}CH_2\!-\!\overset{\ominus}{C}HR \; + \; \underset{O}{\overset{CH_2\!-\!CH_2}{\diagdown\;\diagup}}$$

$$R'\!-\!(\!CH_2\!-\!CH\cdot R\,)_{\overline{n}}CH_2\!-\!CH_2\!-\!O^{\ominus} \; + \; CH_2\!=\!CH\cdot COCl$$

$$R'\!-\!(\!CH_2\!-\!CH\cdot R\,)_{\overline{n}}CH_2\!-\!CH_2\!-\!O\!-\!CO\cdot CH\!=\!CH_2 \tag{11}$$

MACROMER

(2) Copolymerization of macromers with acrylates or methacrylates;
    copolymers usually contain 10–20% macromer;
    copolymer properties are a combination of the properties of main
    polymer chain and of side chain.

When the living polystyrene was allowed to react with p-(chloromethyl) styrene, a macromer with a styryl reactive group suitable for styrene copolymerization was obtained.

When it was desired to prepare a blend of polymers, one of which is insoluble in all solvents and does not melt, such polymer blends had to be made by chemical means, by sequential polymerization of the individual monomers or comonomer mixtures directly into the desired shapes. It was convenient to select monomers which could not be copolymerized but could be homopolymerized by different methods.

Ionic and radical polymerizations can be carried out in sequence in one reaction vessel with or without solvents [31]. Specific examples were demonstrated in the preparation of polymer blends and interpenetrating networks of chloral polymers and addition polymers, for example, styrene, methyl methacrylate, or methyl acrylate polymers. In this polymerization, to a mixture of chloral, styrene, and AIBN initiator was added, above the threshold polymerization temperature of chloral, an anionic initiator for chloral polymerization, for example, triphenylphosphine. Chloral was now polymerized by cryotachensic polymerization, by cooling below the threshold polymerization temperature; polychloral precipitated during this polymerization as a gel, which imbibed the other monomer and acted as a diluent for the chloral polymerization, producing chloral polymer in high conversion. The temperature was now raised from 0 to 60°C, the temperature at which AIBN produces initiating radicals at a desirable rate which caused the styrene to polymerize within the matrix of the chloral polymer. In this way blends were produced from which polystyrene could be extracted essentially quantitatively.

In spite of the presence of the trichloromethyl groups in polychloral, with AIBN as the radical source and styrene or acrylates as the radically polymerized monomers, no grafting on the trichloromethyl groups occurred.

Instead of chloral, a mixture of chloral and a suitable comonomer, for example, an aromatic isocyanate, may be used and instead of styrene, methyl methacrylate or methyl acrylate may be used. When a cross-linking agent was added to the styrene, for example, divinylbenzene in amounts of 5 mole % or ethylene bismethacrylate at a similar level, for the sequential polymerization of methyl acrylate or methacrylate, interpenetrating networks of chloral polymers and cross-linked addition polymers were formed [eq. (12), Fig. 1].

(1)   Anionic polymerization of chloral (cryotachensic)

$$R'^{\ominus} + (n + 1)CCl_3CHO \xrightarrow[\text{cryotachensic}]{} R—C(CCl_3)—O\left[C(CCl_3)—O\right]_n$$
$$\underset{H}{|} \qquad \underset{H}{|}$$

Followed by   (12)

(2)   Radical polymerization of styrene, MMA, etc. (AIBN)

$$R^{\ominus} + CH_2{=}CH·C_6H_5 \xrightarrow{\text{AIBN}} \text{POLYSTYRENE}$$

Head to head (H–H) linkages in conventional head to tail (H–T) polymers and their influence on polymer stability have attracted the attention of scientists for many years. Recently a more concentrated effort to prepare pure H–H polymers was made [32]. Certain H–H polymers can be prepared by alternating copolymerization of two symmetrically substituted olefin comonomers. In this way, nearly pure H–H polypropylene was prepared from ethylene and 2-butene [33] and nearly pure H–H poly(vinylidene fluoride) [34, 35] was prepared from ethylene and tetrafluoroethylene, as in eq. (13):

$$CH_2{=}CH_2 + CH_3CH{=}CHCH_3 \longrightarrow -(CH_2—CH—CH—CH_2)_n-$$
$$\underset{CH_3}{|} \quad \underset{CH_3}{|}$$
(13)

$$CH_2{=}CH_2 + CF_2{=}CF_2 \longrightarrow -(CH_2—CF_2—CF_2—CH_2)_n-$$

For the preparation of H–H polyolefins we found it most convenient to prepare pure 1,4 poly(2,3 disubstituted dienes), and hydrogenate the double bond of the resulting diene polymer [36]. An example is the preparation of H–H polypropylene from the polymerization of 2,3-dimethylbutadiene followed by catalytic

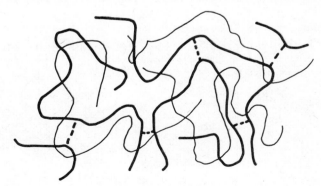

FIG. 1. Blends and IPNs prepared by sequential polymerization.

hydrogenation over Pd/C [eq. (14)]:

$$\text{-----CH}_2\text{---}\underset{\underset{CH_3}{|}}{\overset{\overset{CH_3}{|}}{C}}\text{=C}\text{ CH}_2\text{-----}\xrightarrow{Pd\,5\%/C}\text{-----CH}_2\text{---}\underset{\underset{CH_3}{|}}{CH}\text{---}\underset{\underset{CH_3}{|}}{CH}\text{---}CH_2\text{-----}\quad(14)$$

100% after four hydrogenations

We have recently also been able to prepare H–H polystyrene [36] from 2,3-diphenylbutadiene which was polymerized with **AIBN** as the initiator to a poly(2,3-diphenylbutadiene) which consisted entirely of 1,4 linkages, [eq. (15)]:

$$CH_2\text{=}\underset{\underset{C_6H_5}{|}}{\overset{\overset{C_6H_5}{|}}{C}}\text{---}C\text{=}CH_2\xrightarrow[\substack{60°C,\ 48\ hr\\ quantitative}]{AIBN,\ bulk}\text{-----}CH_2\text{---}\underset{\underset{C_6H_5}{|}}{\overset{\overset{C_6H_5}{|}}{C}}\text{=C}\text{---}CH_2\text{-----}\quad(15)$$

sol. in benzene, etc.
η 2 dl/g

This polymerization of a diene to an entirely 1,4 polymer with a radical initiator is very surprising but it is probably caused by the bulkiness of the phenyl groups in the 2 and 3 position. The polymer was chemically reduced to H–H polystyrene [eq. (16)] and catalytically hydrogenated to H–H poly(vinyl cyclohexane) [eq. (17)]:

$$\text{-----}CH_2\text{---}\underset{\underset{C_6H_5}{|}}{\overset{\overset{C_6H_5}{|}}{C}}\text{=C}\text{---}CH_2\text{-----}\xrightarrow[\substack{K/C_2H_5OH\\ 25\ hr}]{R.T.,\ THF}\text{-----}CH_2\text{---}\underset{\underset{C_6H_5}{|}}{CH}\text{---}CH\text{---}CH_2\text{-----}\quad(16)$$

K/olefin ratio 20:1   1st 93%
K/C$_2$H$_5$OH 1:1   2nd complete

$$\begin{array}{l}H\text{--}H\\ \text{-----}CH_2\text{---}\underset{\underset{C_6H_5}{|}}{\overset{\overset{C_6H_5}{|}}{CH}}\text{---}CH\text{---}CH_2\text{-----}\end{array}\longrightarrow\text{-----}CH_2\text{---}\underset{\underset{C_6H_{11}}{|}}{\overset{\overset{C_6H_{11}}{|}}{CH}}\text{---}CH\text{---}CH_2\text{-----}$$

$$\begin{array}{l}H\text{--}T\\ \text{-----}CH_2\text{---}\underset{\underset{C_6H_5}{|}}{\overset{\overset{C_6H_5}{|}}{CH}}\text{---}CH_2\text{---}CH\text{-----}\end{array}\longrightarrow\text{-----}CH_2\text{---}\underset{\underset{C_6H_{11}}{|}}{\overset{\overset{C_6H_{11}}{|}}{CH}}\text{---}CH_2\text{---}CH\text{-----}\quad(17)$$

Atactic   >97% (UV, NMR)

200°C, 2 days
1200 psi H$_2$
Rh 5%/C

The radical polymerization of 2,3-diphenylbutadiene does not form 1,4-poly(2,3-diphenylbutadiene) of unique stereochemistry of the double bond, but the cis/trans ratio of the double bonds in nearly 1 to 1. It is expected that other dienes with relatively bulky groups in the 2 and 3 position might also polymerize directly to pure 1,4 polymers.

CONCLUSIONS

A number of new developments in the last few years have given new impetus to the study of radical polymerization which is still the method of choice for most industrial processes. It is believed that with the need for more co- and graft polymers and modified polymers in modern polymer technology, some of the subjects and ideas presented in this paper will stimulate further developments.

I would like to express my appreciation to the National Science Foundation for the opportunity to organize the First United States–Romanian Seminar on Polymer Science and to present my views on what I believe are important developments in radical polymerization. This paper was written while I was on sabbatical leave at the C.N.R.S., Centre de Recherches sur les Macromolécules, and Professeur Associé at the Université Louis Pasteur, Strasbourg, France. I would also like to thank L. S. Corley for assistance in the preparation of this manuscript.

REFERENCES

[1] J. C. Bevington, *Acrylonitrile in Macromolecules,* Academic, London/New York, 1961.
[2] G. C. Eastmond, in *Encyclopedia of Polymer Science and Technology,* Vol. 7, 361 (1967).
[3] J. L. Benton and C. A. Brighton, in *Encyclopedia of Polymer Science and Technology,* Vol. 14, 343 (1971).
[4] W. E. Gibbs, Macromolecular Secretariat Symposium on "Toughened Polystyrene," ACS Meeting in New Orleans, 1977, Preprints, ACS Division of Polymer Chemistry, Vol. 18, No. 1, 835 (1977).
[5] E. Pearce, Macromolecular Secretariat Symposium on "Acrylonitrile," ACS Meeting in Chicago, 1973, Preprints, ACS Division of Polymer Chemistry, Vol. 14, No. 2 (1973), and ACS Division of Organic Coatings and Plastics Chemistry (1973).
[6] R. J. Orr and H. L. Williams, *J. Am. Chem. Soc., 78,* 3273 (1956); *ibid., 79,* 3137 (1957).
[7] P. E. M. Allen, J. M. Downer, G. W. Hastings, H. W. Melville, P. Molyneux and J. L. Benton, *Nature (London), 177,* 910 (1956).
[8] J. C. Bevington and T. D. Lewis, *Polymer, 1,* 1 (1960).
[9] J. K. Stille, N. Oguni, D. C. Chung, R. F. Tarvin, S. Aoki and M. Kamachi, *J. Macromol. Sci., Chem., A9(5),* 745 (1975).
[10] M. Kato, *J. Polym. Sci., A-1, 7,* 2175 (1969).
[11] C. G. Overberger and E. Sincich, *J. Polym. Sci., Polym. Chem. Ed., 13,* 1783 (1975).
[12] M. Kato, Y. Takemoto, Y. Nakano and M. Yamazaki, *J. Polym. Sci. Polym. Chem. Ed., 13,* 1901 (1975).
[13] D. Bailey, D. Tirrell and O. Vogl, *J. Polym. Sci. Polym. Chem. Ed., 14,* 2725 (1976).
[14] D. Tirrell, D. Bailey and O. Vogl, Preprints, ACS Division of Polymer Chemistry, *18(1),* 542 (1977).
[15] S. F. Reed, Preprints, ACS Division of Polymer Chemistry, *15(2),* 46 (1974).
[16] A. R. Siebert, IUPAC Symposium on Macromolecules, Aberdeen, 1973, A. R. Siebert, Ger. Offen. 1,924,866 (1970); CA. *72:* P 6743h (1970).
[17] C. H. Bamford and A. D. Jenkins, *Nature, 1976,* 78 (1955).
[18] M. Szwarc, *Makromol. Chem., 35,* 132 (1960).

[19] M. Szwarc, *Carbanions, Living Polymers and Electron Transfer Processes*, Interscience, New York, 1968.

[20] T. Yatsu, S. Moriuchi and H. Fujii, Preprints, ACS Division of Polymer Chemistry, Vol. 16, No. 1, 373 (1975).

[21] G. Bier, Belgian Patent No. 602,153 (1961).

[22] K. Saito and T. Saegusa, *Makromol. Chem., 117*, 86 (1968).

[23] T. Saegusa, T. Tanizu and H. Fujii, Japan. Patent 73-42,468 (1973).

[24] M. Szwarc, M. Levy and R. Milkovich, *J. Am. Chem. Soc., 78*, 2656 (1956).

[25] M. Dreyfuss and P. Dreyfuss, *Polymer, 6*, 93 (1965).

[26] H. Imai, T. Saegusa, S. Matsumoto, T. Tadasa and J. Furukawa, *Makromol. Chem., 102*, 222 (1967).

[27] C. H. Bamford and A. D. Jenkins, *J. Chem. Phys., 56*, 798 (1959).

[28] R. B. Seymour, P. D. Kincaid, D. Owen and J. D. Crump, Preprints, ACS Division of Polymer Chemistry, Vol. 13, No. 1, 522 (1972).

[29] J. van Beulen, Ph.D. thesis, University of Leuven, 1977.

[30] R. Milkovich and M. T. Chang, U.S. Patent No. 3 786 116 (1974).

[31] L. S. Corley, P. Kubisa and O. Vogl, *Polym. J. (Jpn.), 9*, 47 (1977).

[32] T. Tanaka and O. Vogl, *Polym. J. (Jpn.), 6*, 522 (1974).

[33] G. Natta, G. Dall'Asta, G. Mazzanti, I. Pasquon, A. Valvassori and A. Zambelli, *J. Am. Chem. Soc., 83*, 3343 (1961).

[34] W. E. Hanford and J. R. Roland, U.S. Patent No. 2,468,664 (1949); CA. *43:* P. 5410 (1949).

[35] C. W. Wilson and E. R. Santee, *J. Polym. Sci., Polym. Chem., 8*, 97 (1965).

[36] H. Inoue, M. Helbig and O. Vogl, *Macromolecules, 10*, 1331 (1977).

RADICAL RING-OPENING POLYMERIZATION AND COPOLYMERIZATION WITH EXPANSION IN VOLUME

WILLIAM J. BAILEY
Department of Chemistry, University of Maryland, College Park, Maryland 20742

TAKESHI ENDO
Research Laboratory of Resources Utilization, Tokyo Institute of Technology, Tokyo, Japan

SYNOPSIS

Although ring-opening polymerization involving ionic intermediates and Ziegler catalysts are quite extensively used in polymer chemistry, free radical ring-opening polymerization is extremely rare. We have been able to show that unsaturated spiro ortho carbonates will undergo double ring-opening polymerization under radical conditions. In addition these monomers will undergo ready copolymerization with common monomers, such as styrene, methyl methacrylate, and diallyl carbonate.

Since this process involves a double ring-opening, a slight expansion in volume occurs. For a number of industrial applications, such as precision castings, strain-free composites, binders for solid propellants, high strength adhesives, dental fillings, and rock-cracking materials, it is highly desirable to have monomers that expand on polymerization. In the formation of composites, such a monomer might minimize poor adhesion between the matrix and the reinforcing agent or eliminate microcracks. Formation of additional polymers usually involves shrinkage since the monomer undergoes a transition from a van der Waals distance to a covalent distance in the polymeric material. Ring-opening polymerization usually involves somewhat less shrinkage because for every new bond that is formed involving a shift from a van der Waals distance to a covalent distance, another bond is broken involving a shift from a covalent distance to a near van der Waals distance. Thus, if monomers are utilized in which two bonds are broken for every new bond that is formed in the polymerization process, near zero shrinkage or expansion takes place.

Journal of Polymer Science: Polymer Symposium 64, 17–26 (1978)
© 1978 John Wiley & Sons, Inc.
0360-8905/78/0064-0017$01.00

TABLE I
Calculated Shrinkages for Addition Polymerization

Monomer	Shrinkage, %
Ethylene	66.0
Propylene	39.0
Butadiene	36.0
Vinyl chloride	34.4
Acrylonitrile	31.0
Methyl methacrylate	21.2
Vinyl acetate	20.9
Styrene	14.5
Diallyl phthalate	11.8
N-Vinylcarbazole	7.5
1-Vinylpyrene	6.0

For a number of industrial applications, such as strain-free composites, potting resins, and binders for solid propellants, it appeared highly desirable to have monomers that undergo essentially zero shrinkage on polymerization. For many other applications, such as precision castings, high strength adhesives, prestressed castings, dental fillings, and rock cracking materials, it appeared highly desirable to have monomers that would undergo expansion on polymerization. Unfortunately, when commercially available materials polymerize or cure, a pronounced shrinkage takes place. In bulk plastics some of the stresses that are induced because of this shrinkage can be relieved with a total change in the over-all dimensions of the article. However, in a composite the reinforcing material which has a high modulus will often not permit appreciable shrinkage in the over-all dimension of the molded object, which causes enormous stresses to be built up in the composite. These stresses can be relieved either by an adhesive failure, in which the matrix pulls away from the reinforcing fiber, or by a cohesive failure in which a void or a microcrack is formed.

Since water undergoes an expansion in volume of about 4% when it freezes, ice will adhere to almost any surface, including Teflon which it does not even wet, by micromechanical adhesion. If monomers could be prepared that would expand on polymerization, one would expect to produce strong adhesives as well as high strength composites. Similarly, nature cracks rocks by also utilizing the fact that when water freezes it expands. For this reason it should be possible to replace explosives with monomers that expand on polymerization and make possible much safer road building, coal mining, and quarrying.

There are·additional examples that would indicate the high desirability of having materials which would expand. Precision type in the printing industry is made possible because of the existence of alloys that expand when they solidify, whereby the metal is forced into every little crevice and corner. Similarly amalgams are available for dental fillings that expand slightly on solidification which permits a close contact between the filling and the tooth structure.

At the present time all available commercial monomers undergo a pronounced contraction in volume when they polymerize. Chemists are not surprised that during condensation polymerization, substantial shrinkage occurs because small molecules are eliminated during the condensation reaction and the space that they occupied in starting material is vacant in the final structure. Furthermore,

TABLE II
Calculated Shrinkages for Ring-Opening Polymerization

Monomer	Shrinkage, %
Ethylene oxide	23
Isobutylene oxide	20
Cyclobutene	18
Propylene oxide	17
Cyclopentene	15
Cyclopentane	12
Tetrahydrofuran	10
Cyclohexane	9
Styrene oxide	9
Cycloheptane	5
Cyclooctene	5
Bisphenol-A digylcidyl ether and diethylaminopropylamine	5
Cyclooctadiene	3
Cyclododecatriene	3
5-Oxa-1,2-dithiacycloheptane	3
Dimethylsilane oxide cyclic tetramer	2
Cyclooctane	2

it can be shown that the larger the molecule eliminated during the condensation reaction, the larger the shrinkage. However, a considerable shrinkage also takes place during an addition polymerization in which no molecules are eliminated. Table I gives some calculated shrinkages during some representative addition polymerizations.

The shrinkage that occurs during polymerization arises from a number of different causes but the major factor is related to the fact that the monomers are located at a van der Waals' distance from one another, while in the corresponding polymer the monomer units have moved to within a covalent distance of one another. In small atoms the covalent radius is approximately one-third of the van der Waals' radius. From Table I it can be seen that the number of monomer molecules that are transformed to polymer per unit volume is roughly related to the shrinkage that occurs during polymerization. For instance, styrene, which has approximately four times the molecular weight of ethylene, undergoes approximately one-fourth as much shrinkage during polymerization as does ethylene. In addition to this main source of shrinkage the other factors that will influence the amount of shrinkage are the change in entropy going from monomer to polymer, as well as the amount of free volume in the monomer and in the corresponding polymer, that is, the efficiency of packing in the monomer and polymer. For example, if either the monomer or the polymer is crystalline and, therefore, is arranged more compactly, the shrinkage will be greatly influenced.

As indicated in Table II, the shrinkage that occurs during ring-opening polymerization is always somewhat less than that which occurs during simple addition polymerization. For example, ethylene oxide undergoes a shrinkage of 23% on polymerization, but on the basis of molecular weight alone one might expect it to undergo a shrinkage of about 40%. It appears that the expansion that occurs when the covalent carbon–oxygen bond is converted to a near van der Waals' distance during the ring opening partially counteracts the normal shrinkage that occurs during polymerization.

It is also obvious from Table II that the larger the ring involved in ring-opening polymerization, the smaller the shrinkage. In other words, the further that the atoms involved in the ring-opening can get from one another and approach a full van der Waals' distance the smaller the shrinkage. We reasoned, therefore, that if monomers were available that for every van der Waals' distance that was converted to a covalent distance during polymerization, at least two rings opened in which a bond was converted from a covalent distance to a near van der Waals' distance in the polymer, it should be possible to have monomers that would have no shrinkage or undergo expansion upon polymerization. Thus, we were able to show that a series of spiro ortho esters [1–4] and ketal lactones [2, 3] would polymerize with essentially no shrinkage during polymerization, while a number of spiro ortho carbonates [5–10] and bicyclo ortho esters [11] would undergo polymerization with expansion.

In the most impressive example found so far [5], the polymerization of crystalline 1,5,7,11-tetraoxaspiro[5.5]undecane just below its melting point of 141°C produced an expansion of 17%. In this case a very compact crystalline monomer is converted to an amorphous polymer with about 15% of the expansion associated with the melting of the crystalline monomer and only 2% associated with the polymerization. Furthermore, this material copolymerized as a slurry with epoxy resins to produce materials either with slight shrinkage, no change in volume, or slight expansion on polymerization, depending on the amount of the spiro ortho carbonate present in the mixture. Synthesis of bifunctional derivatives related to these materials has made possible the preparation of highly cross-linked resins by homopolymerization and the preparation of elastomers by copolymerization with a large excess of monofunctional monomer [4].

Although the literature contains a large number of examples of ring-opening polymerizations involving ionic intermediates, there are very few examples involving radical ring-opening polymerizations. The few examples that exist in the literature involve the polymerization of vinylcyclopropane derivatives, such as 1,1-dichloro-2-vinylcyclopropane and 1-carbethoxy-2-vinylcyclopropane, or spiro-o-xylylene. Since these examples all contain a highly strained ring, it appeared possible that a number of other strained ring systems containing unsaturation either in or adjacent to the ring could also undergo ring-opening or double ring-opening polymerization by a radical mechanism. For this reason we undertook the synthesis of 3,9-dimethylene-1,5,7,11-tetraoxaspiro[5.5]-

undecane (I) by the following set of reactions [6]:

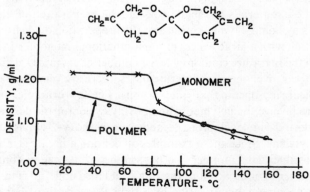

It was found when this monomer was treated with di-*tert*-butyl peroxide at 130°C and the reaction was stopped below 30% conversion, a soluble polymer was obtained having a structure of a polycarbonate with pendant methylene

FIG. 1. Densities of the monomeric spiro ortho carbonate and related polyoxycarbonate versus temperature.

groups. The structure of the polymer was established by elemental analysis as well as infrared and NMR spectroscopy. A very similar polymer could be attained by treatment of the monomer with boron trifluoride etherate at low conversions. The mechanism of the polymerization appeared to involve a radical double ring-opening according to the following mechanism [7]:

$$
\begin{array}{c}
RO^{\cdot} + CH_2{=}C\big\langle\!\!\begin{array}{c}CH_2{-}O\\CH_2{-}O\end{array}\!\!\big\rangle C \big\langle\!\!\begin{array}{c}O{-}CH_2\\O{-}CH_2\end{array}\!\!\big\rangle C{=}CH_2 \longrightarrow
\end{array}
$$

$$
RO{-}CH_2{-}\overset{\cdot}{C}\big\langle\!\!\begin{array}{c}CH_2{-}O\\CH_2{-}O\end{array}\!\!\big\rangle C \big\langle\!\!\begin{array}{c}O{-}CH_2\\O{-}CH_2\end{array}\!\!\big\rangle C{=}CH_2 \longrightarrow
$$

$$
RO{-}CH_2{-}C\big\langle\!\!\begin{array}{c}CH_2\quad O^{\cdot}\\CH_2{-}O\end{array}\!\!\big\rangle C \big\langle\!\!\begin{array}{c}O{-}CH_2\\O{-}CH_2\end{array}\!\!\big\rangle C{=}CH_2 \longrightarrow
$$

$$
RO{-}CH_2{-}C\big\langle\!\!\begin{array}{c}CH_2\quad O\\CH_2{-}O\end{array}\!\!\big\rangle C \big\langle\!\!\begin{array}{c}O{-}CH_2\\\cdot O{-}CH_2\end{array}\!\!\big\rangle C{=}CH_2 \xrightarrow{\text{repeat}}
$$

$$
RO{-}\!\!\left[CH_2{-}\overset{\overset{\textstyle CH_2}{\|}}{C}{-}CH_2{-}O{-}\overset{\overset{\textstyle O}{\|}}{C}{-}O{-}CH_2{-}\overset{\overset{\textstyle CH_2}{\|}}{C}{-}CH_2{-}O \right]_x
$$

The driving force for the double ring-opening polymerization apparently is the relief of the strain at the central spiro atom.

At high conversions this monomer produced a highly cross-linked resin, very similar in appearance to the material produced from the polymerization of diallyl carbonate. Furthermore it was shown that this unsaturated spiro ortho carbonate would readily copolymerize with styrene, methyl methacrylate, and diallyl carbonate, but with slightly lower reactivity than these other monomers. As indicated in Figure 1 the volume change that occurred during homopolymerization was quite unusual. At room temperature a 4.3% expansion in volume occurred, while just below its melting point at 70°C, a 7% expansion in volume occurred; at 85°C a 2% expansion took place and the expansion decreased until at 115°C no change in volume took place during polymerization; above 115°C a slight shrinkage occurred. It is obvious from these data that the large expansion in volume that occurs below the melting point involves not only the increase in volume due to the double ring-opening, but also a change in volume of 3–6% due to the process of going from a crystalline monomer to a liquid monomer. Since the monomer is a crystalline solid, it is difficult to find examples of homopolymerization in which the full 7% expansion in volume can be utilized. However, in copolymerizations it is possible to use a slurry of the crystalline monomer in a liquid monomer so that as copolymerization progresses, the crystalline monomer dissolves with some expansion and also polymerizes with expansion.

A potential use of this monomer is in the area of dental fillings in which a slurry containing 20% of very fine crystals of the unsaturated spiro ortho carbonate (I) in 60% of the adduct of methacrylic acid to bisphenol-A diglycidyl ether (Bis-GMA) plus 20% trimethylolpropane trimethacrylate produces on polymerization a material with essentially no change in volume. An investigation of a bubble test on tooth enamel showed that this copolymer had nearly double the adhesion to the tooth structure that the base resin had without the addition of the unsaturated spiro ortho carbonate. The copolymer also had improved impact strength but yet essentially the same modulus and filled composites appeared to have somewhat improved abrasion resistance.

Since the synthesis of the spiro ortho carbonates through the tin compounds could be modified to produce unsymmetrical materials, we undertook the synthesis of the unsymmetrical 2-methylene-1,5,7,11-tetraoxaspiro[5.5]undecane by the following set of reactions:

$$
\begin{array}{c}
\text{CH}_2\!=\!\text{C}\!\begin{array}{c}\text{CH}_2\!-\!\text{OH}\\[2pt]\text{CH}_2\!-\!\text{OH}\end{array}
\qquad\qquad
\text{HO}\!-\!\text{CH}_2\!-\!\text{CH}_2\!-\!\text{CH}_2\!-\!\text{OH}
\end{array}
$$

$$\downarrow \quad\begin{array}{l}(1)\ \text{Na}\\ (2)\ (n\text{-Bu})_3\text{SnCl}\end{array}$$

$$(n\text{-Bu})_3\!-\!\text{Sn}\!-\!\text{O}\!-\!\text{CH}_2\!-\!\text{CH}_2\!-\!\text{CH}_2\!-\!\text{O}\!-\!\text{Sn}(n\text{-Bu})_3$$

$$\downarrow (n\text{-Bu})_2\text{Sn}\!=\!\text{O} \qquad\qquad \downarrow \text{CS}_2$$

$$
\text{CH}_2\!=\!\text{C}\!\begin{array}{c}\text{CH}_2\!-\!\text{O}\\[2pt]\text{CH}_2\!-\!\text{O}\end{array}\!\!\text{Sn}(n\text{-Bu})_2
\qquad\qquad
\text{CH}_2\!\begin{array}{c}\text{CH}_2\!-\!\text{O}\\[2pt]\text{CH}_2\!-\!\text{O}\end{array}\!\!\text{C}\!=\!\text{S}
$$

$$\downarrow\ 86\%$$

$$
\text{CH}_2\!=\!\text{C}\!\begin{array}{c}\text{CH}_2\!-\!\text{O}\\[2pt]\text{CH}_2\!-\!\text{O}\end{array}\!\!\text{C}\!\begin{array}{c}\text{O}\!-\!\text{CH}_2\\[2pt]\text{O}\!-\!\text{CH}_2\end{array}\!\!\text{CH}_2
$$

mp 61—62°C
II

The resulting monomer was a crystalline solid with a melting point of 61–62°C. When the polymerization was carried out in the presence of di-*tert*-butyl peroxide and the reaction was stopped at low conversion, a linear polycarbonate containing pendant methylene groups was obtained.

The structure of the polymer was established by elemental analysis as well

$$
\text{II}\ \xrightarrow[\substack{\text{di-}\textit{tert}\text{-butyl}\\ \text{peroxide}\\ 43\%}]{130°\text{C}}\
\left[\!\!\begin{array}{c}\text{CH}_2\!-\!\underset{\underset{\text{CH}_2}{\|}}{\text{C}}\!-\!\text{CH}_2\!-\!\text{O}\!-\!\underset{\underset{\text{O}}{\|}}{\text{C}}\!-\!\text{O}\!-\!\text{CH}_2\!-\!\text{CH}_2\!-\!\text{CH}_2\!-\!\text{O}\end{array}\!\!\right]_x
$$

$$[\eta]\ ^{25°\text{C}}_{\text{CHCl}_3} = 0.11$$

as infrared and NMR spectroscopy. The structure of this material was very similar to the polymer that could be obtained by the ionic polymer of this same monomer at low conversions. Bulk polymerization of II with peroxide catalyst gave a material at 25°C with an expansion of 4.5% and at 60°C an expansion of 5.5%; above the melting point of II (61–62°C) the expansion decreased until at 111°C the density of the monomer and the density of the polymer were the same.

When the 3-methylene derivative was mixed with an equal amount of styrene in the presence of di-*tert*-butyl peroxide and the reaction was stopped at below 30% conversion, a soluble copolymer was obtained containing 79% styrene and 21% of the linear polycarbonate units.

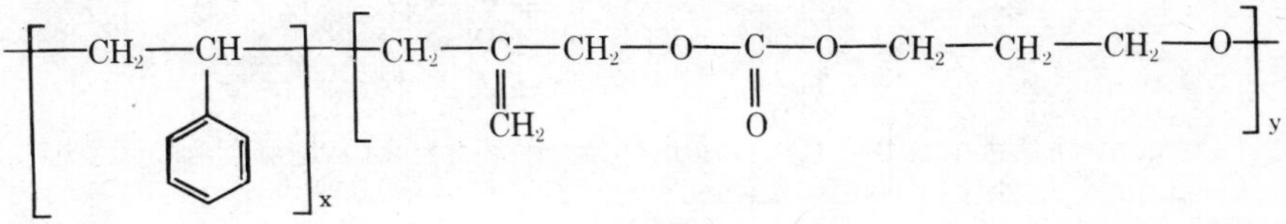

$$\left[-CH_2-CH-\right]_x \left[-CH_2-\underset{\underset{CH_2}{\|}}{C}-CH_2-O-\underset{\underset{O}{\|}}{C}-O-CH_2-CH_2-CH_2-O-\right]_y$$

(with phenyl substituent on the CH)

$$x = 0.79$$
$$y = 0.21$$

By a very similar synthetic scheme, other unsaturated spiro ortho carbonates were prepared.

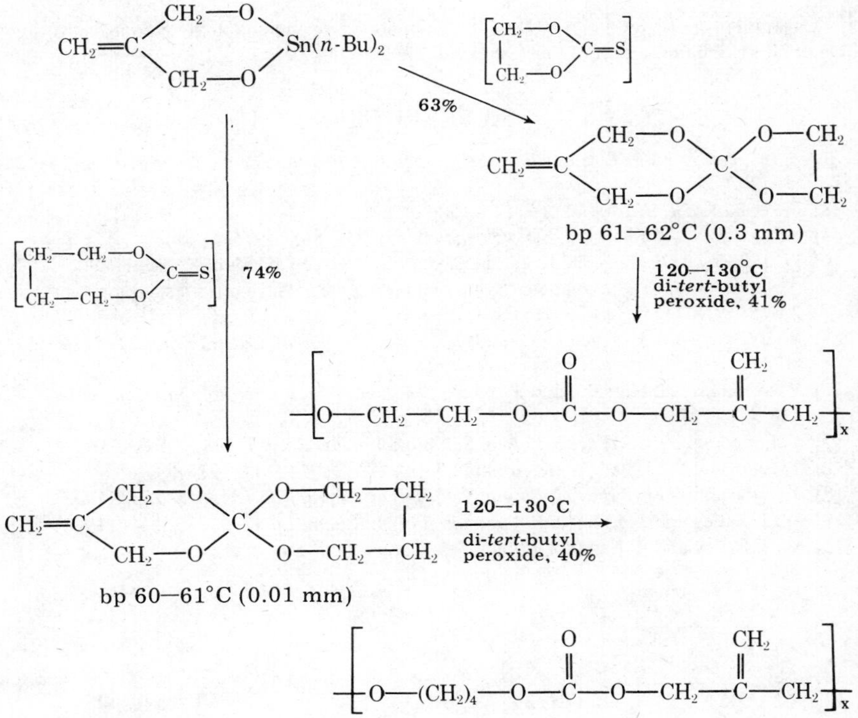

$$CH_2=C \underset{CH_2-O}{\overset{CH_2-O}{<}} Sn(n\text{-}Bu)_2 \quad \xrightarrow{63\%} \quad \left[\underset{CH_2-O}{\overset{CH_2-O}{<}} C=S\right]$$

$$CH_2=C \underset{CH_2-O}{\overset{CH_2-O}{<}} C \underset{O-CH_2}{\overset{O-CH_2}{<}}$$
bp 61–62°C (0.3 mm)

$$\left[\underset{CH_2-CH_2-O}{\overset{CH_2-CH_2-O}{<}} C=S\right] \quad 74\%$$

120–130°C
di-*tert*-butyl
peroxide, 41%

$$\left[-O-CH_2-CH_2-O-\underset{\underset{O}{\|}}{C}-O-CH_2-\underset{\underset{CH_2}{\|}}{C}-CH_2-\right]_x$$

$$CH_2=C \underset{CH_2-O}{\overset{CH_2-O}{<}} C \underset{O-CH_2-CH_2}{\overset{O-CH_2-CH_2}{<}} \quad \xrightarrow[\text{peroxide, 40\%}]{\substack{120-130°C \\ \text{di-}tert\text{-butyl}}}$$
bp 60–61°C (0.01 mm)

$$\left[-O-(CH_2)_4-O-\underset{\underset{O}{\|}}{C}-O-CH_2-\underset{\underset{CH_2}{\|}}{C}-CH_2-\right]_x$$

Bulk polymerization or solution polymerization in chlorobenzene gave soluble polymer if the reaction was stopped at low conversion. Both monomers gave cross-linked resins at high conversions.

A high melting linear polymer was produced by the following set of reactions:

Treatment of this unsaturated monomer with di-*tert*-butyl peroxide at 130°C gave a hard, highly cross-linked resin.

Thus we have demonstrated that unsaturated analogs of the highly strained spiro ortho carbonate monomers will undergo double ring-opening radical polymerization with expansion in volume. Research to determine whether this ring-opening polymerization is a general phenomenon is in progress. Thus we are in the process of preparing unsaturated derivatives of spiro and bicyclo ortho esters to test this hypothesis.

The authors are grateful to the Naval Air Systems Command and to the National Institute of Dental Research for support of this research.

REFERENCES

[1] W. J. Bailey and R. L. Sun, *Am. Chem. Soc., Div. Polym. Chem., Prepr., 13* (1), 281–286 (1972).

[2] W. J. Bailey, *J. Elastoplast., 5,* 142–152 (1973).

[3] W. J. Bailey, *J. Macromol. Sci., Chem., A9* (5), 849–865 (1975).

[4] W. J. Bailey, H. Iwama, and R. Tsushima, *J. Polym. Sci.,* Symposium No. 56, 117–127 (1976); Abstracts of the 4th International Symposium on Cationic Polymerization, Akron, Ohio, June 20–23, 1976.

[5] W. J. Bailey and H. Katsuki, *Am. Chem. Soc., Div. Polym. Chem., Prepr., 14,* 1169–1174 (1973).

[6] W. J. Bailey, H. Katsuki, and T. Endo, *Am. Chem. Soc., Div. Polym. Chem., Prepr., 15,* 445–451 (1974).

[7] T. Endo and W. J. Bailey, *J. Polym. Sci., Polym. Lett. Ed., 13,* 193–195 (1975).

[8] T. Endo and W. J. Bailey, *Makromol. Chem., 176,* 2897–2903 (1975).

[9] T. Endo and W. J. Bailey, *J. Polym. Sci., Polym. Chem. Ed., 13,* 2525–2530 (1975).

[10] W. J. Bailey and T. Endo, *J. Polym. Sci., Polym. Chem. Ed., 14,* 1735–1741 (1976).

[11] W. J. Bailey and K. Saigo, *J. Polym. Sci., Polym. Lett. Ed.,* in press.

SOME RHEOLOGICAL PROPERTIES OF POLYCARBONATE BLENDS

VIORICA DOBRESCU and VALENTIN COBZARU

*Institutul de Cercetări Chimice, ICECHIM, Bucureşti Splaiul
Independenţei Nr.202, Bucureşti, Sector 7, Romania*

SYNOPSIS

The influence of the nature and proportion of components in the blends of polycarbonate (PC) with poly(methyl methacrylate) (PMMA), ABS, polyethylene (PE), and polypropylene (PP) upon the rheological properties in the molten and solid state was studied. It has been observed that the PC–PMMA system is compatible in the molten state, while the other systems are incompatible. For incompatible systems, as blending ratio increases, the melt viscosity of blends goes through a minimum. The experimental results support the idea of reciprocal lubricating effect of polymeric components.

INTRODUCTION

At present one of the most important directions for obtaining new polymeric materials is the formulation of blends of two or more homo- and copolymers.

It is clear now that blending in various proportions of two different polymers may lead to the improvement of the mechanical and physical properties of the finished products.

Thus, good results have been obtained in the field of improving impact strength and stress-cracking, of obtaining materials with higher softening temperatures, or even with qualitatively new properties, such as bicomponent fibers and films.

Inasmuch as the great majority of polymer mixtures are thermodynamically imcompatible in the molten as in the solid state, an intensive study has been made of clarifying the influence of the nature and proportion of components in the blend upon the physical and rheological properties of such multiphase, complex systems [1–6].

The interest in the identification of variables determining the rheological properties of polymer blends is increased by the fact that even the homopolymers are inhomogeneous systems in which the macromolecules have different length and steric defects. In copolymers the inhomogeneity is increased by the dispersion of chemical composition and sequential distribution.

These characteristic features of polymers which make them very similar to blends reveal a new, important way of clarifying the problems connected to the

Journal of Polymer Science: Polymer Symposium 64, 27–42 (1978)
© 1978 John Wiley & Sons, Inc.   0360-8905/78/0064-0027$01.00

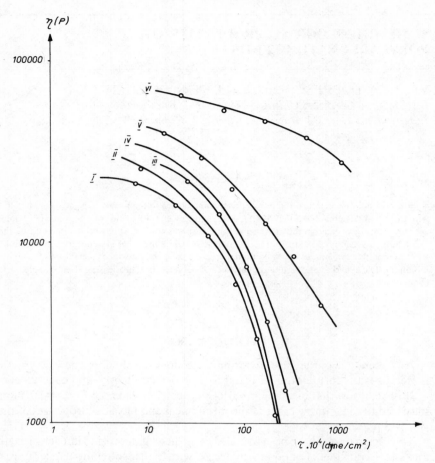

FIG. 1. The dependence of apparent viscosity at 240°C on shear stress for the PC–PMMA system. I, PMMA; II, 10% PC; III, 35% PC; IV, 50% PC; V, 80% PC; VI, PC.

processing and utilization of polymers in general, i.e., the study of the influence of different factors upon the properties of polymer blends.

The purpose of our study was to compare the rheological properties of blends of bisphenol-A polycarbonate with different polymers. In this paper we report the results obtained on such blends with poly(methyl methacrylate), ABS, polypropylene, and polyethylene.

EXPERIMENTAL

Materials investigated were a polycarbonate (PC, Makrolon type 3100), poly(methyl methacrylate) (PMMA, with $\overline{M}_v = 80,000$), an ABS type C, a polypropylene (PP, Daplen AD 10 Natur) and a polyethylene (PE, Argetena 5200 B).

The blends were prepared by tumbling homogenization of pellets and extrusion.

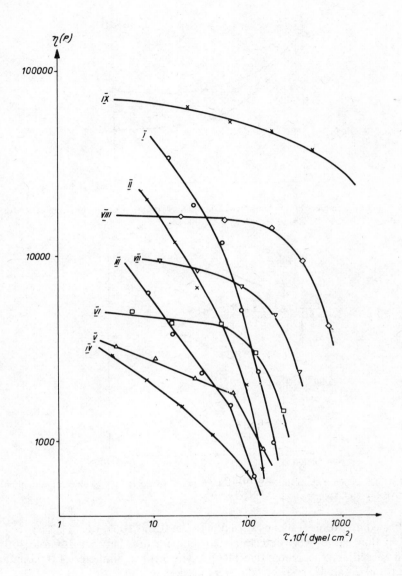

FIG. 2. The dependence of apparent viscosity of 240°C on shear stress for the PC–ABS system. I, ABS; II, 10% PC; III, 20% PC; IV, 35% PC; V, 50% PC; VI, 65% PC; VII, 80% PC; VIII, 90% PC; IX, PC.

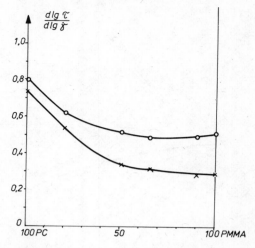

FIG. 3. The dependence of the "non-Newtonian index" ($d \log \tau / d \log \dot\gamma$) on composition for the PC–PMMA system. O, $\dot\gamma = 100$ sec^{-1}; ×, $\dot\gamma = 1000$ sec^{-1}.

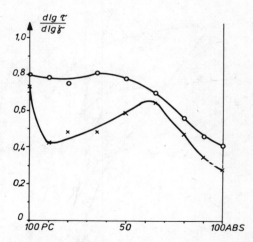

FIG. 4. The dependence of the "non-Newtonian index" ($d \log \tau / d \log \dot\gamma$) on composition for the PC–ABS system. O, $\dot\gamma = 100$ sec^{-1}; ×, $\dot\gamma = 1000$ sec^{-1}.

The composition of blends varied between 0 and 100% PC.

The flow curves were determined at temperatures between 160 and 300°C with a capillary rheometer INSTRON type 3211; a diameter of 0.05 in. and lengths of 2 and 4″, respectively, were used.

The correction for end effects and the Rabinowitsch correction was applied.

Before the measurements the samples were dried for 2 hr at 120–130°C.

We have also determined the flow curves of concentrated solutions of PC–ABS blends in chloroform at 20°C with the coaxial cylinder system SV-1 of a Rotovisko rheometer.

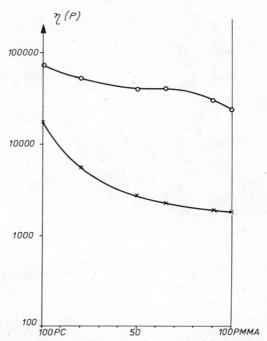

FIG. 5. The dependence of apparent viscosity at 240°C on composition for the PC–PMMA system. O, $\dot{\gamma} = 1\ \text{sec}^{-1}$; ×, $\dot{\gamma} = 1000\ \text{sec}^{-1}$.

The thermomechanical behavior of PC blends was studied using a modified Höppler-consistometer [7] on samples of cylindrical form with $\phi = 10$ mm and $h = 10$ mm cut from injection molded rods.

The compression stress of 31.5 Kgf/cm² was applied periodically for 10 sec at intervals of 5°C, the heating rate being 1°C/min. The deformation developed in 10 sec of stress application was plotted versus the corresponding temperature.

RESULTS AND DISCUSSION

First we shall compare the results obtained for blends of PC with PMMA and ABS.

Figures 1 and 2 show plots of viscosity versus shear stress at 240°C for the PC–PMMA and PC-ABS blends, respectively.

It is seen from these figures that the lower Newtonian viscosity of PMMA is lower than that of PC, while the lower Newtonian viscosity of ABS is higher than that of PC, the flow curve indicating a very pronounced pseudoplastic character of ABS melt.

The essentially different rheological behavior of the two systems is apparent. While the curves for PC–PMMA blends are intermediate between those of the individual components, the curves of PC–ABS blends can be separated in two distinct groups.

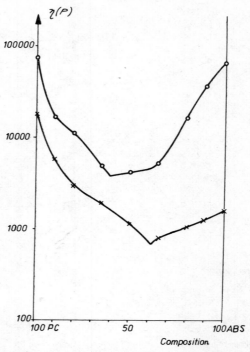

FIG. 6. The dependence of apparent viscosity at 240°C on composition for the PC–ABS system. $O, \dot{\gamma} = 1 \text{ sec}^{-1}; \times, \dot{\gamma} = 1000 \text{ sec}^{-1}$.

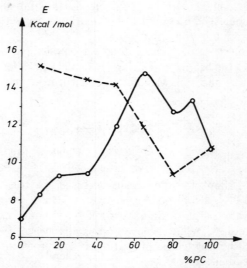

FIG. 7. The dependence of the activation energy of viscous flow at constant shear stress (10^6 dyne/cm^2) on composition. $\times$, PC–PMMA system; O, PC–ABS system.

Thus, blends with up to 60–65% ABS have viscosity–shear stress curves of a form similar to that of pure PC, but considerably lower viscosities. Blends with

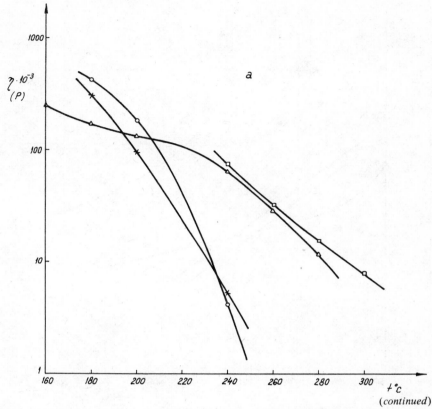

FIG. 8. The variation of viscosity with temperature for the PC–ABS system. (a) $\dot{\gamma} = 1$ sec^{-1}; (b) $\dot{\gamma} = 1000$ sec^{-1}. □, PC; △, ABS; ×, blend with 35% PC; ○, blend with 50% PC.

more than 60–65% ABS have viscosity–shear stress curves similar to that of pure ABS but still lower viscosities.

Figures 3 and 4 show the dependence on composition of a non-Newtonian index ($d \log \tau / d \log \dot{\gamma}$), at two shear rates, for the two systems, PC–PMMA and PC–ABS. The difference between the blends of PC with PMMA and ABS is apparent. The increase of the amount of PC in PMMA gives rise, for both shear rates, to a continuous increase of the value of the non-Newtonian index up to that observed for PC, the more Newtonian component, while the behavior of PC–ABS blends is different at low and high shear rates. At shear rates below 1.5 sec^{-1} the behavior of blends with less than 60–65% ABS is nearly Newtonian. For a shear rate of 10^2 sec^{-1} the non-Newtonian index increases with increasing amount of PC in the blend from the value observed for ABS to that of PC, reaching this value for the blends with 40–45% ABS. For higher shear rates the dependence has a maximum for the blends with 60–65% ABS.

A more illustrative picture of the above mentioned effects is given in Figures 5 and 6 where the apparent viscosity at 240°C is plotted as a function of composition for two shear rates.

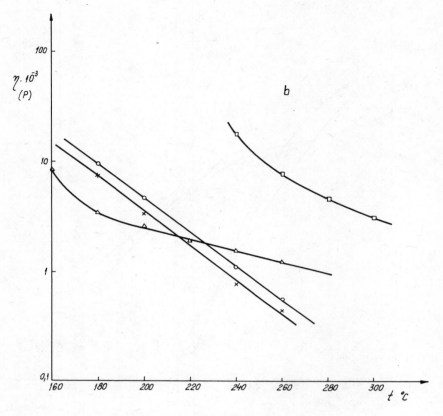

FIG. 8. (*continued from previous page*)

For PC–PMMA system the viscosity of blends increases continuously with increasing PC content, while for PC–ABS blends, in the indicated range of blending ratios, a minimum is observed. The ratio of the viscosity of PC to that of the blend with 65% ABS, corresponding to the minimum on the curve, for a shear rate of 1 sec^{-1}, is about 18.

Of interest is the dependence of the blend viscosity on temperature, given in Figure 7 as a plot of activation energy of viscous flow versus composition, for a constant value of the shear stress, 10^6 dyn/cm^2.

For the temperature range 240–260°C the maximum on the plot for PC–ABS blends is located at an amount of 65% ABS in the blend, indicating that, besides the lowest viscosity, this blend is characterized by the highest dependence of viscosity on temperature.

This observation is illustrated in Figure 8(a) and (b) where the variation of viscosity as a function of temperature is presented for PC, ABS, and the blends with 50% and 65% ABS.

The comparative thermomechanical behavior of the two systems is also interesting.

From the curves given in Figures 9 and 10 and from Figures 11 and 12, where

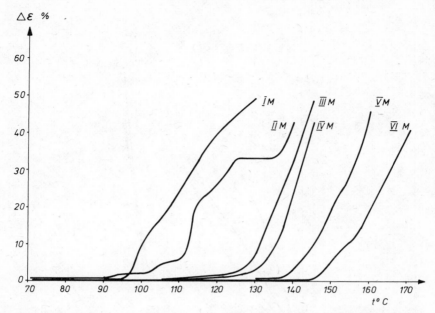

FIG. 9. Thermomechanical curves for the PC–PMMA system. IM, PMMA; IIM, 10% PC; IIIM, 35% PC; IVM, 50% PC; VM, 80% PC; VIM, PC.

the temperatures of equal deformability are plotted versus composition it is seen that for both types of blends the "softening" temperatures increase with increasing PC content.

The plot for PC–ABS blends in Figure 12 has however a somewhat different aspect, with an inflection located again in the range of 40–65% ABS, indicating a sharp change of the nature of the continuous phase of the system.

On the basis of the experimental results the following conclusions may be drawn. In the blends of PC with ABS, PC appears to act as a lubricant for ABS up to an amount of 35–40% PC in the blend, even though in the range of shear rates used the PC has a higher viscosity. The lower values of the glass transition temperatures of the blends of ABS with up to 65% PC support this conclusion (Table I).

In the blends with a higher content of PC, ABS acts as a lubricant for PC. The lubricating effect of the ABS is supported by the lower glass transition and melting temperatures of the PC in the blends, as shown by the data in Table I and by the thermomechanical curves in Figure 10.

We have tried to make evident a similar behavior for concentrated solutions of PC–ABS blends. From Figure 13 it is clear that the "lubricating" effect of the solvent is much stronger than that of the polymeric components.

It is also seen that, similar to the melts, the concentrated solution of ABS has, at low shear rates, a higher viscosity than that of the PC solution of the same concentration.

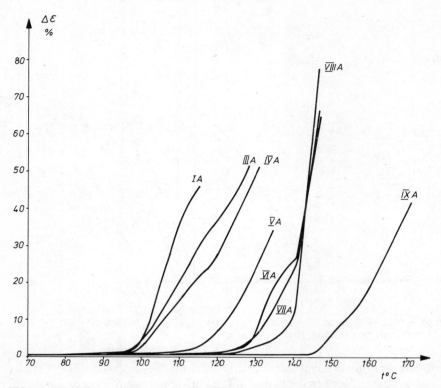

FIG. 10. Thermomechanical curves for the PC–ABS system. IA, ABS; IIIA, 20% PC; IVA, 35% PC; VA, 50% PC; VIA, 65% PC; VIIA, 80% PC; VIIIA, 90% PC; IXA, PC.

TABLE I
The Values of Glass Transition Temperatures Obtained by DTA for the Blends of PC with ABS at a Heating Rate of 4°C/min

Composition	T_g (°C)	
ABS	100	125
10% PC	100	122
20% PC	100	122
35% PC	100	122
50% PC	100	123
65% PC	100	124
80% PC	100	128
90% PC	—	141
PC	—	146

For the PC–PMMA blends it appears that for the entire range of compositions the lubricating role belongs to PMMA, the lower viscosity component, this behavior indicating the compatibility of the two polymers in the molten state.

The other two types of blends studied were the blends of PC with PP and PE,

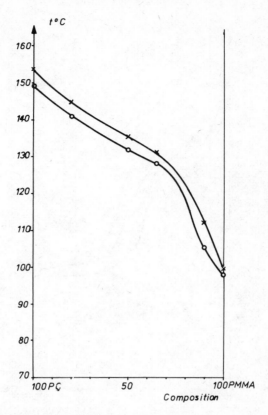

FIG. 11. The dependence of the temperature of equal deformability ($\Delta\epsilon$) on composition for the PC–PMMA system. ×, $\Delta\epsilon$ = 5%, O, $\Delta\epsilon$ = 10%.

two polymers which differ from PC not only chemically but also morphologically.

In Figure 14 the plots of the apparent viscosity at 240°C as a function of shear stress for pure PC and PP and of their blends are compared. The lower Newtonian viscosity of PP is much higher than that of PC and the pseudoplastic character of the PP melt is very pronounced.

Similar to the blends with ABS the curves for the blends of PC with PP are not intermediate between those of pure components, but the viscosity of blends is generally lower than that of PC and PP. An interesting feature is that the blends with 0.5 and 99% PP have higher viscosities than pure PC and pure PP, respectively.

This is apparent from the plot of viscosity at 240°C as a function of composition, in Figure 15. We can see that the blend with 50% PP has, at a shear rate of 1 sec^{-1}, a viscosity about three times lower than that of PP and PC, while the

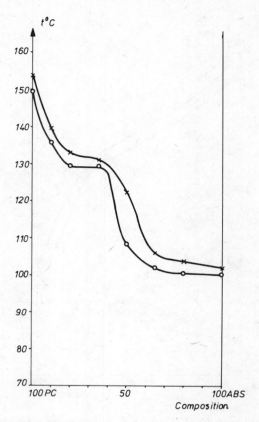

FIG. 12. The dependence of the temperature of equal deformability ($\Delta\epsilon$) on composition for the PC–ABS system. $\times$, $\Delta\epsilon = 5\%$; $\bigcirc$, $\Delta\epsilon = 10\%$.

blend with 0.5% PP has a viscosity of 82,000 P in comparison with 74,000 P, the viscosity of PC. Similarly, the blend with 1% PC in PP has a viscosity of 105,200 P in comparison with 99,100 P for PP.

The same was observed for the blends of PC with PE (Fig. 16).

Owing to the different chemical nature of the components we could make evident the distribution of polyolefins in the blends. From the extruded rods, obtained from the rheometer capillary, we have removed the PC by extraction with methylene chloride, and have found that the polyolefins are distributed as thin threads, oriented in the direction of flow. The threads are better separated in the blends with low polyolefin contents. In the blends with high polyolefin content the PP and PE form feltlike mass formed of thin, short threads.

In Figures 17 and 18 are presented the photographs of the distorted PP threads extracted from the blends with 1% and 5% PP, respectively.

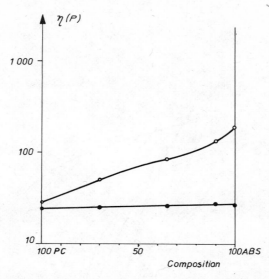

FIG. 13. The dependence of the apparent viscosity of 20% solutions at 20°C on composition for the PC–ABS system. O, $\dot{\gamma} = 1$ sec^{-1}; ●, $\dot{\gamma} = 1000$ sec^{-1}.

CONCLUSIONS

As we have already seen, the examined blends, except those of PC with PMMA, are incompatible in the molten state.

In the shear field the unmiscible components distribute themselves in a manner which is specific to the nature of the components and to the flow rate employed, giving rise to a pronounced anisotropic character of the melt.

For incompatible systems this fact is easily observable but the anisotropy exists even in homopolymers, due to the differences in the length of the macromolecules and to the orientation of the long chains in the shear field.

The anisotropic character of polymers in the molten state is a very important problem from the rheological point of view.

Of course, not only the apparent viscosity is obtained assuming the isotropy of the fluid, but the different degree of anisotropy, as a function of shear rate, will lead to different magnitude errors in the calculus of viscosity.

In addition, a question may arise connected to the flow behavior of polymer melts, namely that of the anisotropy and especially of the variation of the anisotropy as a function of flow rate, as one of the causes of the observed pseudoplastic character of polymer melts.

To support this statement we must recall the well known fact that the broadening of the molecular weight distribution causes the pseudoplasticity to be higher, this being the basis of the use of NNI values for the polydispersity evaluation.

Of course, the problem of the evaluation of anisotropy and its shear rate dependence is not a simple one, not only from the mathematical point of view but also from the experimental one, and we consider that the study of the blends of

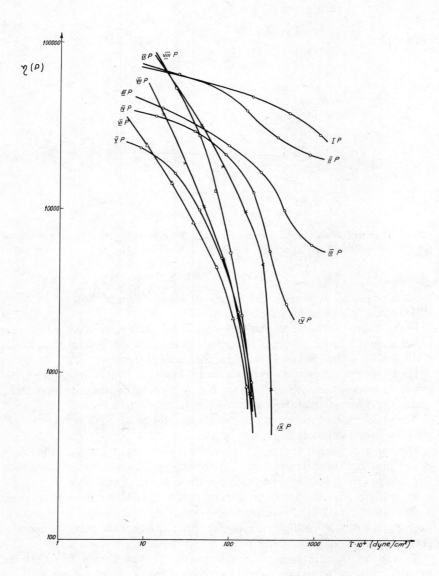

FIG. 14. The dependence of the apparent viscosity of 240°C on shear stress for the PC–PP system. IP, PC; IIP, 0.5% PP; IIIP, 1% PP; IVP, 5% PP; VP, 50% PP; VIP, 75% PP; VIIP, 95% PP; VIIIP, 99% PP; IXP, PP.

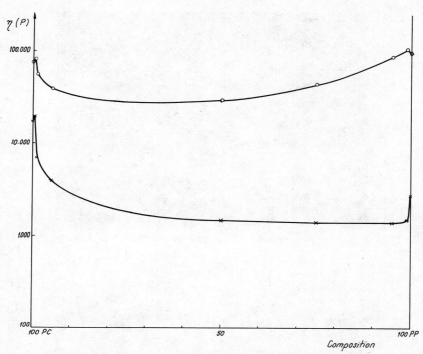

FIG. 15. The dependence of apparent viscosity at 240°C on composition for the PC–PP system. O, $\dot{\gamma} = 1$ sec^{-1}; ×, $\dot{\gamma} = 1000$ sec^{-1}.

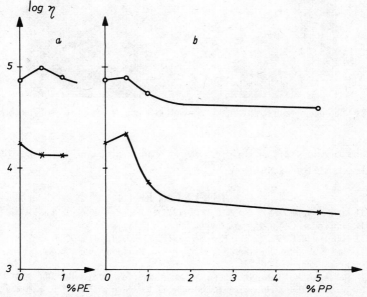

FIG. 16. The dependence of apparent viscosity at 240°C on composition of (a) PC–PE system, (b) PC–PP system. O, $\dot{\gamma} = 1$ sec^{-1}; ×, $\dot{\gamma} = 1000$ sec^{-1}.

FIG. 17. Microphotograph of distorted PP threads extracted from an extruded rod (blend with 1% PP).

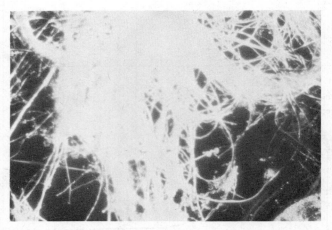

FIG. 18. Microphotograph of distorted PP threads extracted from an extruded rod (blend with 5% PP).

polymers of the same or different types, compatible and incompatible, will provide a basis for its solution.

REFERENCES

[1] C. D. Han, *J. Appl. Polym. Sci., 15,* 2579 (1971).
[2] C. D. Han and T. C. Yu, *J. Appl. Polym. Sci., 15,* 1163 (1971).
[3] C. D. Han and T. C. Yu, *Polym. Eng. Sci., 12,* 77 (1972); *12,* 81 (1972).
[4] C. D. Han, K. U. Kim, J. Parker, N. Siskovic and C. R. Huang, *Appl. Polym. Symp., 20,* 191 (1973).
[5] L. K. Beliakova, V. V. Guzeev and Yu. M. Malinsky, *Plast. Massy, 12,* 39 (1973).
[6] T. I. Ablazova, M. B. Tsebrenko, A. B. V. Yudin, G. V. Vinogradov and B. V. Yarlykov, *J. Appl. Polym. Sci., 19,* 1781 (1975).
[7] V. Dobrescu, *Materiale Plastice, 8,* 631 (1971).

INORGANIC FRAGMENTS IN GRAFT AND BLOCK COPOLYMERS: A REVIEW OF SOLVED AND UNSOLVED PROBLEMS

IONEL HAIDUC*

Chemistry Department, Babes–Bolyai University, R-3400 Cluj-Napoca, Romania

SYNOPSIS

In the first part of the paper some problems related to the formation of inorganic chains in polymeric backbones are discussed and the difficulties in the use of polymerization reactions in inorganic polymer chemistry are reviewed: the lack of suitable inorganic monomers; the tendency of inorganic ring formation; and the tendency of structure unit redistribution. In the second part several possibilities of introducing inorganic fragments in block and graft copolymers are reviewed.

INTRODUCTION

Among the chemical elements, carbon is rather unique in many respects. This is based mostly upon two properties of this element: (a) The ability of carbon atoms to concatenate, which results in the possibility of building most varied structures, with linear, branched, cyclic, low and high molecular carbon–carbon backbones; and (b) the ability of carbon atoms to form interconvertible single, double or triple bonds, involving p_π–p_π orbital overlap.

Polymer chemistry has made use of these two properties in two ways: (a) By building a large variety of high molecular structures; and (b) by using organic monomers, i.e., p_π–p_π carbon–carbon double bond compounds able to polymerize, as starting materials for the synthesis of polymeric structures. The organic monomers are isolatable, stable but reactive compounds, which can be polymerized by various mechanisms. The ability of chemically different monomers to copolymerize greatly increases the possibilities of building polymeric structures of most varied types.

The organic polymers satisfy many technical needs, but not all of them. For certain special purposes, especially for extremely low and high temperature uses, the organic polymers fail to meet all the requirements. Since many inorganic compounds, e.g., silicates, borates, and phosphates, are much more stable thermally than the organic compounds, the logical conclusion was that building polymers with inorganic backbones or introducing inorganic fragments in a polymeric backbone, might improve the thermal stability, or, in a more general manner, might change the properties of a polymer in a useful way. These ideas,

Journal of Polymer Science: Polymer Symposium 64, 43–55 (1978)
© 1978 John Wiley & Sons, Inc.
0360-8905/78/0064-0043$01.00

which emerged in the late thirties, resulted eventually in the development of silicon polymer chemistry and of other types of organo–element polymers. These polymers have met considerable commercial success in several application fields, which stimulated new research.

INORGANIC POLYMERIZATION

The introduction of inorganic building units or fragments in a polymer would be a fairly simple matter if inorganic monomers, i.e., compounds containing polymerizable double bonds between atoms other than carbon, were available. Unfortunately, this is not so, for several reasons.

The first problem arises in connection with the very limited number of available monomers with p_π–p_π double bonds, which can be isolated. Two examples are the monomeric phosphazenes [1–3] and the monomeric borazenes [4]:

$$Cl_3P{=}N{-}R \qquad C_6F_5{-}B{=}N{-}R$$

which can be isolated in certain cases. However, these monomers are of little use in polymer chemistry, since their polymerization does not go further than the step of cyclic dimer and trimer, respectively [1]:

It is rather difficult to find more examples of isolatable inorganic monomers of the type cited.

Some monomers, especially organosilicon types, e.g., silicon–oxygen and silicon–nitrogen double bonded species:

have been identified [5] only as unstable, transient species and are not isolatable as compounds. Therefore, in this respect there is no hope for having Si$=$O and Si$=$N monomers, which could be used in the synthesis of polymers. This difficulty can be explained in a simple way, by taking into account the hybridization which can be achieved by the orbitals of various nonmetal atoms. Thus, boron achieves only sp^2 and sp^3 stable hybridizations, in which the atom is three- and four-coordinated, respectively, and forms only single bonds; therefore, with the exception mentioned, boron does not form monomers of the R—B$=$X type, where X would be O, N, S. Similarly, for silicon only sp^3 hybridization is stable, which corresponds to four single bonds with other substituents. The formation of a double bonded monomer would require sp^2 hybridization, which is unstable for this element. The reason why p_π–p_π double bonding, i.e., sp hybridization for boron, and sp^2 hybridization for silicon, cannot be achieved, is not yet fully understood.

In contrast to the examples mentioned above, there are many phosphorus compounds (as well as arsenic, sulfur, and selenium compounds) which contain double bonds between elements other than carbon and are perfectly stable as monomers. These include, for example, triorganylphosphine oxides [6], imides [7] and sulfides [8], or sulfoxides [9] and sulfones [10]:

$$O{=}P\diagdown^{R}_{\diagdown R}{-}R \qquad HN{=}P\diagdown^{R}_{\diagdown R}{-}R \qquad S{=}P\diagdown^{R}_{\diagdown R}{-}R \qquad O{=}S\diagup^{R}_{\diagdown R} \qquad {}^{O}_{O}{\gtrless}S\diagup^{R}_{\diagdown R}$$

In such monomers p_π–d_π double bonds are involved, and the hybridization of the four-coordinated phosphorus atom is of the sp^3 type. This is a very stable hybridization state, which makes possible the isolation of the corresponding monomers. Polymerization of these double bond monomers would require the transition to a five-coordinate, sp^3d hybridized phosphorus atom. Such a type of bonding, although not forbidden, is less stable, and as a result, the monomers mentioned do not polymerize. The only exception, already quoted, involves the trichlorophosphazenes, $Cl_3P{=}NR$, which tend to dimerize [1]. Therefore, the phosphorus-containing double bonded monomers are again of little use in polymer chemistry, and cannot be used as starting materials in the synthesis of high molecular compounds containing P–O, P–N or P–S building units. Similarly, the sulfur(IV) and sulfur(VI) derivatives do not polymerize.

A particular feature of inorganic chemistry is the tendency of ring formation, in many cases when double or triple bonded monomers might be expected. This is similar to the tendency of aldehydes, particularly formaldehyde, to form cyclic trimers. Thus, in many reactions in which compounds are formed whose composition (elemental analysis) would suggest a double or triple bond monomer, actually they are cyclic dimers, trimers, and tetramers. Thus, cyclic rather than monomeric species are formed in these cases, as illustrated for siloxanes, silazanes, borazines, phosphazenes and many other types [1]:

$$R_2SiX_2 + H_2O \rightarrow (R_2SiO)_n + 2\,HX \qquad n = 4, 5, 6. . .$$

$$R_2SiCl_2 + 2NH_3 \rightarrow (R_2SiNH)_n + NH_4Cl \qquad n = 3\ \&\ 4$$

$$RBX_2 + 2\,NH_3 \rightarrow (RBNH)_3 + NH_4Cl$$

$$PCl_5 + NH_4Cl \rightarrow (Cl_2PN)_n + 4\,HCl \qquad n = 3, 4. . .$$

Silicon, the element next to carbon in Group IV, is also unable to form double bond monomers of the type $R_2Si{=}SiR_2$ and only cyclic oligomers are obtained instead [11]. Similarly, no monomers such as $R_2Ge{=}GeR_2$, $RP{=}PR$, or $RAs{=}AsR$ can be isolated and only their cyclic counterparts have been obtained [1].

The inorganic ring systems are, however, interesting for polymer chemistry, in two ways:

TABLE I
Ring-Chain Equilibrium Constants[a]

Repeating Unit	K_6^0	Temperature
—Me$_2$SiS—	3000	25°C
—CH$_2$CH$_2$—	1200	25°
—Me$_2$SiNMe—	300	120°
—Me$_2$GeS—	50	120°
—ClAsNMe—	15	27°
—CH$_2$O—	4	120°
—Me$_2$GeO—	0.9	35°
—(CF$_3$CH$_2$CH$_2$)MeSiO—	0.5	50–150°
—Cl$_2$PN—	0.1	146°
—O$_2$SO—	0.1	24°
—(NaO)PO$_2$—	0.04	500°
—Me$_2$SiO—	0.02	80°
—FAsO—	0.05	r.t.
—FAsNMe—	0.05	r.t.
—S—S—	0.05	

[a] The equilibrium constants refer to the conversion of six-membered rings into a linear polymer.

(a) Some of them, e.g., elemental sulfur (cyclooctasulfur), cyclosiloxanes, and cyclophosphazenes, can be converted into linear high polymers, by catalytic or thermal (in certain instances even radiation induced) ring-opening polymerizations [1]. This is similar to the ring polymerization known for certain organic heterocyclic systems, e.g., paraformaldehyde.

(b) The inorganic rings can be incorporated as building blocks in the structures of several types of polymers, thus affording the synthesis of block polymers containing cyclic units, interconnected in a high molecular network by different organic or inorganic bridging chains.

The ability to polymerize is, however, not general, and some rings are known, which cannot be converted into linear polymers by ring-opening polymerization reactions, e.g., borazines (B$_3$N$_3$), boroxines (B$_3$O$_3$), cyclosilazanes (Si$_3$N$_3$), and cyclosilthianes (Si$_3$S$_3$). Such rings may only serve as building blocks for cyclolinear or ring-block copolymers.

The ability of a ring system to polymerize is described by an appropriate ring-chain equilibrium constant. The ring-opening polymerization reactions mentioned are equilibrium controlled [1], and the corresponding equilibrium constants have been determined in certain cases [12], as shown in Table I.

Ring systems with high ring-chain equilibrium constants do not polymerize readily. The equilibrium is strongly shifted towards the cyclic species. The corresponding linear polymers are broken down to cyclic molecules. The rings with small ring-chain equilibrium constants, such as cyclosiloxanes, cyclophosphazenes, sulfur trioxide, and sodium metaphosphate, are readily polymerized by ring opening, to form linear polymers. These polymerizations occur via thermodynamically controlled exchange of parts between molecules (redistribution reactions) rather than kinetically controlled reactions (the latter being typical for organic polymerizations). Such equilibrium controlled reactions are also possible for those rings which do not polymerize readily, such as cyclosilthianes [13], and results in an interconversion of rings of different sizes, without building a high polymeric chain:

$$2 \ (R_2SiS)_3 \rightleftarrows 3 \ (R_2SiS)_2 \qquad K_{3,2}^{00} = 7200 \text{ (at } 25°C)$$
$$3.5 \text{ (at } 200°C)$$

It can also be achieved for cyclogermoxanes [14] and other cyclic species of different ring sizes [1].

$$3 \ (Me_2GeO)_4 \rightleftarrows 4 \ (Me_2GeO)_3 \qquad K_{4,3}^{00} = 6.0 \text{ (at } 35°C)$$

Even rings made up of different building units may exchange parts, to undergo ring–ring equilibria, involving a redistribution of structural moieties, but no linear polymers are formed in the process. Such examples are known for the cyclo-silazane–cyclosilthiane or cyclogermathiane–cyclosilthiane pairs of rings [13] (R = Me):

$$(R_2SiNR)_3 \ + \ (R_2SiS)_3 \ \rightleftarrows \ (R_2SiNR)_2(R_2SiS) \ + \ (R_2SiNR)(R_2SiS)_2$$

$$(R_2GeS)_3 \ + \ (R_2SiS) \ \rightleftarrows \ (R_2SiS)_2(R_2GeS) \ + \ (R_2SiS)(R_2GeS)_2$$

To summarize the discussion presented above, it can be concluded that monomer polymerization or copolymerization seems to be of little use in building inorganic polymers, or in introducing inorganic fragments into block and graft copolymers. Inorganic ring systems may play, to some extent, a role similar to that of organic monomers (by ring-opening polymerizations), but the inability of certain inorganic rings to polymerize and their equilibrium-controlled exchange of parts (redistribution reactions) have also to be taken into account. Under these circumstances, which could be the ways of introducing inorganic fragments in graft and block copolymers? There are several possibilities:

(a) *Polymerization reactions* may serve this purpose in two ways: to build a polymeric carbon–carbon backbone, by polymerization of vinyl derivatives of inorganic fragments (cyclic or linear), followed by further grafting on the inorganic side chains; to build an inorganic polymeric backbone, by ring-opening polymerization of some inorganic ring systems, bearing adequate substituents for further grafting of other chains on the inorganic main chain. Radiation processes are of particular interest for the synthesis of graft polymers, containing inorganic fragments.

(b) *Addition reactions,* e.g., of hydridosiloxanes to vinyl derivatives of silicon or possibly phosphorus or other elements, can also be used for the synthesis of graft and block copolymers.

(c) *Polycondensation reactions* are of much use in the synthesis of graft and block copolymers containing inorganic fragments. They have been applied to the synthesis of copolymers containing two different types of chains or blocks, as well as for grafting siloxane chains to polyalkylenes or other organic polymers.

GRAFT COPOLYMERS VIA POLYMERIZATION REACTIONS

The most direct way to prepare graft polymers is based upon the polymerization of a vinyl group-terminated polymer (macromer), containing inorganic fragments. Thus, vinyl-terminated siloxane macromers, containing *p*-vinylphenyl (styryl) groups, can be polymerized catalytically, to give a polystyrene, grafted with polysiloxane chains [15]:

$$CH{=}CH_2 \qquad \xrightarrow{\text{catalyst}} \qquad \cdots{-}CH{-}CH_2{-}CH{-}CH_2{-}\cdots$$

(polysiloxane-grafted polystyrene structure: each aromatic ring bearing $-SiMe_2-O-SiMe_2-O-\vdots$ chains)

A second approach is to use appropriate transformations of a preformed organic polymer, the inorganic fragment being introduced after the polymerization step. Thus, *p*-iodostyrene is first polymerized to poly(*p*-iodostyrene), which can be readily metallated in the aromatic nucleus with lithium metal or organolithium reagents (via halogen–metal exchange); the lithiated polystyrene reacts with trimethyl-chlorosilane, to give a silylated polymer [16]. There is no reason why this reaction could not be used with long-chain chlorine-terminated polysiloxanes, to give a silylated graft copolymer. Such graft polymers have been indeed obtained by the reaction of a metallated polystyrene with a poly(dimethylcyclosiloxane) [17].

$$\cdots{-}CH{-}CH_2{-}\cdots \quad \xrightarrow{Na^+C_{10}H_8^-} \quad \cdots{-}CH{-}CH_2{-}\cdots \quad \xrightarrow{(Me_2SiO)_8} \quad \cdots{-}CH{-}CH_2{-}\cdots$$

(I)   (M)   ($-SiMe_2-O-SiMe_2-O-\vdots$)

Vinylsilanes containing reactive groups can be polymerized catalytically, to give a polyethylene, substituted with functional organosilicon groups. These can be further converted into polymeric chains, thus affording grafting. An interesting example is the synthesis of a ladder polymer, containing two parallel chains, one organic (polyethylene), the other inorganic (siloxane) [18]:

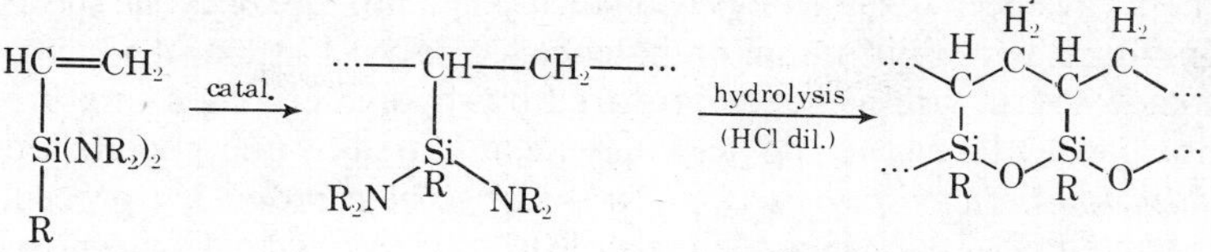

$$HC{=}CH_2 \quad \xrightarrow{\text{catal.}} \quad \cdots{-}CH{-}CH_2{-}\cdots \quad \xrightarrow[\text{(HCl dil.)}]{\text{hydrolysis}}$$

(with $Si(NR_2)_2$ and R substituents on the left vinyl; Si with R_2N, R, NR_2 on the middle; and the hydrolyzed siloxane ladder structure on the right)

If suitable catalysts for the polymerization of vinyl-chlorosilanes were found, it could be possible to prepare silicon–nitrogen analogues:

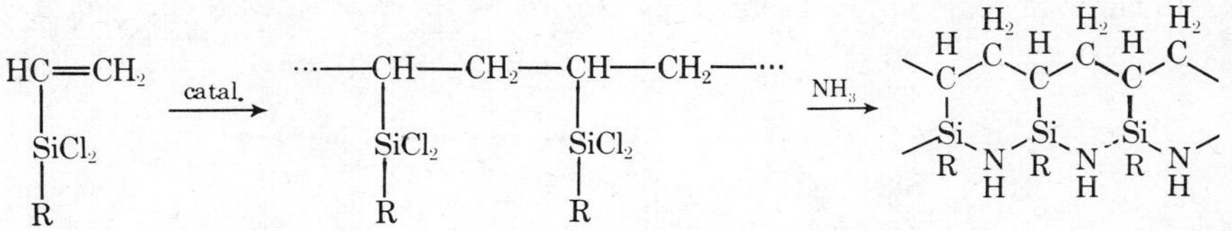

$$HC{=}CH_2 \quad \xrightarrow{\text{catal.}} \quad \cdots{-}CH{-}CH_2{-}CH{-}CH_2{-}\cdots \quad \xrightarrow{NH_3}$$

(with $SiCl_2$, R on left; $SiCl_2$, R groups on middle; and the Si–N ladder structure on the right)

Such Si–N ladder polymers would be of considerable interest, since parallel grafting to a polyethylene chain should stabilize the polysilazane chain, which otherwise tends to break down to cyclosilazanes, and thus cannot exist independently.

For building inorganic chains capable of further grafting, the vinylcyclosiloxanes are of interest. They can be polymerized or copolymerized via ring-opening reactions [19, 20], to give a siloxane chain bearing vinyl groups, and the latter can be used for grafting of organic chains:

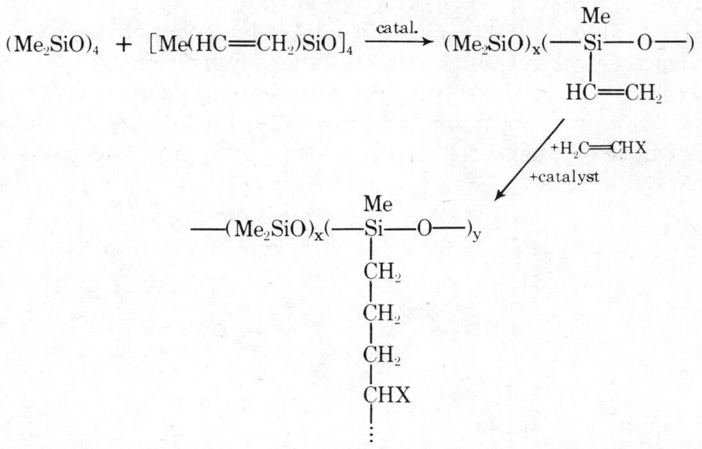

$$(Me_2SiO)_4 + [Me(HC{=}CH_2)SiO]_4 \xrightarrow{\text{catal.}} (Me_2SiO)_x\Big({-}\underset{\underset{HC=CH_2}{|}}{\overset{\overset{Me}{|}}{Si}}{-}O{-}\Big)$$

$$\xrightarrow[\substack{+H_2C=CHX \\ +\text{catalyst}}]{}$$

$${-}(Me_2SiO)_x\Big({-}\underset{\underset{\underset{\underset{CHX}{|}}{\overset{\vdots}{CH_2}}}{\underset{\overset{CH_2}{|}}{\overset{\overset{CH_2}{|}}{}}}}{\overset{\overset{Me}{|}}{Si}}{-}O{-}\Big)_y$$

An interesting recent development in the field of copolymerization is based upon the fact that both vinylic groups and cyclosiloxanes are sensitive to similar polymerization catalysts, e.g., organolithium compounds. This affords simultaneous polymerization by ambident initiators, processing two anionic sites of unequal reactivity (a vinylic and a siloxane site). Thus, dilithiobenzophenone or 1,1-diphenylhexyllithium can convert a mixture of hexamethylcyclotrisiloxane and methyl methacrylate into a block copolymer [21]. Similarly, block co-polymers have been obtained by copolymerizing a mixture of cyclic siloxane and vinylpyridine with dilithiumnaphthalene as catalyst [22]. A block copolymer of unknown structure has been prepared by copolymerizing a mixture of cy-closiloxanes with styrene and acrylonitrile, initiated by cumyl peroxide [23].

Radiation induced polymerization seems to be a promising procedure for preparing block and graft copolymers of the type discussed. Cyclosiloxanes, as well as cyclophosphazenes, can be polymerized under the action of ionizing radiations, especially at temperatures near their melting point, in the solid state [1]. It seems that both ring-opening polymerization and coupling via proton abstraction occurs in this process [24–26].

The radiation-induced polymerization of vinylcyclosiloxanes is of particular interest [27]. Thus, in the liquid state vinyl polymerization occurs, to give a polyethylene chain substituted with cyclosiloxane moieties; near the melting point, in the solid state, ring-opening polymerization occurs, to give a siloxane polymeric chain, bearing vinylic groups as substituents. Both these types of polymers can be of interest for further grafting by appropriate reactions.

The vinyl-substituted cyclosiloxanes, required for these reactions are readily prepared by hydrolysis [28–30] or cohydrolysis [31] of vinylchlorosilanes.

Gamma radiation-induced grafting of methyl methacrylate to a siloxane polymer was reported [32]. Probably the polymer thus obtained contains polyacrylate chains interconnected by siloxane groups. In a similar fashion, poly(dichlorophosphazene) polymers, treated with styrene monomer, under the action of gamma rays gave a phosphazene–polystyrene graft copolymer [33].

Thermomechanical treatment may have in certain instances an effect similar to that of ionizing radiations. Thus, polydimethylsiloxane polymers mixed with poly(tetrafluoroethylene) and subjected to heat and mechanical treatment, gave a graft copolymer with improved mechanical and chemical properties [34–36].

Another possible source of excitation in grafting could be the *cold plasma (glow discharge)*. It was used for the polymerization of octamethylcyclotetrasiloxane [37], as well as of many vinylic monomers, which suggests its possible use in the simultaneous polymerization of cyclosiloxane rings and their vinylic substituents (in vinylcyclosiloxanes) to yield graft copolymers.

GRAFT COPOLYMERS VIA ADDITION REACTIONS

Interesting copolymers were obtained by using the catalytic addition of olefins to hydrido-substituted silanes or siloxanes. In a first example, a polysiloxane, containing R-Si-H groups (prepared by copolymerization of two cyclosiloxanes), when treated with olefins and a polymerization catalyst, gave a graft copolymer, in which polyethylene chains were attached to the main siloxane chain [38, 39].

Similarly, a hydrido-substituted polysiloxane reacted with a vinyl group terminated siloxane in the presence of H_2PtCl_6 to give a graft copolymer with a basic polysiloxane network [40].

$$
\begin{array}{c}
\quad\quad\quad\quad\quad\quad\quad HC{=\!=}CH_2 \\
\quad\quad\quad R\quad\quad Ph \quad\quad\quad\quad | \quad\quad\quad\quad\quad\quad\quad\quad R\quad\quad Ph \\
R(Si\!-\!O)_x(Si\!-\!O)_y SiR_3 \ + \ SiR_2 \ \xrightarrow{\ H_2PtCl_6\ } \ R(Si\!-\!O)_x{}'(Si\!-\!O)_y SiR_3 \\
\quad\quad H\quad\quad Ph \quad\quad\quad\quad | \quad\quad\quad\quad\quad\quad\quad\quad |\quad\quad\quad Ph \\
\quad\quad\quad\quad R = CH_3 \quad\quad\quad\ O \quad\quad\quad\quad\quad\quad\quad\quad CH_2 \\
\quad\quad\quad\quad\quad\quad\quad\quad\quad\quad\quad\quad | \quad\quad\quad\quad\quad\quad\quad\quad\quad | \\
\quad\quad\quad\quad\quad\quad\quad\quad\quad\quad\quad SiR_2 \quad\quad\quad\quad\quad\quad\quad CH_2 \\
\quad\quad\quad\quad\quad\quad\quad\quad\quad\quad\quad\quad | \quad\quad\quad\quad\quad\quad\quad\quad\quad | \\
\quad\quad\quad\quad\quad\quad\quad\quad\quad\quad\quad\ O \quad\quad\quad\quad\quad\quad\quad\quad SiR_2 \\
\quad\quad\quad\quad\quad\quad\quad\quad\quad\quad\quad\quad \vdots \quad\quad\quad\quad\quad\quad\quad\quad\quad | \\
\quad\quad\quad\quad\quad\quad\quad\quad\quad\quad\quad\quad\quad\quad\quad\quad\quad\quad\quad\quad\ O \\
\quad\quad\quad\quad\quad\quad\quad\quad\quad\quad\quad\quad\quad\quad\quad\quad\quad\quad\quad\quad\quad | \\
\quad\quad\quad\quad\quad\quad\quad\quad\quad\quad\quad\quad\quad\quad\quad\quad\quad\quad\quad SiR_2 \\
\quad\quad\quad\quad\quad\quad\quad\quad\quad\quad\quad\quad\quad\quad\quad\quad\quad\quad\quad\quad\quad | \\
\quad\quad\quad\quad\quad\quad\quad\quad\quad\quad\quad\quad\quad\quad\quad\quad\quad\quad\quad\quad\ O \\
\quad\quad\quad\quad\quad\quad\quad\quad\quad\quad\quad\quad\quad\quad\quad\quad\quad\quad\quad\quad\quad \vdots
\end{array}
$$

The linear vinylsiloxanes (macromers), required for such reactions, are readily prepared by cohydrolysis of α,ω-dichlorosiloxanes with vinylchlorosilanes [41]. Therefore, reactions of this type could be much extended. Indeed, some other interesting examples have been described, involving the grafting of vinyl-terminated polyoxyethylenes to hydridosiloxanes [42], addition of vinylcyclosiloxanes to hydridocyclosiloxanes [43], synthesis of block copolymers by interconnecting cyclosiloxanes (bearing vinyl groups) with linear polysiloxane chains (end-terminated with Si-H groups) [44] or by interconnecting vinylcyclosiloxanes via silazane chains [45].

It would be of interest to extend such reactions to vinyl-substituted cyclosilazanes, which are readily available by ammnolysis or coammonolysis of various vinylchlorosilanes [46–48]. However, so far very few such attempts have been made.

BLOCK AND GRAFT COPOLYMERS VIA POLYCONDENSATION REACTIONS

The polycondensation reaction is probably the major synthetic method in the preparation of inorganic polymeric structures, and it can be used for the synthesis of block and graft copolymers in several ways, affording rather varied types. Probably the most commercially successful copolymers, obtained by such reactions, are the Dexsil copolymers, made of carborane and siloxane blocks [49, 50].

$$
\begin{array}{c}
\ R\quad\quad\quad\quad\quad\quad R \quad\quad\quad\quad\quad R\quad\quad\quad R \\
RO\!-\!Si\!-\!CB_{10}H_{10}C\!-\!Si\!-\!OR \ + \ ClSi\!-\!(O\!-\!Si)_nCl \ \xrightarrow[\substack{heat \\ -RCl}]{FeCl} \\
\ R\quad\quad\quad\quad\quad\quad R \quad\quad\quad\quad\quad R\quad\quad\quad R \\[1.2em]
\quad\quad\quad\quad\quad\quad R\quad\quad\quad\quad\quad R\quad\quad R\quad\quad\quad R \\
\quad\quad\quad [-\!O\!-\!Si\!-\!CB_{10}H_{10}C\!-\!Si\!-\!O\!-\!Si\!-\!(O\!-\!Si)_n\!-\!]_x \\
\quad\quad\quad\quad\quad\quad R\quad\quad\quad\quad\quad R\quad\quad R\quad\quad\quad R
\end{array}
$$

Other carborane-containing block copolymers were made by the condensation of a cyclosilazoxane (with elimination of ammonia) with hydroxymethyl-derivatives of carboranes [51].

$$R_2Si\!-\!O\!-\!SiR_2$$
$$| \qquad |$$
$$O \qquad O \quad + \quad HO\!-\!CH_2\!-\!\overset{R}{\underset{R}{Si}}\!-\!\overset{}{\underset{B_{10}H_{10}}{C\!-\!C}}\!-\!\overset{R}{\underset{R}{Si}}\!-\!CH_2\!-\!OH \quad \xrightarrow{-NH_3}$$
$$| \qquad |$$
$$R_2Si\!-\!O\!-\!SiR_2$$

$$[-(\overset{R}{\underset{R}{Si}}\!-\!O)_4CH_2\!-\!SiR_2\!-\!\overset{}{\underset{B_{10}H_{10}}{C\!-\!C}}\!-\!SiR_2\!-\!CH_2\!-\!O\!-\!]_x$$

Other polycondensation reactions used in the synthesis of block copolymers involve the condensation of bicyclic (or polycyclic) chlorosiloxanes with α,ω-dihydroxysiloxanes [52, 53],

$$\begin{array}{c} O\!-\!SiR_2\!-\!O \\ ClSiR_2O\!-\!Si\overset{\nearrow}{\underset{\searrow}{\quad}}O\!-\!SiR_2\!-\!O\overset{\quad}{\underset{\quad}{\quad}}Si\!-\!O\!-\!SiR_2Cl \\ O\!-\!SiR_2\!-\!O \end{array}$$

$$+ \quad HO\!-\!SiR_2\!-\!(OSiR_2)_x\!-\!OSiR_2\!-\!OH \quad \xrightarrow[-HCl]{} \quad \text{Block Copolymer}$$

condensation of α,ω-diethoxysiloxanes with polyethylene terephthalate [54], α,ω-bis(dialkylamino)polysiloxanes with oxyalkylene polymers [55–57], silarylene diols with siloxanes [58], or borazine derivatives with various linking reagents [59, 60], as shown in the following two examples:

$$\overset{Ph}{\underset{OEt}{EtO\!-\!Si\!-\!O}}\!-\!(\overset{R}{\underset{R}{Si}}\!-\!O)_x\overset{Ph}{\underset{OEt}{Si}}\!-\!OEt$$

$$+ \quad HO(CH_2CH_2O\overset{}{\underset{O}{C}}\!-\!\bigcirc\!-\!\overset{}{\underset{O}{CO}}\!-\!)_yCH_2CH_2OH \quad \xrightarrow[-EtOH]{} \quad \text{Block Copolymer}$$

$$\overset{R}{\underset{R}{R_2N\!-\!(Si}}\!-\!O)_{\bar{x}}\overset{R}{\underset{R}{Si}}\!-\!NR_2 \quad + \quad HO(CH_2CH_2O)_yH \quad \xrightarrow[-HNR_2]{} \quad \text{Block Copolymer}$$

The examples given deal mainly with block copolymers, but the reaction can be adapted for the synthesis of similar graft copolymers.

CONCLUSIONS

This review, by no means exhaustive, was meant to illustrate a few points related to the synthesis of graft and block copolymers containing inorganic fragments. Although the traditional types of reactions known in the chemistry of organic polymers cannot always be used (e.g., the copolymerization of double bond monomers) there are adequate means of introducing inorganic fragments into a polymeric structure. Space limitations did not afford any comment about the properties of the resulting polymers, to illustrate the way in which they were modified by building organic–inorganic polymeric hybrids, but it can be stated that considerable modification can be achieved. Therefore, this seems to be a field which deserves further exploration.

REFERENCES

[1] I. Haiduc, *The Chemistry of Inorganic Ring Systems*, Wiley, New York, 1970.
[2] I. N. Zhmurova and A. V. Kirsanov, *Zh. Obshch. Khim., 30*, 3044 (1960); *J. Gen. Chem. USSR (Engl. transl.), 30*, 3018 (1960).
[3] M. Bermann, *Ad. Inorg. Chem. Radiochem., 14*, 1 (1972).
[4] P. I. Paetzold and W. M. Simpson, *Angew. Chem., 78*, 825 (1966).
[5] L. H. Sommer et al., *J. Am. Chem. Soc., 98*, 618 (1976); *97*, 7371 (1975).
[6] H. R. Rays and D. J. Peterson, in *Organic Phosphorus Compounds*, G. M. Kosolapoff and L. Maier, Eds., Wiley-Interscience, New York, 1972, Vol. 3, p. 341.
[7] E. Fluck and W. Haubold, in *Organic Phosphorus Compounds*, G. M. Kosolapoff and L. Maier, Eds., Wiley-Interscience, New York, 1973, Vol. 6, p. 579.
[8] L. Maier, in *Organic Phosphorus Compounds*, G. M. Kosolapoff and L. Maier, Eds., Wiley-Interscience, New York, 1972, Vol. 4, p. 1.
[9] K. J. Wynne and I. Haiduc, in *Methodicum Chimicum*, H. Zimmer and K. Niedenzu, Eds., Thieme, Stuttgart, 1976, Band 7, S. 683.
[10] I. Haiduc and K. J. Wynne, in *Methodicum Chimicum*, H. Zimmer and K. Niedenzu, Eds., Thieme, Stuttgart, 1976, Band 7, S. 699.
[11] E. Hengge, *Topics Curr. Chem., (Fortschr. Chem. Forsch.), 51*, 1 (1974).
[12] R. M. Levy and J. R. Van Wazer, *J. Chem. Phys., 45*, 1824 (1966).
[13] K. Moedritzer and J. R. Van Wazer, *J. Phys. Chem., 70*, 2030 (1966).
[14] K. Moedritzer and J. R. Van Wazer, *Inorg. Chem., 4*, 1753 (1965).
[15] G. Greber and E. Roese, *Makromol. Chem., 55*, 96 (1962).
[16] B. Houel, *C.R., 248*, 800 (1959); *250*, 2209 (1960).
[17] A. Zilhka, A. Eisenstadt and A. Ottolenghi, *Bull. Res. Counc. Isr., 1962* (5), 159.
[18] J. L. Speier, *Australian Patent*, 443,991 (1973).
[19] A. R. Gilbert and S. W. Kantor, *J. Polym. Sci., 40*, 35 (1959).
[20] S. W. Kantor, R. C. Osthoff and D. T. Hurd, *J. Am. Chem. Soc., 77*, 1685 (1955).
[21] P. C. Juliano, D. E. Florian, R. W. Hand and D. D. Karthunen, *Proc. Sagamore Army Mater. Res. Conf.*, 19th, 1972, pp. 61–83; *Chem. Abstr., 82*, 31561 (1975).
[22] J. W. Dean, U.S. Patent, 3,875,254 (1975).
[23] J. R. Hilliard, U.S. Patent, 3,898,300 (1975).
[24] W. J. Burlant and C. Taylor, *J. Polym. Sci., 41*, 547 (1959).
[25] E. J. Lawton, W. T. Grubb, and J. S. Balwit, *J. Polym. Sci., 19*, 455 (1956).
[26] Y. Tabata, H. Kimura, and H. Sobue, *J. Polym. Sci., Polym. Lett., 2B*, 23 (1964).
[27] Y. Tabata, H. Kimura, and H. Sobue, *Kogyo Kagaku Zasshi, 67*, 827 (1964); *Chem. Abstr., 61*, 8417 (1964).
[28] K. A. Andrianov, L. M. Khananashvili, and Yu. F. Konopchenko, *Vysokomol. Soedin., 2*, 719 (1960).
[29] S. W. Kantor, R. C. Osthoff, and D. T. Hurd, *J. Am. Chem. Soc., 77*, 1685 (1955).
[30] M. Momonoi and N. Suzuki, *Nippon Kagaku Zasshi, 78*, 581 (1957); *Chem. Abstr., 53*, 5159 (1959).
[31] A. N. Pines, R. N. Mixer, D. L. Bailey, and W. T. Black, German Patent, 1,060,862 (1959); British Patent, 847,082 (1960).
[32] S. Pinner and V. Wycherley, *Plastics, 23*, 27 (1958).
[33] W. W. Spindler and R. L. Vale, *Makromol. Chem., 43*, 231, 237 (1961).
[34] A. Iray, *Mater. Methods, 1955*, 42.
[35] G. S. Irby, *Rubber Plast. Age, 37*, 105 (1956).

[36] K. A. Andrianov and A. I. Glukhova, *Khim. Prom., 1957*, 347.

[37] F. Denes, C. Ungurenasu and I. Haiduc, *Eur. Polym. J., 6*, 1155 (1970).

[38] I. Z. Speier, I. A. Webster, and G. H. Barnes, *J. Am. Chem. Soc., 79*, 974 (1957).

[39] V. O. Reichsfeld and A. G. Ivanova, *Vysokomol. Soedin., 4*, 30 (1962).

[40] A. E. Mink and D. D. Mitchell, U.S. Patent, 3,801,544 (1974).

[41] H. Niebergale, *Makromol. Chem., 52*, 218 (1962).

[42] E. A. Rick and G. Omietanski, U.S. Patent, 3,798,253 (1974).

[43] A. A. Zhdanov, K. A. Andrianov, and A. P. Malyukhin, *Vysokomol. Soedin., A, 16*, 2345 (1974).

[44] K. A. Andrianov, A. A. Zhdanov, and A. P. Malyukhin, *Vysokomol. Soedin., A 16*, 1765 (1974).

[45] K. A. Andrianov, V. N. Kopylov, M. I. Shkolnik, and E. A. Koroleva, USSR Patent, 472,952 (1974).

[46] K. A. Andrianov, I. Haiduc and L. M. Khananashvili, *Dokl., 150*, 92 (1963).

[47] K. A. Andrianov, I. Haiduc, and L. M. Khananashvili, *Izv. Akad. Nauk SSSR, Ser. Khim., 1963*, 948.

[48] D. Ya. Zhinkin, V. N. Markova, and M. V. Sobolevskii, *Zh. Obshch. Khim., 33*, 1293 (1963).

[49] S. Papetti, B. B. Schaefer, A. P. Grey, and T. L. Heying, *J. Polym. Sci., 4*, 1623 (1966).

[50] H. A. Schroeder, *Inorg. Macromol. Rev., 1*, 45 (1970).

[51] K. A. Andrianov, B. A. Astapov, and N. M. Orlova, USSR Patent, 401,693 (1974).

[52] K. A. Andrianov, N. N. Makarova et al., *Doklady, 223*, 861 (1975).

[53] K. A. Andrianov, V. N. Tsvetkov et al., *Vysokomol. Soedin., A 18*, 890 (1976).

[54] K. A. Andrianov, O. I. Grivanova et al., *Vysokomol. Soedin., 2*, 521 (1960).

[55] B. Prokai and B. Kanner, U.S. Patent, 3,792,073 (1974).

[56] J. R. Walsh and C. J. Litteral, U.S. Patent, 3,821,122 (1974).

[57] G. Rosmy, *Ger. Appl.*, 2,353,161 (1975).

[58] W. R. Dunnavant, *Inorg. Macromol. Rev., 1*, 165 (1971).

[59] V. V. Korshak, V. A. Zamyatina et al., *Vysokomol. Soedin., 5*, 1127 (1963).

[60] I. B. Atkins and B. R. Currell, *Inorg. Macromol. Rev., 1*, 203 (1971).

SOME UNSOLVED PROBLEMS CONCERNED WITH RADIATION GRAFTING TO NATURAL AND SYNTHETIC POLYMERS

V. STANNETT and T. MEMETEA*

Department of Chemical Engineering, North Carolina State University, Raleigh, North Carolina 27607

SYNOPSIS

The radiation grafting behavior of styrene to cellulose acetate, polyethylene terephthalate, and wool was examined in some detail. A combination of ESR and an analysis of the grafted products was used. Each system exhibited a certain feature which was difficult to explain. The preirradiation grafting of styrene to cellulose acetate gave a bimodal molecular weight distribution of the grafted side chains, the bulk consisting of an extremely narrow distribution high molecular weight polystyrene. Some possible explanations are presented. The yields of grafted side chains in the case of polyethylene terephthalate were considerably greater than the radical yields determined by ESR. Finally in the case of radiation grafting of styrene to wool, numerous low molecular weight branches were obtained. A backbiting type of chain transfer to polymer mechanism was postulated to explain these results and may also be present in the grafting of styrene to polyethylene terephthalate.

INTRODUCTION

The development of graft copolymerization was first started in 1944 with the use of chemical means of initiation [1]. The use of high energy radiation both from low dose rate isotope sources and from electron accelerators began about 10 years later. Extensive research has been carried out in numerous laboratories since then, and the subject has been reviewed in detail in a number of publications [2–4]. In spite of the large body of research, little commercialization has developed, notable exceptions being the combined grafting and cross-linking processes successfully practiced by the Millikan Corporation [5] and the Bell Telephone Laboratories [6] for cotton and poly(vinyl chloride) cables, respectively. In the past few years however, activity has expanded and there appear to be reasonable prospects for the increased industrial use of high energy radiation, particularly from electron accelerators for grafting. The successful development of reliable high power but low penetration electron accelerators, capable of treating films and webs at high speeds, is largely responsible for this optimism.

* N.S.F. Visiting Fellow from Institute of Physical Chemistry, Bucarest, Romania.

Journal of Polymer Science: Polymer Symposium 64, 57–69 (1978)
© 1978 John Wiley & Sons, Inc.   0360-8905/78/0064-0057$01.00

The Radiation Grafting Process

Many different methods of grafting have been developed including the joining of two preexisting polymers. The most common method, however, is

Polymer → Polymer* (activation step)

Polymer* plus monomer → Grafted Polymer (grafting step)

The asterisk implies either a free radical or ion; in most cases however free radicals are involved. The "activation" step can be achieved by numerous chemical routes including chain transfer and direct oxidation. High energy radiation however directly forms free radicals in the polymer without catalysts or pretreatments which gives it a unique advantage as a method of initiation. The free radicals are formed by the breakdown of excited species formed directly or after electron-positive ion recombinations. Usually the product is a macroradial plus a small fragment such as a hydrogen or methyl radical; sometimes, however, chain breaking occurs and a block copolymer is eventually produced. The small radicals can either abstract a hydrogen from a polymer chain and disappear as hydrogen or methane gas, for example, or initiate the formation of homopolymer. General discussions of these various processes have been presented by Chapiro [2].

Three different methods of grafting using high energy radiation have been developed and these will be briefly discussed together with their advantages and disadvantages. The presence of a swelling agent to facilitate the diffusion of the monomer to the free radicals is often necessary.

(1) *The Mutual Method*. The polymer is irradiated directly in the presence of the monomer. The monomer can be in the liquid or vapor phase. It is clear that with this method, homopolymer is also formed, initiated by the direct radiolysis of the monomer and swelling agent. This can be minimized by either the judicious choice of the monomer–polymer combination, which obviously gives a limited source of systems, or by having the monomer in the vapor phase. The success of the latter method is due to the low degree of energy absorption in low density vapors.

An example of this method is the grafting of acrylic acid to polyethylene terephthalate. If the polyester is irradiated in the presence of the liquid monomer, homopolymer is formed in abundance and almost no grafting occurs. Grafting can be satisfactorily achieved, however, by irradiating the polymer in the presence of, for example, acrylic acid vapor. In some cases, where only a small amount of grafting is needed, the monomer can be applied alone or in solution and irradiated without too much homopolymer being produced. If only surface grafting is needed, this method works quite well; the homopolymer can be rendered inextractable by the addition of a small amount of a difunctional monomer, thus causing cross-linking which ties the homopolymer into the graft copolymer. It is clear that a simple addition and irradiation procedure using an electron accelerator can be a very practical and attractive method.

(2) *The Preirradiation Method*. The polymer is irradiated and then treated with the monomer or monomer solution. This technique almost completely

eliminates the production of homopolymer. It has a disadvantage for polymers such as cellulose which degrade with radiation, since the protective effect of the monomer is missing. Furthermore, preirradiation grafting is more sensitive to inhibition by oxygen than is the mutual method and more extensive deaeration procedures are necessary.

When electron accelerators are used, the high intensity of the radiation, resulting in high radical fluxes, can lead to very short grafted chains with the mutual method but not with preirradiation due to the different kinetics. In practice, however, the mutual method becomes in reality a combination of mutual and preirradiation grafting kinetics in that when the polymer leaves the radiation field a considerable population of trapped radicals remain. These continue to grow leading to high grafting yields.

(3) *The Peroxide Method.* The fibers are irradiated in air, leading to the production of peroxides and hydroperoxides. These can later be decomposed in the presence of monomer leading to grafting and homopolymer. This method has not been explored in great depth. In many cases a mixture of trapped radicals and peroxides are believed to be responsible for the grafting in heterogeneous systems [7]. The method is quite interesting and well worthy of further study.

In principle, grafting by the mutual and peroxide methods can be carried out heterogeneously or in solution. Most reported studies, however, have been made with the solid polymer suspended in the monomer or monomer solution. By the judicious choice of timing, swelling conditions, and degree of penetration of the radiation, grafting can be localized near the surfaces or throughout the solid polymer. With the use of suitable inhibitors, including air, the grafting can be confined to the interior of the polymer substrate.

Fundamental studies of radiation grafting are difficult, mainly because of the difficulty of completely characterizing the products. In a typical system the product contains unmodified base polymer, true graft polymer, and homopolymer from the monomer used. However, some investigations have resulted in, often elaborate, procedures to bring about satisfactory separation of the three components. The characterization of the purified graft copolymer presents a further problem. However, if a polymer substrate (backbone) is chosen which can be destroyed, leaving the grafted side chains intact, their number, molecular weight, and molecular weight distributions can be determined. Dene rubber backbones have been used because of the great industrial interest in grafted rubbers for impact-resistant plastics. Ozone and other oxidizing agents have been used to destroy the backbone polymers but these often result in degradation of the side chains. More elegant systems have involved hydrolyzable backbones such as cellulosics, keratin, polyamides, and polyesters. This paper will summarize some of the results of such studies with cellulose acetate, polyethylene terephthalate and, more briefly, wool as the substrate polymers. In each case, styrene was chosen as a model monomer since polystyrene proved to be quite inert to the hydrolytic destruction of the backbone. Some of the results obtained with each study are still difficult to interpret in an unequivocal way and further work is clearly needed.

EXPERIMENTAL

The cellulose acetate used had a D.S. of 1.84 and a viscosity average molecular weight of 55,000. (Eastman Kodak Co.) and was cast into thin (0.002 cm) films from acetone solution and dried completely under vacuum at 50°C.

The polyethylene terephthalate (Monsanto Co.) was in the form of 800 denier monofilaments, the crystallinity was about 50%, and the viscosity average molecular weight 52,000. The finish was removed by washing with carbon tetrachloride and drying under vacuum. The wool used was Beltsville 56's topping with an average fiber diameter of 22 μ. It was extracted with diethyl ether followed by ethanol.

The styrene and solvents were obtained from the Fisher Scientific Co. and were freshly distilled before use.

The mutual grafting experiments were carried out under vacuum in sealed pyrex tubes. Air was removed by several freeze–thaw cycles at 10^{-5} torr before sealing. Cobalt 60 irradiation was used at the stated temperatures and dose rates.

In the preirradiation experiments, the polymers and monomer solutions were degassed under vacuum. After irradiating the polymer separately, the degassed monomer was added through a break seal and the grafting reaction allowed to proceed at the stated time and temperature.

The separation and hydrolysis procedures varied with each system and these will be described separately. The molecular weights of the polystyrene were determined either by measuring the intrinsic viscosity in benzene at 30°C and calculated by the formula of Fox and Flory [8],

$$(\eta) = 9.77 \times 10^{-5} \, Mn^{0.73}$$

or by osmometry using a high speed Mechrolab membrane osmometer.

The ESR measurements were carried out with a Varian V.4502-10 spectrometer (cellulose acetate and wool experiments) and with a JEOL Model JES-ME.1X spectrometer (polyethylene terephthalate experiments). In all experiments quartz tubes were used and any color centers removed by flaming in the conventional manner.

RESULTS AND DISCUSSIONS

The Cellulose Acetate–Styrene Systems

After grafting, the thin films were freed from both homopolymers by alternate extractions with benzene and a 70–30 acetone–water mixture. The purity of the graft copolymer was checked by density gradient centrifugation. The graft copolymer was then hydrolyzed with sulfuric acid. The polystyrene isolated in this manner was fractionated and the molecular weights and molecular weight distributions measured. Details of the experimental procedures have been presented elsewhere [9].

The procedures and results obtained with both the mutual and preirradiation grafting methods are presented in Table I. It was interesting that with the

TABLE I

Properties of Isolated Graft Copolymers for the Styrene–Cellulose Acetate System[a]

Method	Percent Combined Polystyrene in Isolated Graft Copolymer	Isolated Polystyrene Side Chains		Mw/Mn
		Mw	Mn	
Mutual	56.7	830,000	90,000	9.2
Prirradiation	56.8	1,630.000	57,000	28.6

[a] A total dose of 10.0 Mrads was used in both cases, dose rate 30,000 rads per hour. The mutual irradiation was conducted at 25°C. The preirradiation was carried out at −78°C and the subsequent polymerization at 25°C for 72 hr. In both cases a 80/20 styrene–pyridine solution was used.

preirradiation procedure only 5.2% of homopolystyrene was produced, indicating that any small radicals produced by the radiolysis must have extracted hydrogens from the substrate leaving mainly macroradicals to initiate the subsequent polymerization.

The molecular weight distributions were very broad reflecting the diffusion-controlled nature of the hetrogeneous grafting method. The polystyrene obtained from the mutual grafting had a unimodal distribution but the preirradiation sample showed a clear bimodal distribution. About 20% of the sample had a number average molecular weight of only 13,200 and a Mw/Mn ratio of 15.0. However, the remaining 80% had an Mn of 1.83 million and a Mw/Mn ratio of only 1.09. If the number of grafted chain ends (one end per chain) is calculated, it is found that 97% of the trapped radicals produce the low molecular weight broad distribution sample and the remaining 3% are responsible for 80% of the grafted side chains with the narrow distribution.

The reasons for this unexpected behavior remain largely a mystery. It is tempting to speculate that the bulk of trapped radicals rapidly initiate and terminate as the monomer and swelling agent reach them, leading to the low molecular weight grafts. A very small number of radicals could be in the crystalline regions, which are in a small amount in cast unoriented cellulose acetate films. These would initiate at relatively the same time and continue to grow to the very high chain lengths of narrow distribution actually found with 80% of the grafted side chains. Two further sets of experiments were conducted to try to help establish this simple picture. In one series of separate experiments, preirradiation grafting was carried out under similar conditions but the yield of pure graft and the viscosity average molecular weight of the isolated polystyrene side chains determined as a function of time. The results are shown in Table II. The rate of grafting was found to be very rapid at first and then to slow down; the molecular weights continued to increase with time. These results are reasonably consistent with the ideas presented. However the rapid reaction continues up to about 40% rather than the 20% of low molecular weight material, although the molecular weight data are reasonably consistent. In a second set of experiments, the radical yields were determined as a function of radiation dose up to the 10 Mrads used for the grafting experiments. A degassed mixture of styrene and pyridine was then added to the irradiated sample in the ESR tube through

TABLE II
The Yields and Side Chain Molecular Weights for Preirradiation Grafting versus Time at 25°C
for the Styrene–Cellulose Acetate System

Time (hours)	Percent Graft	Molecular Weight of Grafted Side Chains	Percent of Radical Population Remaining
1.0	14.0	180,000	88.2
3.0	40.0	222,000	64.7
8.8	56.0	709,000	58.8
9.8	81.0	997,000	23.5
185	102.00	1,630,000	11.8

Conditions the same as described in Table I.

a break seal and the radical concentration measured as a function of time. The actual curves are presented elsewhere [10] but the percentage of the original population of radicals remaining after different reaction times are included in Table II. Even after 185 hr, 11.8% remained, after 72 hr, the time of the original study, about 30% remained. This is in contradiction to the simple picture of the grafting process. Even more surprising was the fact that the ESR spectrum itself remained the same and did not change to that of the growing polystyryl radical. It can be speculated that the growing chains rapidly terminate, leaving the normal cellulose acetate radical spectrum intact and leading to the low molecular weight portion of the grafted side chains. The long lived radicals leading to the very high molecular weight portion would involve too few radicals to confound the observed signal. Why the reaction takes place in the way it does is not clear; a knowledge of the distribution pattern of the trapped radicals would be helpful but no way is known at the present time to determine this. Additional and more ingenious experiments are clearly needed.

The Polyethylene Terephthalate (PET)–Styrene System [11, 12]

PET is one of the most difficult polymers on which to graft. The reasons for this difficulty are essentially twofold: (1) the polymer, particularly in the form of drawn fibers, absorbs organic liquids, such as the vinyl monomers, to only a small degree and the rate of diffusion is very low; (2) in the case of radiation grafting, the yield of free radicals on radiolysis is extremely low. Nevertheless radiation grafting can be effected, particularly in the presence of chlorinated solvents. The studies to be described try to find the reasons for the low radical yields and also to learn about the actual yield of grafted chain ends and other features of the polystyrene grafts. Techniques somewhat similar to the previous investigation were used. The grafted fibers, after extraction with benzene, were dissolved in tetrachlorethane and precipitated in methanol, ground to a fine powder and extracted with benzene, redissolved, and the whole process repeated several times. A very small amount of occluded polystyrene was still present as revealed by column chromatography. The grafted polystyrene side chains were isolated by hydrolysis with $N/4$ KOH in benzyl alcohol according to the method of Sakurada et al. [13].

TABLE III
Yields and Compositions for Styrene Grafted PET Fibers

Method	Total Dose (Mrads)	Percent Apparent Grafting	Percent Occluded Polystyrene	Mn of Isolated Polystyrene*		Fraction of PET Grafted	G branches per 100 eV
				Occluded Homopolymer	Grafted Branches		
Mutual	5.75	25.4	3.6	78,800	63,600	0.24	0.57
Pre-irradiation	9.0	64.0	3.5	662,000	662,000	0.11	0.10

* Mn by osmometry.

With 50–50 by volume of styrene in methylene chloride solution good yields were obtained by both methods. At 0.38 Mrads per hour and 9.8 Mrads total dose at 25°C, 48 and 62.5% grafting was obtained by the mutual and preirradiation methods, respectively. Without the methylene chloride only about 6% grafting was achieved by both methods. It was interesting that less than 4% of homopolystyrene was obtained with both methods.

Details of the graft copolymers are presented in Table III. A number of interesting features emerge. The molecular weights of the occluded homopolystyrene are similar to those of the isolated side chains. Since their growth occurs in the same environment, this is understandable. The molecular weights are an order of magnitude smaller with mutual grafting compared with the preirradiation method. Presumably, the constant production of radicals during the mutual grafting process leads to chain termination. With the preirradiation method, on the other hand, the initial population of radicals continue to grow in an unhindered fashion. The lack of homopolymer in the mutual method is gratifying but surprising. It would seem that the growing chains abstract a hydrogen from the PET, leading to additional branches and almost no homopolymer. An extreme example of this type of behavior appears to be found with the grafting of styrene to wool fibers, to be described in the next section. Preirradiation often leads to little homopolymer as discussed in the styrene–cellulose acetate example presented above. Only a small amount of the PET is grafted; this probably reflects the diffusion-controlled nature of the reaction—perhaps the grafting is localized to the surface regions, for example. This, coupled with the ~50% crystallinity, means that the grafting reactions are concentrated into rather limited regions, leading to the backbiting type of termination discussed above. Finally, the G values for the grafted side chains are interesting. The value of 0.57 for the mutual method is close to that expected with an aromatic system, but far more than the ESR value of 0.025 for trapped radicals. It could be that the radicals are produced in normal yields and are captured by the styrene monomer, whereas they terminate before they can be measured by the ESR technique. This, however, does not explain the lack of homopolymer. The comparatively small amounts of solvent and monomer present in the swollen PET fibers, however, could account for, at least, some of the lack of homopolymer. This would reduce the effective yield of radicals somewhat but could still be in a reasonable range. In the case of the preirradiation method, the G

TABLE IV

The Molecular Weights of the Polystyrene Side Chains in the Mutual Radiation Grafting of Styrene to Wool

Dose Rate Mrads per hour	Total Dose Mrads	Percent Graft	Intrinsic Viscosty	Mv
0.003	5.0	467	0.32	65,400
0.01	1.0	111	0.19	32,100
0.01	2.5	245	0.30	59,800
0.01	5.0	558	0.36	78,000
0.05	5.0	206	0.23	41,600
0.10	2.6	50.1	0.22	39,100
0.10	5.0	159	0.22	39,100
0.20	5.0	85.2	0.21	36,700

Grafting solution contained 30% styrene, 52% Dioxane and 18% methanol. 25°C.

value is only 0.1. This is a factor of four greater than the actual value for trapped radicals. The very high molecular weight of the side chains and the long reaction time of 29 days make it quite feasible that some chain transfer to the substrate PET polymer is eventually taking place.

In a series of ESR experiments, attempts were made to study the effect of the methylene chloride and temperature on the radical yields [12]. In the presence of methylene chloride almost no radicals could be detected, presumably due to their rapid termination in the "plasticized" PET. On adding methylene chloride to the irradiated PET fibers, the rate of sorption of the solvent corresponded closely to the disappearance of the trapped radicals; this implies also the rapid termination of the radicals, once plasticized, by the solvent. Irradiation of the PET at $-196°C$ gave a yield of trapped ionic species of about 0.7 per 100 eV. It was interesting that on warming, the signal changed to that found at room

TABLE V

The Molecular Weights of the Polystyrene Side Chains in the Preirradiation Grafting of Styrene to Wool[a]

Total Dose Mrads	Percent Graft	Intrinsic Viscosity	Mv
4.3	13.5	0.041	3,900
4.3	14.9	0.046	4,600
4.3	19.1	0.040	3,800
4.3	21.0	0.040	3,800
7.5	13.5	0.040	3,800

Grafting solution contained 30% Styrene, 52% Dioxane and 18% Methanol. 25°C. Reaction time was 3 days.

[a] Dose rate 0.1 Mrads per hour.

TABLE VI

The Molecular Weight of the Polystyrene Side Chains in the Mutual Radiation Grafting of Styrene to Wool

Dose Rate Mrads per hour	Total Dose Mrads	Percent Graft	Intrinsic Viscosity	Mv
0.005	0.67	3151	3.03	1,300,000
0.050	0.52	105	0.74	206,000
"	1.23	331	1.18	367,000
"	1.37	1375	2.34	911,000
"	2.14	3251	2.39	943,000
o.1	4.1	2584	1.76	620,000

Grafting solution contained 80% Styrene, 20% Formic Acid

temperature, normally attributed to trapped radicals, at about $-130°C$ with a G value of 0.42. A transition point has been noted at $-100°C$. Perhaps the reaction $PET^+ + e \rightarrow PET* \rightarrow PET$. begins to occur at this temperature, followed by the rapid recombination of radicals of higher temperature. In the presence of styrene monomer, the radicals can add, leading to grafting before they terminate by mutual recombination.

These are largely hypotheses, and much further work is needed to rationalize the low G (radical) values with the structure of the polymer and the comparatively high grafting yields.

The Wool–Styrene System

An extensive series of grafting studies have been carried out with the wool-styrene system. The results of these studies have been published, including the experimental details [14, 15, 16]. One additional and important aspect will be considered here. This is concerned with the molecular weights of the grafted side chains.

It is not easy to separate the grafted polystyrene side chains from the wool backbone. However, it was found that two separate treatments of the grafted wool with a two-phase system of toluene and 5% aqueous sodium hypochlorite, each for 24 hr, completely removed the wool with no degradation of the polystyrene. A considerable number of grafted samples were treated in this way and their viscosity average molecular weights determined. It was interesting that distinct evidence of amino acid residues was found with infrared and the ninhydrin color test showing that grafting had indeed occurred.

The molecular weights obtained with the mutual grafting method are presented in Table IV, and those with the preirradiation method in Table V. Both were conducted in similar composition styrene solutions. The low molecular weights found were quite unexpected and unique. Similar studies with cellulose acetate and polyethylene terephthalate gave viscosity average molecular weights in the range of 150,000–1,500,000; see Tables I–III for example. A very con-

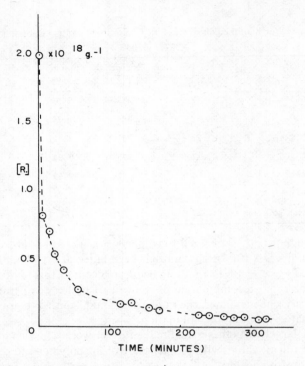

FIG. 1. Decay of free radicals in wool after introduction of methanol–dioxane solution.

servative estimate of the G (branches) gave values of 20–50. This compares with a G (radical) value for dry wool of only 0.8 [15]. It is clear that considerable chain transfer is taking place but with regeneration of growing chains. No other radiation grafting system could be found in the literature showing this effect, although evidence of a tendency of this kind could be inferred from the PET grafting studies discussed above and from the mutual grafting of styrene to cellulose acetate in the presence of large excesses of carbon tetrachloride [17]. The latter compound appeared to act both as a radiation sensitizer and a chain transfer agent. In principle, many short grafted branches could be produced by using very high dose rates such as those obtained with an electron accelerator. Normally, however, the gel effect and residual long lived radicals, always present in the swollen systems used, leads to high molecular weights. In any case, the yields would be very low except at very high total doses. This is because the yield of graft copolymer is given by the product of the number of chains and their molecular weight. Since the chain lengths are low, many more chains need to be started, leading, in turn, to a need for very high doses. These considerations result in a considerable limitation to the versatility of radiation grafting. The discovery that with wool many short chains can be produced at reasonably low total doses is, therefore, of great significance. It is believed that the growing polystyrene radicals chains transfer with certain moieties present in the wool

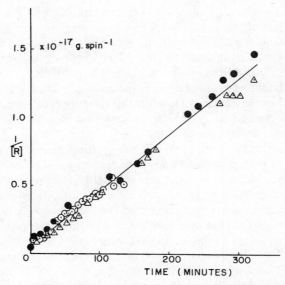

FIG. 2. Second order decay plots for wool and monomer solutions. ●, Methanol/dioxane; △, methanol/styrene/dioxane; ○, methanol/2-ethylhexyl acrylate.

by a kind of backbiting mechanism, leading to a new wool radical which again acts as a grafting side. The introduction of readily transferable groups, such as sulfhydryls, into a polymer backbone would therefore provide a method of forming grafts with numerous short branches. The wool molecule contains numerous transfer groups including disulfides, sulfhydryl, and imines. The most frequently asked question to researchers in this field is how to produce grafts containing numerous short branches and the possible effects on their properties.

Considering the data presented in Table IV and some additional data, already published, a number of facts emerge. There is a tendency towards higher molecular weights at higher grafting yields at low dose rates; also earlier data obtained with lower swelling (10% methanol) gave somewhat lower chain lengths [15]. There appeared to be little effect of dose rate other than that at very low dose rates, higher degrees of grafting were obtained. These results are quite consistent with the chain transfer mechanism of termination. The molecular weights would be proportional to the ratio of the concentrations of monomer and wool. As the swelling is increased, the monomer content should become greater and the wool concentration decrease leading to the observed increase in molecular weight. As the degree of grafting increases, the grafted polystyrene will cause further swelling and a higher monomer content, again increasing the chain length, as actually observed. In another series of experiments, an 80% styrene: 20% formic acid was used. This combination resulted in very substantial swelling; the results are given in Table VI. The grafting yields were found to be very high and the molecular weights in the range of 200,000 to more than one million. Again it can be seen that the molecular weight increases steadily with increasing grafting yield. Two parallel experiments, one with 5% water and one with 20% water, were conducted in styrene–dioxane solution. At similar grafting yields, the molecular weights were 134,000 and 507,000, again emphasizing the importance of swelling in determination of the length of the grafted branches. It is clear that other effects are involved, including the rate of diffusion of the monomer to the active sites and perhaps the gel effect. The latter would be very prevalent in the case of water. However the evidence presented, including the lack of dependence on dose rate over a seventyfold range, is all in favor of the backbiting chain transfer mechanism.

The data obtained with the preirradiation grafting method are given in Table V. The yields of graft copolymer and the molecular weights are extremely low. In an attempt to understand this, some parallel ESR studies were conducted. Dry wool was irradiated to 4.0 Mrads, under vacuum, and degassed grafting solution of the same composition introduced through a break seal and the radical concentration was then followed as a function of time. The results are plotted in Figure 1. The trapped radicals decayed very rapidly and only about 10% remained after 1 hr. The same data is shown plotted as a second order reaction in Figure 2. Included in this plot are similar data, but where the styrene has been substituted, in the same volume concentration, by dioxane and by 2-ethyl hexyl acrylate. The rate of decay can be seen to be essentially the same. It is clear that the rate of mutual termination of the radicals is governed by the methanol content only. Presumably the rapid diffusion of the methanol causes plasticization of the wool giving rise to growth and decay. The diffusion constant of methanol into wool, for example, was determined to be 15 times larger than acrylonitrile; styrene would be a larger penetrant and even more slowly diffusing. This would account for both the low yields and the low molecular weights.

The grafting results obtained with wool are interesting and open up a number of interesting possibilities of both yield and molecular weight control. The heterogeneous grafting process is, however, very complex and much more work is needed to understand it in detail.

REFERENCES

[1] T. Barr and J. B. Speakman, *J. Soc. Dyers Colour.*, *60*, 238 (1944).
[2] A. Chapiro, *Radiation Chemistry of Polymeric Systems*, Chap. 12, Interscience, New York, 1962.
[3] R. D. Gilbert and V. Stannett, *Isot. Radiat. Technol.*, *4*, 403 (1967).
[4] V. Stannett and A. S. Hoffman, *Am. Dyest. Rep.*, *57*, 998 (1968).
[5] M. Kasten, *Daily News Record*, Jan 10, 1968.
[6] L. D. Loan, *Proceedings of the First International Radiation Processing Conference*, Puerto Rico, May 1976, Pergamon, New York, 1977, p. 253.
[7] R. E. Kesting and V. Stannett, *Makromol. Chem.*, *55*, 1 (1962).
[8] T. G. Fox and P. J. Flory, *J. Am. Chem. Soc.*, *73*, 1915 (1951).
[9] J. D. Wellons, A. Schindler and V. Stannett, *Polymer, 5*, 499 (1964).
[10] D. Campbell, J. L. Williams and V. Stannett, *J. Polym. Sci.*, *A1*, 7429 (1969).
[11] T. Memetea and V. Stannett, *A.C.S. Polymer Preprints, 18* (1), 783 (1977).
[12] T. Memetea and V. Stannett, *ibid.*, 777 (1977).
[13] I. Sakurada, Y. Ikata and T. Kawahara *J. Polym. Sci., 11*, 2329 (1973).

[14] V. Stannett, K. Araki, J. A. Gervasi and S. W. McLeskey, *J. Polym. Sci., A3,* 3763 (1965).

[15] D. Campbell, J. L. Williams and V. Stannett, *A.C.S. Adv. in Chem. Series., 66,* 221 (1967).

[16] P. Ingram, J. L. Williams, V. Stannett and M. W. Andrews, *J. Polym. Sci., A1* (6), 1895 (1968).

[17] J. M. Bentvelzen, F. Kimura-Yeh, H. B. Hopfenberg and V. Stannett *J. Appl. Polym. Sci., 17,* 809 (1973).

RADICAL CYCLO- AND CYCLOCOPOLYMERIZATION

GEORGE B. BUTLER

Department of Chemistry and Center for Macromolecular Science,
University of Florida, Gainesville, Florida 32611

SYNOPSIS

Although the principles of cyclopolymerization have been well established, certain aspects of the originally proposed mechanism still remain to be justified in terms of experimental observations. It is the purpose of this paper to discuss these conflicting aspects, to present new information, and to point out those areas where no single mechanistic explanation is presently sufficient. The areas to be considered are: (1) much evidence is available that the cyclization step proceeds via a fast concerted reaction, while kinetic studies on a limited number of monomers indicate that the activation energy for cyclization is greater than that for linear propagation; and (2) although six-membered rings are predicted in radical cyclopolymerization of 1,6-dienes, based upon radical stabilities, numerous examples of such polymerizations are now known to lead to large fractions of five-membered rings; in addition, extensive stuides on model compounds leading to small molecules have shown that the five-membered ring is predominant. New evidence indicates that a driving force for cyclization may be that the cyclized polymer is in a lower free energy level than the monomer; also, that radical stability exerts a marked influence on the ring size. This paper also includes the results of a study designed to determine the effect of the nature of a substituent at the 5-position of 5,6-unsaturated radicals on the ring size of the cyclic product. Thus, it was shown that 2-methallyloxy ethyl radical yielded five- to six-membered rings in the ratio of 30:1, while the 2-(2-phenylallyloxy)ethyl radical yielded products in the ratio of 1:2. The latter structure which simulates cyclopolymerization of di(2-phenylallyl)ether reflects the stabilizing influence of the phenyl group on the intermediate radical. The rate constants and energies of activation for the competing reactions were determined.

INTRODUCTION

Cyclopolymerization, more accurately defined as an alternating intramolecular–intermolecular chain propagation of a suitable nonconjugated diene, has now become well established. Numerous dienes have been investigated following the first observation of this phenomenon [1] and the original proposal [2] to explain the failure of certain 1,6-dienes to function as cross-linking agents as accepted theory predicted [3]. Several reviews dealing with cyclopolymerization phenomena have been published [4–7]. It is the purpose of this paper to reconsider certain aspects of the mechanism of cyclopolymerization which appear to still be open to some question and to present new evidence which may help clarify some of these uncertainties. Although the general course of cyclopolymerization as shown in eq. (1),

$$(1)$$

Journal of Polymer Science: Polymer Symposium 64, 71–93 (1978)
© 1978 John Wiley & Sons, Inc.   0360-8905/78/0064-0071$01.00

(propagation may also produce five membered ring at B) is well supported by numerous examples and is not open to question, there does not at present appear to be a totally satisfactory, single mechanism which can adequately account for all of the experimental observations.

Essentially two similar but somewhat divergent mechanistic explanations have been offered for cyclopolymerization. The first, the "electronic interaction" theory [8], suggests that there may be an interaction between the neighboring double bonds of a 1,6-diene as implied in eq. (2):

$$\tag{2}$$

If such interactions occur, bathochromic shifts in the ultraviolet (UV) spectra of such molecules are predicted which may lower the activation energy for cyclization. Facile cyclizations should occur, and this proposal is synonymous with a concerted process from attack on the diene to generation of the cyclic radical. The second mechanistic explanation, the "steric interaction" or "steric control" theory, was first formalized by Gibbs and Barton [5] although others [2] had earlier recognized the importance of steric factors in cyclopolymerization or related cyclizations. This theory proposes that steric factors alone control cyclopolymerization and that cyclopolymerization of a 1,6-diene occurs because the second double bond will frequently be presented to the reactive site in a conformation favorable for reaction. Also, the intermediate existing prior to cyclization is sterically shielded from an incoming monomer molecule, thus decreasing the probability for linear propagation. The cyclized intermediate will have less steric hindrance than the uncyclized structure thus causing it to react faster. As was pointed out by the authors [5], according to this theory, there should be no difference between the activation energies for the two steps.

The Electronic Interaction Theory—Bathochromic Shifts in UV Spectra of 1,6-Dienes

By formal analogy to conjugated dienes, in which an electronic interaction between the adjacent double bonds is generally accepted to account for the bathochromic shift in the UV absorption spectra of these compounds relative to their monoolefinic counterparts, such an "across-space" interaction can be envisioned in nonconjugated dienes. Such interactions should be observable as bathochromic shifts in the UV spectra of these dienes. Bathochromic shifts have been reported for a variety of monomers [8–10]; however, others failed to show such shifts [11, 12]. Also, the bathochromic shifts exhibited by certain monomers have an alternative explanation [13].

Kinetic Studies

The kinetic relationship between intramolecular and intermolecular propagation for radical cyclopolymerization has been derived [14]. These equations permit the ratio of rates and the differences in energy of activation requirements for the intramolecular and intermolecular propagations to be determined. The ratio k_c/k_i, defined as the ratio of the cyclization rate constant to the intermolecular propagation rate constant for acrylic anhydride, was found to be 5.9 M/l. at 45°C in cyclohexanone. The difference in the activation energies $(E_c - E_i)$ was 2.4 kcal/M while the value for the ratio of the frequency factors for the two steps, A_c/A_i, was found to be 167 M/l. For radiation-induced polymerization of acrylic anhydride in toluene at 30°C, $k_c/k_i = 11.1$ M/l. and $E_c - E_i$ was 2.3 kcal/M [15].

For radical polymerization of o-divinylbenzene in benzene, k_c/k_i values of 2.1, 2.3, 2.8, 3.3, and 4.0 mole/l. at 20, 30, 50, 70, and 90°C, respectively, were found, and $E_c - E_i$ was 1.9 kcal/M; cyclization was favored by a frequency factor of 10 at the monomer concentration where the probability of finding a double bond of another molecule around the uncyclized radical is the same as that of finding the intramolecular double bond [16].

Similar studies on divinylformal by radical polymerization in benzene at 50°C indicate the k_c/k_i ratio to be 200 M/l. and $E_c - E_i$ to be 2.6 kcal/M [17]. For methacrylic anhydride in cyclohexanone at 37°C with radical initiators, k_c/k_i was found to be 46 M/l. [18]. However, a radical polymerization study of the same monomer in dimethylformamide at 0°C, yielded a value of k_c/k_i of 2.4 M/l.; $E_c - E_i$ was found to be 2.6 kcal/M and A_c/A_i was 256 M/l. [13]. The value of k_c/k_i was 6.2 M/l. at 80°C. The marked difference in the k_c/k_i ratio from that reported in cyclohexanone was attributed to a solvent effect. A previous study of methacrylic anhydride [19] in dimethylformamide led to an estimate of 11.4 M/l. for k_c/k_i.

Kinetic studies have been done on several unsymmetrical monomers. In a free radical cyclopolymerization study of vinyl trans-cinnamate [20], $k_c/k_i = 13.3$ M/l. at 70°C in benzene, and $E_c - E_i \simeq 0$. In a similar study of allyl ethenesulfonate, $k_c/k_i = 45$ M/l. at 45°C in benzene [21]. Under the same conditions, allyl ethanesulfonate did not polymerize and propyl ethenesulfonate polymerized at about the same rate as allyl ethenesulfonate. Copolymerization of allyl ethanesulfonate and propyl ethenesulfonate yielded a polymer consisting of more than 90% of the latter. It was suggested that the high degree of cyclization in poly(allyl ethenesulfonate), 85% for $[M] = 0.29$ mol/l., corresponded to an enhanced reactivity of the allyl group in the monomer compared to the allyl group in allyl ethanesulfonate.

In unsymmetrical monomers, where the double bonds differ significantly in reactivity, it is possible to neglect initiation or intermolecular propagation reaction involving the less reactive double bond. This permits a simplification of the kinetic expressions which describe these systems, and such have been applied to vinyl trans-cinnamate [20], allyl ethenesulfonate [21], N-allylmethacrylamide [27], N-allylacrylamide [23], allyl methacrylate [24], allyl acrylate [25], and vinyl trans-crotonate [26]. For these monomers, only one type of repeating unit

TABLE I
Activation Parameters in Cyclopolymerization of o-Divinylbenzene

Catalyst	Solvent	A_c/A_i $(M/l.)$	$E_c - E_i$ $(kcal/M)$	Ref.
$BF_3O(C_2H_5)_2$	Toluene	2.2×10^4	5.3	29
$(C_6H_5)_3CBF_4$	Methylene chloride	5×10^4	6.2	30
Sodium naphthalene	Tetrahydrofuran	15	2.2	28
Free radical	Benzene	50	1.9	16, 27

is formed, and the simplified kinetic expressions could be used. On this basis, it has been suggested [25] that in free radical cyclopolymerization of nonconjugated dienes the intermolecular chain propagation reactions must be slow to allow suitable orientation of the intramolecular double bond in order to have the formation of a large fraction of cyclic structural units.

In a cationic polymerization study of o-divinylbenzene [27], there was a large variation of k_c/k_i (0.1–5 $M/l.$) with various Friedel–Crafts catalysts. The value of k_c/k_i also decreased with decrease in the dielectric constant of the solvent.

In an anionic polymerization study of the same monomer [28], reaction conditions which give higher k_c/k_i values also gave higher rates of propagation in styrene polymerization. It was concluded that k_c/k_i increased with the reactivity of the growing anion.

A comparison of the activation parameters in cyclopolymerization of o-divinylbenzene, with different initiating systems is shown in Table I.

These results are in accord with the respective driving forces which would be predicted for the relative interactions of neighboring double bonds with cation, free radical, and anion centers [31]. Apparently inconsistent with these concepts, however, is the fact that 2,6-diphenyl-1,6-heptadiene undergoes essentially complete cyclization at moderate monomer concentrations by anionic initiation [32, 33]. The following resonance stabilized anions (3) may be important:

$$(3)$$

Until recently, ground state interactions between the neighboring double bonds of 1,6-heptadienes had little or no experimental support. It has been reported [34a] that according to wave-mechanical methods the predicted delocalization energy for bicycloheptadiene in the ground state is zero. Heats of hydrogenation studies also gave no indication of homoconjugative stabilization of bicycloheptadiene [34b]. However, recent evidence supports the possible existence of an electronic interaction in 1,6-hexadiene. This evidence is based upon a photoelectron spectroscopic study [35] in which it was shown that a conformation of 1,5-hexadiene in which the two double bonds are crossed (4I) is more stable by

2.3 kcal/M than the open chain conformation (4H).

$$\text{(4)}$$

There is, however, a more reasonable basis, from both spectral and chemical evidence, for assuming that interaction can exist between monomer double bonds in the excited state and between the ion or radical and the pendent double bond. Bicycloheptadiene, which shows no interaction in the ground state, is described as possessing excited state π-electron interactions [36]. Chemical abnormalities appear in the reactivity of bicycloheptadienes and other compounds towards electrophilic [37], free radical [38], and thermal additions [39] and towards irradiation [40]. More recently, Miller has provided evidence for the resonance stabilization of carbanions [41]. While π-orbital interactions certainly occur with both polar and photochemical activation, the results of radical experiments require the existence of at least two intermediate radicals [38]. However, they do not rule out the possibility that one of these may be nonclassical or that a third, presumably nonclassical, radical exists in addition to the two classical ones.

For those open-chain systems [9, 10] however, which do show bathochromic shifts in their UV spectra, it must be assumed in accordance with the Franck–Condon principle [42] that the molecule is preoriented in a conformation favorable for the electronic transition to occur. However, it must be restated here that many monomers undergo cyclopolymerization in which there is presently no evidence for an excited state interaction. Further, until a satisfactory explanation can be offered for the observed difference in activation energies between the cyclization step and the intermolecular propagation step, use of the electronic interaction as the only explanation for the preference for cyclization is not justified. It is quite likely that the additional energy requirement for cyclization comes about as the result of the introduction of angular and torsional strain during ring formation [43].

Small Molecule Radical Cyclization

The results of monomeric radical cyclizations also support the electronic interaction theory in that the less thermodynamically favored cyclopentane adduct, with the proposed development of an intermediate primary radical, appears to predominate in many of these studies [44]. These results may be explained on the basis of the nonclassical radical (5), consistent with the electronic interaction theory under discussion:

$$\text{(5)}$$

The chain transfer step could lead to a five- or six-membered ring depending upon the steric requirements of the diene and the incoming chain transfer agent, an explanation which avoids the necessity of developing the less stable primary radical. Studies on the cyclization of 4-(1-cyclohexenyl) butyl radical, however, fail to support such an intermediate [45].

Cation Initiated Cyclopolymerization

Cationic initiated intramolecular cyclizations [46–51] and polymerizations [32, 33] support the electronic interaction theory. Monomeric cyclization studies show that concerted ring closures occur in some cases while in others an allylic cation is apparently present as an intermediate. Cationic initiated cyclopolymerization of 2,6-diphenyl-1,6-heptadienes [32, 33] led to high conversion to cyclopolymer (6) having less than 5% residual unsaturation. The nonclassical carbonium ion may be a factor:

$$\text{(6)}$$

Metal alkyl coordination-initiated cyclopolymerizations have been reported which support the electronic interaction theory. Such polymerization attempts on β,β-disubstituted alkenes have been notably unsuccessful. However, 2,5-dimethyl-1,5-hexadiene was converted to a reasonably high polymer containing very little residual unsaturation by this technique while 2-methyl-1-pentene was converted only to a pentamer under the same conditions [52]. This remarkable result was attributed to a "driving force" inherent in the cyclopolymerization process.

Relative Rates of Competing Steps and Degree of Polymerization

Two quite unique and characteristic traits of the cyclopolymerization process which must be recognized in mechanistic considerations are the facts that the intramolecular step is highly favored over the intermolecular step [52, 53] and that the degree of polymerization is higher than that for the corresponding mono-unsaturated derivatives [2, 32, 52, 53, 54]. As illustrative examples, neither α-methylstyrene nor allyltrimethylammonium bromide undergo radical initiated polymerization while their diene counterparts, 2,6-diphenyl-1,6-heptadiene [32, 33] and diallyldimethylammonium bromide [2], respectively, yield high polymers under radical conditions.

Enhanced Rate of Cyclopolymerization

A third aspect of cyclopolymerization to be considered is that the rate of cyclopolymerization is significantly greater than that of homopolymerization of the mono-olefinic counterpart of the diene [17, 18, 20, 55]. Relative rates up to 590 have been reported [43] in a comparison between 2-cyano-1-heptene and 2,6-dicyano-1,6-heptadiene; however, methacrylonitrile is reported to polymerize at about the same rate as the diene [56]. Also, in a telomerization study of allyl ethyl ether and diallyl ether with bromo-trichloromethane, it was reported that the rate of the reaction with allyl ethyl ether is greater than that with diallyl ether [57].

Statistical Consideration in Cyclopolymerization (Fig. 1)

The probability of cyclopolymerization, assuming a freely rotating model [43], has been calculated [58]. The results indicate that from statistical considerations alone, 1,6-diene monomer concentrations must be reduced to <0.10 mole/liter in order to attain >95% cyclization, whereas in many studies, monomer concentrations of 5–8 M have been reported to lead to essentially quantitative cyclization [59, 60]. Consistent with these results and their implications are the results of calculations which assume pseudosecond order kinetics for reaction of the intermediate radical with its intramolecular double bond in order to determine the "apparent" double bond concentration required to account for the observed rates. This value is >20 moles/l. for cyclopolymerization of 2,6-dicyano-1,6-heptadiene [43], and has been calculated to be 40–100 moles/l. for small molecule cyclizations, and the absolute unimolecular rate constant to be 10^5 sec^{-1} [61]. The absolute rate constant for addition of the radicals to alkenes has been found to be 1×10^3. However, there is evidence available that a unimolecular process is generally favored by the frequency factor over a bimolecular process in the same system by a factor of about 100 [62].

Low Volume Shrinkage in Cyclopolymerization

Polymerization of methacrylic anhydride has been shown to be accompanied by a volume shrinkage of only 13.7 ml/M of unsaturation as compared to 22–23 ml/M for methyl, ethyl, and n-propyl methacrylates [63]. The low volume shrinkage was interpreted to indicate a preorientation of the molecule to a cyclic conformation, thus requiring little volume change on polymerization. This

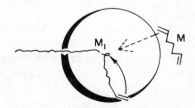

FIG. 1. Statistical model for cyclopolymerization

preorientation was also suggested to be the driving force for cyclization of the molecule. However, as has been pointed out in a recent paper [13], low volume shrinkage may be associated with polymers containing cyclic units in or near the chain as the volume shrinkages of phenyl, benzyl, and cyclohexyl methacrylates are all very close to 15 ml/mole [64].

Preorientation in Solid State Polymerization

In an effort to establish whether preorientation of the diene molecule in a conformation favorable for cyclopolymerization may occur, several solid state polymerization studies of characteristic 1,6-dienes have been undertaken. N-Phenyldimethacrylamide [65, 66] and N-methyldimethacrylamide [66] have been polymerized in solid state by use of suitable irradiation. N-Methyldimethacrylamide [66] via γ-ray initiated solid state polymerization led to cyclopolymer containing 70% five-membered rings and 30% six-membered rings. N-Phenyldimethacrylamide [66] led to a polymer containing only six-membered rings. These monomers were also converted to similar polymers [66] by radical initiation. N-Methyl- and N-phenyl-N-isobutyrylmethacrylamide did not undergo radical initiated polymerization as did the dimethacrylamides [66]. The facile polymerization of these monomers in solid state supports the preorientation theory, particularly in the crystal; however, no satisfactory explanation is yet available to account for the distribution of five- and six-membered rings in the polymers.

Ring-Strain versus Propagation Rate

In an effort to evaluate the effect of ring strain on propagation rate, the following monomers (7) were studied [12]:

$$\text{(7)}$$

The cyclic lactone arising from the first monomer would possess higher ring strain than the tetrahydropyran ring. These factors were reflected in the higher propagation rate for the second monomer.

The Steric Control Theory

It has been pointed out that the nature of an unsaturated acyclic radical may, even in the absence of electronic effects, favor the cyclization reaction over intermolecular propagation [5]. A pendent group on the β carbon atom of a monovinyl monomer has a significant effect on the absolute rate constants for propagation and termination. Values reported for methyl acrylate and butyl acrylate are 720 versus 14 and 2.2×10^6 versus 9×10^3 l./M/sec for propagation and termination, respectively. Thus the rate of intermolecular reaction of an initiated monomer molecule would be expected to be lower than that of a similar monomer with less than five atoms in the pendent group. However, the pendent double bond will frequently be presented to the reactive species in a conformation which is favorable for reaction. Further the cyclic radical formed will be less sterically hindered than its noncyclic counterpart, thus facilitating the approach and reaction of the next monomer unit. This effect may be important in explaining the high degrees of polymerization obtained through the cyclopolymerization of otherwise poorly polymerizable structures. However, values of the propagation rate constant reported for methyl acrylate at 25°C vary from 580 to 1580 while the corresponding termination rate constants reported vary from 6.5×10^6 to 55×10^6. Also, the propagation rate constants reported for n-butyl acrylate at the same temperature vary from 13 to 2100 while the corresponding termination rate constants vary from 1.8×10^4 to 3.3×10^8. Also it has been stated [67] that because of the pronounced autoacceleration experienced with methyl acrylate and the accompanying necessity for extrapolating measurements to a conversion of less than 1%, the data on this monomer are less accurate than for most other monomers investigated. A comparison between methyl methacrylate and n-butyl methacrylate would have been more realistic and when these comparisons are made, the above argument cannot be supported. The propagation and termination rate constants of methyl methacrylate at 30°C, determined by the rotating sector method, are 143 and 12.2×10^6, respectively. Values for n-butyl methacrylate at the same temperature and by the same method are 369 and 10.2×10^6 [68]. Thus, it appears that from this comparison the length of the pendent group on the β carbon has little effect on the absolute rate constants for propagation and termination. However, it should be pointed out that while N-methyldimethacrylamide readily undergoes free radical-initiated cyclopolymerization, N-methylisobutyryl-methacrylamide yields no polymer under identical conditions. It is also significant that 2,6-dicyano-1,6-heptadiene undergoes polymerization up to 590 times as fast as 2-cyano-1-heptene [43].

The decrease in entropy for a cyclization step would perhaps be expected to be smaller than that for addition of a new monomer unit. Only rotational motion will be lost in cyclization, however, on adding a new monomer molecule to the chain, the loss of translational and rotational degrees of freedom will result. Therefore, as far as entropy is important, cyclization would be favored over intermolecular propagation. It has been pointed out that a mechanism advocating a complete steric control of cyclopolymerization would require $E_c - E_i$ to be zero, since the double bonds are considered to be identical throughout and the cyclization and intermolecular activation energies would be the same. Such a mechanism is not consistent with the fact that $E_c - E_i$ is, experimentally, most often greater than zero.

A comparison between the most probable propagation mechanism of cyclopolymerization of o-divinylbenzene, cis-1,2-divinylcyclohexane, o-phthaldehyde, cis-cyclohexene-4,5-dicarboxaldehyde, and o-vinylbenzaldehyde by cationic initiators has been made [69]. o-Divinylbenzene is proposed to propagate by a stepwise process and an intermediate, while the divinyl cyclohexane and the aldehydes are proposed to proceed by a concerted process. o-Vinylbenzaldehyde shows the existence of both reaction schemes, depending upon the type of cation produced.

Further consideration of the original premise that in the analogy between 1,6-nonconjugated dienes and conjugated dienes such as 1,3-butadiene, the assumption must be made that the activation energy of cyclization would be lowered may be open to question, particularly in the case of cationic-initiated cyclopolymerization. It is well known that addition of hydrogen bromide to 1,3-butadiene [70] yields both the 1,2- and the 1,4-addition products, and that the proportions in which they are obtained are considerably affected by the temperature of the reaction. Reaction at −78°C produces a mixture containing about 20% of the 1,4-product and about 80% of the 1,2-product. However, reaction at high temperature (about 40°C) yields a mixture of about 80% of the 1,4-product and about 20% of the 1,2-product. Also, the fact that either compound is converted to the same mixture as obtained by reaction at the higher temperature when equilibrated at that temperature leads to the conclusion that the mixture is the result of equilibrium between the two. Since the 1,4-adduct predominates (80:20) at the elevated temperature it is indicated to be the more stable isomer. Since the reaction at −78°C leads to a preponderance of the 1,2-adduct (80:20), it is indicated to form faster than the 1,4-product. According to transition state theory, for two competing reactions, that one having the higher activation energy is favored by an increase in temperature. In addition, as the temperature is raised there is faster conversion of the initially formed products into the equilibrium mixture. This has been used as a classical example of kinetic control of product ratio at the lower temperature and thermodynamic control of product ratio at the higher temperature. These conclusions have been justified in terms of molecular orbital theory [71].

To the extent that the analogy can be drawn, it is not too surprising that the activation energy for the 1,6-addition step in cyclopolymerization, the cyclization step which, by analogy is related to 1,4-addition in 1,3-butadiene, is higher than that for the 1,2-addition step, the intermolecular propagation step, which, by analogy to the butadiene case, corresponds to 1,2-addition. If, for example, in the case of 2,6-diphenyl-1,6-heptadiene, the nonclassical carbonium ion contributes to stabilization of the product of initial attack on monomer by initiator, the products of the next step of the propagation could conceivably be governed by the relative stabilities of the products. Attack from above the plane by another monomer molecule could occur with approximately equal probability at C-2 as at C-6. Furthermore, formation of the product via attack at C-2 would probably require less energy than that resulting from attack at C-6, owing to the angular and torsional strain introduced in ring formation. On the other hand, the cyclized product should be favored from an enthalpy standpoint since two π-systems are converted to σ-systems by attack at C-6 whereas only one π-system is consumed by attack at C-2.

On the basis of the present evidence it appears that, in one case of cyclopolymerization or another, all of the factors considered such as statistical probability, thermodynamic stability, entropy effects, nonconjugated interactions, and polymerization conditions are important in understanding the mechanism of cyclopolymerization. However, in most cases the exact extent to which each of these factors is important remains unclear.

Ring Size in Cyclopolymerization

Radical cyclization via intramolecular attack of a radical species on an olefinic double bond have been known and studied since the initial observation was made by Butler and Bunch [1] and later reported by Butler and Ingley [1b]. These authors reported that certain 1,6-dienes polymerized via free radical initiators to yield soluble, linear, and saturated polymers, results which could be accounted for reasonably only on the basis of an alternating intra–intermolecular chain propagation [72]. The proposed mechanism was confirmed by structural and degradation studies on the products [72d] although this study did not unequivocably establish the ring size of the cyclic unit produced. It is obvious that in an intramolecular radical attack such as may occur in eq. (8), two intermediates are possible; however, the evidence

$$\tag{8}$$

available in the literature was overwhelming in favor of formation of the more stable secondary radical [73]. This proposal is consistent with "head to tail" enchainment in radical-initiated polymerization of vinyl monomers, an arrangement which has been unquestionably established as the major mode of enchainment in a wide variety of vinyl monomers [74].

1,6-Dienes then would be expected to lead to six-membered rings. By Hofmann degradation reactions on suitable polymers, ring structures were shown to be present [72d]. By oxidation of a suitable polymer with potassium permanganate [72d] under conditions analogous to that reported for ring opening of benzoylpiperidine, ring opening occurred. It is apparent, however, that the corresponding five-membered rings would have responded to these degradative procedures in an approximately analogous manner. In other work, dehydrogenation of a suitable polymer [75] by potassium perchlorate led to evidence for the formation of aromatic rings, a result which could reasonably have come about only by dehydrogenation of six-membered alicyclic structures.

However, in 1961, Miyake [76] studied the structure of the cyclic polymer obtained from divinyl formal via hydrolysis and determination of the 1,2-glycol content. The 1,2-glycol structure could reasonably arise only through five-membered cyclic structures, produced via "head to head" enchainment. He found that the ratio of 1,2-glycol content was much higher than that found in polyvinyl alcohol derived in the normal fashion.

Sultanov and Arbuzova [77] in a quantitative study, reported the ratio of six- to five-membered rings to be 77:23, in this cyclic polymer. Following this original work numerous studies have demonstrated that in many cases of cyclopolymerization the *less* stable intermediate radical is the predominant propagating species [78]. A study of cyclopolymerization of dimethacrylamide and *N*-substituted dimethacrylamide led to the conclusion that considerable five-membered ring was present in these polymers [79]. These results have been confirmed by other workers [65, 66]. *N*-Phenyldimethacrylamide was found to produce polymer by γ-radiation containing 90% five-membered ring and 10% six-membered ring [66]. In contrast, it had been shown that *N*-methyldimethacrylamide yields polymer which consists exclusively of six-membered rings [80]. Other examples of cyclopolymerization of 1,6-dienes to polymers containing five-membered rings along with six-membered rings have been reported [6]. An explanation based upon solvent polarity has been offered [81] to account for ring size in cyclopolymerization of dimethyacrylamides, an increase in polarity favoring the five-membered ring. Recent electron spin resonance evidence has been presented [82] in support of the five-membered cyclized radical (9) as the intermediate in the rate-controlling step in free radical cyclopolymerization of *N*-(*n*-propyl)-dimethacrylamide.

$$
\begin{array}{c}
\text{CH}_3 \ \ \text{CH}_3 \\
| \quad\quad | \\
-\text{CH}_2-\text{C}-\!\!-\!\!-\text{C}-\text{CH}_2\text{·} \quad\quad -\text{P} = -n-\text{C}_3\text{H}_7 \\
| \quad\quad | \\
\text{C} \quad\quad \text{C} \\
\text{O}^{\nwarrow}\ \ \ _\text{N}\ \ \ ^{\nearrow}\text{O} \\
| \\
\text{P}
\end{array}
\tag{9}
$$

The amount of information which can be obtained concerning the microstructures of these cyclopolymers is usually quite limited and depends to a large extent upon the nature of the monomer. Most of the cyclopolymers which have been prepared do not offer such distinguishing features to serve as a probe of the cyclic structures as poly-(divinylformal). Another approach to the study of the cyclic units formed in these radical cyclization reactions is through the study of the radical cyclization reactions of selected model compounds.

In 1958 Friedlander [83] reported that radical-initiated addition of various active chain transfer agents to 1,5-diolefins provides a new synthetic tool for preparing six-membered heterocyclic and carbocyclic compounds. This was in agreement with the earlier polymerization studies [1, 2] where the polymers were postulated to have six-membered rings as the major cyclic repeating unit. In 1964, however, Brace [84] reported that products obtained from the addition of several active chain transfer agents to certain 1,6-dienes initiated exclusive formation of five-membered cyclic products, in contradiction to the earlier results of Friedlander [83]. However, Cadogan et al. reported a predominance of six-membered rings via use of other telogens [85].

Although radical cyclization via the less stable intermediate cyclic radical had been unquestionably demonstrated in cyclopolymerization in 1961 [4], a remarkable observation by Lamb et al. [86] in 1963 was published. These authors reported that 6-heptenoyl peroxide decomposed to give methylcyclopentane as the major product. Only a very small amount of cyclohexane was formed. This result is hard to justify. Cyclohexane had been shown to be more stable than methylcyclopentane at the reaction temperature [87]. Cyclohexaneformyl peroxide decomposes 34 times as fast as cyclopentaneacetyl peroxide [88] indicating that the cyclohexyl radical is more stable than the cyclopentylmethyl radical. These authors [86] also decomposed these two peroxides to determine that formation of the cyclic radicals from the 5-hexenyl radical are both irreversible processes. The irreversibility of the conversion of 5-hexenyl radical to cyclopentylmethyl radical was also demonstrated by Walling and Pearson [89], who observed almost exclusive five-membered ring formation in the cyclization of 5-hexenyl mercaptan with triethyl phosphite, and more recently by the ESR studies of Kochi and Krusic [90].

The model compound study was the first of many such radical cyclization reactions leading to small molecules containing five-membered rings or mixtures of five- and six-membered rings. Justification of this apparent preference for five-membered ring formation has been based on both electronic [86, 91] and steric [89, 58] factors. An early suggestion that there may be an electronic interaction [91] between the initially formed radical and the neighboring double bond of the diene has received considerable attention [7]. The formation of methylcyclopentane from the reaction of 5-hexenyl mercaptan with triethyl phosphite was explained by Walling and Pearson [89] as arising from the attack of the radical at the more accessible end of the double bond, the process being irreversible. Butler and Raymond [58] noted that for approach of the radical to the double bond with the p-orbitals in a common plane formation of five-membered rings would be less sterically hindered. One of the terminal hydrogens lies in the nodal plane directly between the radical carbon and the carbon on the terminal end of the double bond, thus hindering six-membered ring formation. No such steric interference exists for approach of the radical to the other end of the double bond, leading to five-membered rings. Much of the work in this area has been summarized in a review article by Julia [92]. In a study of cyclization of 1-substituted 4-hexenyl radicals, it was shown that as substitutions in the 1-position by radical stabilizing groups (one or two —CN or —C(O)OEt groups) gradually increased, the mixture of cyclized products changed from nearly pure cyclopentane to nearly pure cyclohexane derivatives. Higher temperatures favored cyclohexane formation. Also, it was shown that the cyclization reactions were reversible when C—1 was disubstituted, although it had been shown [86] earlier that in the case of the primary hexenyl radical, the cyclization step is irreversible. On this basis, it was proposed that the cyclopentane product is the kinetically preferred product while the cyclohexane product is preferred via thermodynamic control, with the energy of activation for cyclization being higher for the cyclohexane derivatives.

The cyclization of unsaturated radicals is of interest since the direction of ring closure varies strikingly with changes in radical structure. With highly substituted radicals the formation of cyclohexyl derivatives predominates while the simple 5-hexenyl radical cyclizes almost exclusively to the cyclopentylmethyl radical and five-membered ring products are also preferred in many other cases. Although the available evidence does not allow a satisfactory explanation of the results, it seems that they could be interpreted by assuming the existence of a nonclassical radical, the choice of product being determined by the steric requirements of the system.

Other methods of radical generation on the δ position with respect to a double bond have been investigated and likewise these studies have led to the conclusion that large fractions of five-membered ring are formed. Gas phase studies [93, 94] were followed by studies based upon decomposition of 6-heptenoyl peroxide [86], the result of which was a predominance of the five-membered ring over the six-membered ring to the extent of 94:6. No rate enhancement for decomposition of the peroxide was observed. However, some "concerted cyclization" was implicated by a later study involving generation of 5-hexen-1-yl radical from 6-bromo-1-hexene with tri-n-butylstannane in presence of azobisisobutyronitrile as initiator.

The majority of the radical cyclizations reported have been accomplished by permitting alkyl radicals to react with an unsubstituted double bond thus producing the intermediate, nonresonance stabilized primary or secondary radical [95]. As stabilization of the generated radical increased, the ratio of six-membered ring increased. A more recent study [96] has shown that when the radical generated after cyclization of the six-membered ring can be stabilized by resonance the ratio of six-membered to five-membered ring is markedly increased. The ratio of five-membered to six-membered ring formed from 2-methallyloxyethyl radical was 30:1 but this ratio dropped to 1:2 in the 2-(2-phenylallyloxy)ethyl radical case. The results of this study are summarized and discussed in this paper.

RESULTS AND DISCUSSION

Although alkenes such as propene undergo reaction with radical species, it is well known that only those alkenes bearing substitutents which stabilize the incipient radical lead to high polymers via radical initiation. Likewise, dienes such as 1,6-heptadiene and 2,6-dimethyl-1,6-heptadiene fail to yield high polymers via radical initiation; however, 2,6-diphenyl-1,6-heptadiene [32] readily leads to a high molecular weight cyclic polymer containing little or no residual unsaturation. It was a major purpose of this investigation to generate 5-substituted-5-hexenyl radicals in which the 5-substituent was selected from among those highly radical-stabilizing groups present in well known vinyl monomers, and to determine the ratio of five- to six-membered ring formed. Thus, 2-phenylallyl 2-bromoethyl ether, 2-methylallyl 2-bromoethyl ether, and the products which could arise from generation of the corresponding radicals in the presence of tributyltin hydride were synthesized. Each ether was reduced in presence of a radical initiator and tributyltin hydride, and the ratios of products produced under a variety of conditions were determined. Thus, it was possible to conduct a kinetic study of the reactions of the radicals produced, and to calculate activation energies for the competing reactions. The results obtained with 2-methylallyl 2-bromoethyl ether are consistent with those subsequently obtained from 2-methyl-6-bromo-1-hexene [97].

Synthesis of Reactants and Predicted Products

2-Methallyl-2-bromoethyl ether (I), ethyl methallyl ether (Ia), 3,3-dimethyltetrahydrofuran (Ib), 3-methyltetrahydropyran (Ic), 2-phenylallyl 2-bromoethyl ether (II), ethyl 2-phenylallyl ether (IIa), 3-methyl-3-phenyl-tetrahydrofuran (IIb), 3-phenyltetrahydropyran (IIc), and 3-phenyl-3-(bromomethyl)-tetrahydrofuran (III) were synthesized for this study by standard synthetic processes. Tributyltin hydride (Bu₃SnH) was prepared according to the procedure of Kuivila and Buemel [98] by reaction of tributyltin chloride with lithium aluminum hydride.

Generation and Reactions of Radicals

Scheme I shows the generation and reactions of the two radical systems studied. It is analogous to the reaction scheme described for the reactions of the 4-(1-cyclohexenyl)-butyl radical by Struble et al. [45].

The 2-Methallyloxyethyl Radical

The techniques involved in generating the radicals and analyzing the resulting products were modeled after previously published procedures [100]. In all cases the molar ratio of the bromide to Bu₃SnH in the reaction mixtures was about three to one. The amount of AIBN was 1.5 mole % based on Bu₃SnH concentration.

Quantitative determinations of the concentration of products resulting from the reactions of the radicals with Bu₃SnH were made by gas chromatography (GC). A number of standard solutions were prepared containing authentic samples of the expected products in the range of concentrations in which they were obtained during the radical reactions. A standard calibration curve was prepared for the peak area versus concentration for each of the expected products.

The ratio of 3,3-dimethyltetrahydrofuran [Ib] to 3-methyltetrahydropyran [Ic] formed at each temperature was approximately the same for all concentrations of reactants. The ratios observed at the three reaction temperatures are listed in Table II.

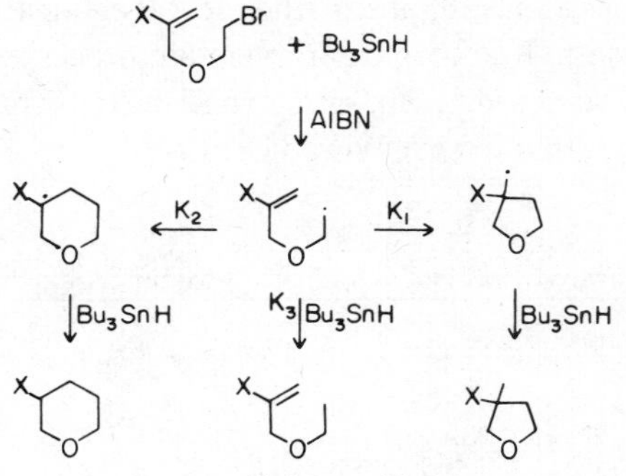

Scheme I

TABLE II
Ratios of 3,3-Dimethyltetrahydrofuran [Ib] to 3-Methyltetrahydropyran Produced [Ic]

Temperature	[Ib]:[Ic]
40	43 ± 1.5
90	30 ± 1.0
125	24 ± 1.0

The 2-(2-Phenylallyloxy)Ethyl Radical

The procedure, molar ratios of reactants, and methods of analysis were analogous to those used for the previous system.

The ratio of 3-phenyltetrahydropyran [IIc] to 3-phenyl-3-methyltetrahydrofuran [IIb] formed at each temperature was approximately the same for all three concentrations of reactants. The ratios observed at the different reaction temperatures are listed in Table III.

The 3-(3-Phenyltetrahydrofuranyl)Methyl Radical

The 3-(3-phenyltetrahydrofuranyl)methyl radical was generated from 3-phenyl-3-bromomethyltetrahydrofuran under the usual reaction conditions at 90°C to check the reversibility of its formation from the 2-(2-phenylallyloxy)-ethyl radical (10):

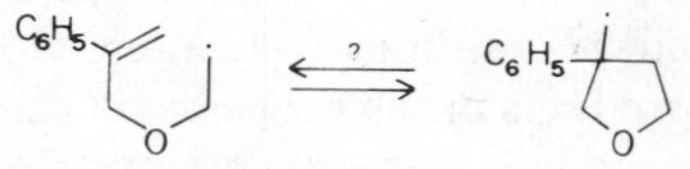

$$C_6H_5 \quad \overset{?}{\rightleftharpoons} \quad C_6H_5 \qquad\qquad (10)$$

It was reasoned that if either of the cyclization reactions of the 2-(2-phenylallyloxy)ethyl radical was reversible, the one shown above should be since the other radical leading to 3-phenyltetrahydropyran should be a more stable benzylic radical. The only product obtained when this bromide was treated with Bu₃SnH under the usual conditions at 90°C was 3-phenyl-3-methyltetrahydrofuran, which proves the stability of the 3-(2-phenyltetrahydrofuranyl)-methyl radical under these conditions. Thus it appears the radical cyclization reactions of the 2-(2-phenylallyloxy)ethyl radical are irreversible processes, just as they were shown to be for the cyclization reactions of the 5-hexenyl radical [86] and subsequently for the 5-methyl-5-hexenyl radical [97].

TABLE III
Ratios of 3-Phenyltetrahydropyran [IIc] to 3-Phenyl-3-Methyltetrahydrofuran Produced [IIb]

Temperature	[IIc]:[IIb]
40	1.82 ± 0.04
90	1.92 ± 0.04
130	2.00 ± 0.05

Kinetics of the Radical Reactions

The groundwork for studying the kinetics of the reactions of radical systems of the type studied in this work was established by Carlsson and Ingold [61] and by Walling et al. [99]. Carlsson and Ingold established the fact that for the reaction of an alkyl bromide with tributyltin hydride in the presence of a radical initiator, the rate controlling step is abstraction of hydrogen from the tributyltin hydride by the alkyl radical. Thus, the competing reactions of the alkyl radical can be studied kinetically by this method. Walling et al. reported a kinetic treatment for the competing reactions of the 5-hexenyl radical in the presence of Bu_3SnH.

Equation (11) quite satisfactorily fit the data obtained for the reaction products obtained from the 2-methallyloxyethyl radical and the 2-(2-phenylallyloxy)ethyl radical. The equation was integrated between the initial and final reaction conditions to yield eq. (12).

$$\frac{d[Bu_3SnH]}{d[\text{I or II Radical}]} = \frac{k_1 + k_2 + k_3[Bu_3SNH]}{k_3[Bu_3SnH]} \tag{11}$$

$$[\text{I or II Radical}]_{Fin} = [Bu_3SnH]_{In} - \frac{k_1 + k_2 \ln k_1 + k_2 + k_3[Bu_3SnH]_{In}}{k_3 \qquad k_1 + k_2} \tag{12}$$

A constant ratio of tetrahydrofuran derivatives to tetrahydropyran derivative over a fourfold change in tributyltin hydride concentration at each temperature for both of the radical systems studied was obtained.

Since the cyclization of the 2-(2-phenylallyloxy)ethyl radical to the 3-(3-phenyltetrahydrofuranyl)methyl radical was shown to be an irreversible process, [eq. (13)],

$$\tag{13}$$

it would seem reasonable to assume that all four cyclic radicals formed in the two radical systems studied were formed irreversibly. The methyl analog of the five-membered cyclic radical should be no more likely to undergo the reverse reaction than the phenyl system, while the two six-membered cyclic radicals, one being tertiary and the other benzylic, should be less likely to undergo the reverse reaction.

The lack of reversibility of the radical cyclizations coupled with the constant ratios of cyclic products at each temperature supports the proposed reaction scheme, whereby the cyclic products arise from irreversible radical cyclizations whose rates are independent of the tributyltin hydride concentration. Thus the ratio of k_1 to k_2 is just the constant ratio of tetrahydrofuran derivative to tetrahydropyran derivative formed at that temperature (14):

$$\frac{k_1}{k_2} = \frac{[Ib]}{[Ic]} \text{ or } \frac{[IIb]}{[IIc]} \tag{14}$$

With this relationship between k_1 and k_2 established, eq. (12) was solved by trial and error for values of k_3/k_1 or k_3/k_2 which fit each set of data. The values which gave the best fit for the products obtained from the 2-methallyloxethyl radical and the 2-(2-phenylallyloxy)ethyl radical, respectively, were used to calculate values of [Ia] and [IIa] the concentration of uncyclized products.

The values of k_1/k_2, k_3/k_1, and k_3/k_2 which were obtained for the two radical systems at each of the three reaction temperatures were utilized by applying the Arrhenius equation to give eqs. (15) and (16).

$$\frac{k_1}{k_2} = \left(\frac{A}{A_2}\right) \exp\left(\frac{E_{A_2} - E_{A_1}}{RT}\right) \tag{15}$$

$$\ln\left(\frac{k_1}{k_2}\right) = \ln\left(\frac{A_1}{A_2}\right) + \frac{E_{A_2} - E_{A_1}}{RT} \tag{16}$$

Plotting $\ln(k_1/k_2)$ versus $1/T$ gave straight lines with slopes of $E_{A_2} - E_{A_1}/R$ and intercepts of $\ln(A_1/A_2)$. The values of the various activation energy differences and frequency factor ratios obtained from the slopes and intercepts are listed in Table IV.

The product distributions from the 2-methallyloxyethyl radical show that up to 3% of 3-methyltetrahydropyran [Ic] was formed in addition to the two major products. In the case of the 2-(2-phenylallyloxy)ethyl radical, 3-phenyltetrahydropyran [IIc] was the predominant cyclic product, indicating that the increased stability of the benzylic radical intermediate is the major controlling factor in the product distribution.

The kinetic results show that Scheme I quite adequately describes the reactions of the two radicals studied in this research. Equation (7), which was derived from this scheme, gave a good fit to the data for both radical systems at all three temperatures. The formation of the cyclic products can be accounted for by irreversible cyclizations of the initially formed radicals—processes whose rates are independent of the Bu_3SnH concentrations.

From the kinetic results it was possible to get an estimate of the activation energies for the cyclization reactions relative to the activation energy for the abstraction of hydrogen from Bu_3SnH by a primary alkyl radical. Wilt et al. [100] have calculated the activation energy for this hydrogen abstraction process to be between 6.8 and 8.2 kcal/mol using rate constants reported by Carlsson

TABLE IV
Activation Energy Differences and Frequency Factor Ratios from Figures 1 and 2

Property	2-Methallyloxyethyl radical	2-(2-Phenylallyloxy)-ethyl radical
$E_{A_2} - E_{A_1}$ (kcal/mole)	1.7 ± 0.05	0.26 ± 0.05
$E_{A_1} - E_{A_3}$ (kcal/mole)	2.2 ± 0.15	1.3 ± 0.25
$E_{A_2} - E_{A_3}$ (kcal/mole)	3.9 ± 0.3	1.6 ± 0.25
A_2/A_1	0.35 ± 0.08	2.8 ± 0.2
A_3/A_1 (liter/mole)	0.44 ± 0.07	2.3 ± 0.8
A_3/A_2 (liter/mole)	1.3 ± 0.55	0.85 ± 0.25

and Ingold [61]. The comparisons of the rate constants for the cyclization reactions to that for the hydrogen abstraction from Bu_3SnH by the initially formed primary alkyl radicals are presented in Table IV. The rate constants for hydrogen abstraction by the 2-methallyloxyethyl and the 2-(2-phenylallyloxy)-ethyl radical should be very nearly the same. Thus it is possible to compare the activation energies for the cyclization reactions leading to the four cyclic products in the two radical systems. The order of activation energies for the four cyclizations are presented in relation to $E(H)$, the activation energy for the hydrogen abstraction by the initially formed primary radicals:

Product	$E_A - E_A(H)$	(kcal/mole)
3-Phenyl-3-methyltetrahydrofuran	1.3	0.25
3-Phenyltetrahydropyran	1.6	0.25
3,3-Dimethyltetrahydrofuran	2.2	0.15
3-Methyltetrahydropyran	3.9	0.3

These values show that the activation energy for the cyclization of the 2-(2-phenylallyloxy)ethyl radical to form 3-phenyltetrahydropyran is about 2.3 kcal/mole less than the activation energy for the cyclization of the 2-methallyloxyethyl radical to form 3-methyltetrahydropyran. This helps to explain why formation of the six-membered cyclic product competes so much more favorably in the phenyl system than in the methyl system.

The relationships between the three rate constants in both of the radical systems were established. Approximate values of the rate constants for the cyclization reactions of the two radicals were obtained from the rate constant relationships and the reported value [61] of the rate constant for abstraction of hydrogen from tributyltin hydride by the 1-hexyl radical ($k = 1.0 \times 10^6 M^{-1}$ sec^{-1} at 25°C). Comparison of this value to the ratios reported earlier gives the following values for the rate constants of the four cyclization reactions at 40°C:

Product	k_c (sec^{-1}) at 40°C
3-Phenyltetrahydropyran	9.6×10^4
3,3-Dimethyltetrahydrofuran	6.1×10^4
3-Phenyl-3-methyltetrahydrofuran	5.3×10^4
3-Methyltetrahydropyran	1.4×10^3

The total rate constant for cyclization in either of the two radical systems would be simply $k_1 + k_2$, the sum of the two rate constants for the cyclization reactions. The approximate values for the methyl and phenyl systems studied in this work are 6.2×10^4 and 1.5×10^5 sec^{-1}, respectively. The fact that this total rate constant for cyclization is more than twice as great for the phenyl system as for the methyl system appears to arise from the differences in rate constants for cyclization to the tetrahydropyran derivatives. The rate constants for formation of the tetrahydrofuran derivatives are nearly the same. This big difference is likely due to the greater stability of the benzylic radical leading to 3-phenyltetrahydropyran compared to the tertiary radical leading to 3-methyltetrahydropyran.

Carlsson and Ingold [61] determined the rate constants for cyclization at 40°C for the 1-hexenyl radical studied by Walling et al. [99], and for the 4-(1-cyclo-hexenyl)butyl radical reported by Struble et al. [45] by comparing the reported rate constant ratios with the rate constant for hydrogen abstraction from Bu_3SnH by the 1-hexyl radical at 25°C. The values of k_c obtained were 1×10^5 sec^{-1} for the 4-(1-cyclohexenyl)butyl radical. They compared these constants with the rate constant for addition of an ethyl radical to 1-heptene at 40°C, $1 \times 10^3 M^{-1}$ sec^{-1}, and stated that the "effective double bond concentration" for the intramolecular cyclization of such radicals is about 40–100 M. Such a comparison with the cyclization rate constant for the 2-methallyloxethyl radical yields a value of 62 M. For the 2-(2-phenylallyloxy)ethyl radical the value would be 150 M but a more accurate basis for comparison would be with the rate constant for addition of a primary radical to an α-alkyl styrene, which should have a value greater than $1 \times 10^3 M^{-1}$ sec^{-1} at 40°C.

This enhanced "effective double bond concentration" for reaction of the initial radical with the double bond to give a cyclic radical, coupled with the increased steric hindrance to approach of another monomer molecule when the initial radical is part of a growing polymer chain, should be sufficient to explain why so many 1,6-dienes undergo free radical polymerization to yield polymers composed entirely of cyclic repeating units. Propagation by reaction of the initial radical with another molecule of monomer simply cannot compete with the cyclization processes in most cases.

The products obtained from the reactions of the two radicals studied in this work were consistent with a reaction scheme whereby the cyclic products arise from irreversible radical cyclizations. The kinetic treatment further supports this explanation. Equation (7), when applied to the data of Struble et al. [45] for the products from reactions of the 4-(1-cyclohexenyl)butyl radical with tributyltin hydride at 40°C gave a good fit for the data reported by Beckwith and Gara [101] for the reactions of the 2-(3-butenyl)phenyl radical with Bu_3SnH with $k_3/k_c = 2.0$. The data of Walling et al. [99] for the reactions of the 5-hexenyl radical with Bu_3SnH at 40°C were satisfied by eq. (7) with $k_3/k_c = 10$. Only Walling's data at 130°C, which showed no distinct trend in product distribution, failed to satisfactorily fit eq. (7). A scheme involving competition between the simple irreversible radical reactions leading to the cyclic and non-cyclic products satisfactorily explains the observed product distributions.

The proportions of five-membered and six-membered cyclic product in the two radical systems studied appeared to be dependent upon the activation energies for the competing cyclization reactions and also upon steric factors concerning the ease of approach of the radical to the two ends of the double bond. These effects would also be important in determining the ring size of the repeating units of cyclopolymers. When the initial radicals are stable enough that the radical cyclization reactions are partially or totally reversible, as in the case of some of the radical systems studied by Julia [92], the stabilities of the cyclized radicals become an additional factor in determining the proportions of five-membered and six-membered cyclic repeating units in cyclopolymers.

It is apparent from these new data and these discussions that considerably more investigation will be necessary before definite conclusions can be drawn with respect to the ratio of five- to six-membered rings in the many cyclopolymers already synthesized, and a satisfactory explanation for these extensive variations from one system to another is available.

REFERENCES

[1] (a) G. B. Butler and R. L. Bunch, *J. Am. Chem. Soc., 71*, 3120 (1949); (b) G. B. Butler and F. L. Ingley, *J. Am. Chem. Soc., 73*, 895 (1951).
[2] G. B. Butler and R. J. Angelo, *J. Am. Chem. Soc., 79*, 3128 (1957).
[3] H. Staudinger and W. Heuer, *Ber., 67*, 1159 (1934).
[4] G. B. Butler, *Encyclo. Polym. Sci. Tech., 4*, 588 (1966).
[5] W. E. Gibbs and J. M. Barton, in *Kinetics and Mechanism of Polymerization, Vol. 1, Vinyl Polymerizations, Part 1*, G. E. Ham, Ed., Marcel Dekker, Inc., New York, 1967, p. 69.
[6] G. C. Corfield, C. Aso, and G. B. Butler in *Progress in Polymer Science*, Vol. 7, A. D. Jenkins, Ed., Pergamon, Oxford, pp. 71–207.
[7] G. B. Butler, *Proc. International Symp. on Macromolecules*, Rio de Janerio, July 26–31, 1974, E. B. Mano, Ed., Elsevier, Amsterdam, 1975, pp. 57–76.
[8] G. B. Butler and T. W. Brooks, *J. Org. Chem., 28*, 2699 (1963).
[9] G. B. Butler and B. Iachia, *J. Macromol. Sci. Chem., A3*, 803 (1969).
[10] G. B. Butler and M. A. Raymond, *J. Org. Chem., 30*, 2410 (1965).
[11] G. B. Butler and J. J. Van Heiningen, *J. Macromol. Sci. Chem., A8*, 1139, 1175 (1974).
[12] K. B. Baucom and G. B. Butler, *J. Macromol. Sci. Chem., A8*, 1205, 1239 (1974).
[13] T. F. Gray, Jr. and G. B. Butler, *J. Macromol. Sci. Chem., A9*, 45 (1975).
[14] J. Mercier and G. Smets, *J. Polym. Sci., A1*, 1491 (1963).
[15] M. Okada, K. Hayashi, and S. Okamura, *Kobunshi Kagaku, 22*, 135 (1965).
[16] C. Aso, T. Nawata and H. Kamao, *Makromol. Chem., 68*, 1 (1963).
[17] Y. Minoura and M. Mitch, *J. Polym. Sci., A3*, 2149 (1965).
[18] G. Smets, P. Haus, and N. Deval, *J. Polym. Sci., A2*, 4825 (1964).
[19] W. E. Gibbs and J. T. Murray, *J. Polym. Sci., 58*, 1211 (1962).
[20] J. Roovers and G. Smets, *Makromol. Chem., 60*, 89 (1963).
[21] E. De Witte and E. J. Goethals, *Makromol. Chem., 115*, 234 (1968).
[22] L. Trossarelli, M. Guita, and A. Priola, *Ric. Sci., 35*(11-A), 429 (1965).
[23] L. Trossarelli, M. Guita, and A. Priola, *Macromol. Chem., 100*, 147 (1967).
[24] L. Trossarelli, M. Guita, and A. Priola, *Ann. Chim., 56*, 1065 (1966).
[25] L. Trossarelli, M. Guita, and A. Priola, *Ric. Sci., 36*, 993 (1966).
[26] L. Trossarelli and M. Guita, *Polymer, 9*, 233 (1968).
[27] C. Aso, T. Kunitake, and R. Kita, *Makromol. Chem., 97*, 31 (1966).
[28] C. Aso, T. Kunitake, and Y. Imaizumi, *Makromol. Chem., 116*, 14 (1968).
[29] C. Aso and R. Kita, *Kogyo Kagaku Zasshi, 68*, 707 (1965).
[30] (a) C. Aso, *Mem. Fac. Eng., Kyushu Univ., 22*, 119 (1968); (b) C. Aso, T. Kunitake, Y. Matsugama, and Y. Imaizumi, *J. Polym. Sci., A1*, 2049 (1968).
[31] C. L. McCormick and G. B. Butler, *J. Macromol. Sci. Rev., C8*, 201 (1972).
[32] N. D. Field, *J. Org. Chem., 25*, 1006 (1960).
[33] C. S. Marvel and E. J. Gall, *J. Org. Chem., 25*, 1784 (1960).
[34] R. B. Turner, W. R. Meador, and R. E. Winkler, *J. Am. Chem. Soc., 79*, 4116 (1957).
[35] J. C. Bunzli, A. J. Burak and D. C. Frost, *Tetrahedron, 29*, 3735 (1973).
[36] C. F. Wilcox, Jr., S. Winstein, and W. G. McMullen, *J. Am. Chem. Soc., 82*, 5450 (1960).
[37] S. Winstein and M. Shatavsky, *J. Am. Chem. Soc., 78*, 592 (1956).
[38] H. G. Kuivila, *Accnts. Chem. Res., 1*, 299 (1968).
[39] E. F. Ullman, *Chem. Ind. (London)*, 1173 (1958).
[40] S. J. Cristol and R. L. Snell, *J. Am. Chem. Soc., 79*, 1950 (1958).
[41] B. Miller, *J. Am. Chem. Soc., 91*, 751 (1969).
[42] W. J. Moore, *Physical Chemistry*, 3rd Ed., Prentice-Hall, Englewood Cliffs, 1962, p. 598.

[43] G. B. Butler and S. Kimura, *J. Macromol. Sci. Chem.*, *A5*, 181 (1971).

[44] J. J. G. Cadogan, D. H. Hey, and A. D. S. Hock, *Chem. Ind.*, 735 (1964).

[45] D. L. Struble, A. L. J. Beckwith, and D. E. Gream, *Tetrahedron Lett.*, No. 34, 3701 (1968).

[46] P. D. Bartlett and S. Bank, *J. Am. Chem. Soc.*, *83*, 2591 (1961).

[47] P. D. Bartlett, *Ann. Chem.*, *653*, 45 (1962).

[48] W. S. Johnson, D. M. Bailey, R. Owyang, R. A. Bell, B. Jaques, and J. K. Crandall, *J. Am. Chem. Soc.*, *86*, 1959 (1964).

[49] W. S. Johnson, S. L. Gray, J. K. Crandall and D. M. Bailey, *J. Am. Chem. Soc.*, *86*, 1966 (1964).

[50] W. S. Johnson, W. H. Lunn, and K. Fitzo, *J. Am. Chem. Soc.*, *86*, 1972 (1964).

[51] H. E. Ulery and J. H. Richards, *J. Am. Chem. Soc.*, *86*, 3113 (1964).

[52] C. S. Marvel and J. K. Stille, *J. Am. Chem. Soc.*, *80*, 1740 (1958).

[53] G. B. Butler and R. W. Stackman, *J. Org. Chem.*, *25*, 1643 (1960).

[54] J. K. Stille and D. A. Frey, *J. Am. Chem. Soc.*, *83*, 1697 (1961).

[55] J. Mercier and G. Smets, *J. Polym. Sci.*, *57*, 763 (1962).

[56] N. Grassie and E. Vance, *Trans. Faraday Soc.*, *52*, 727 (1956).

[57] R. Breslow, E. Barrett, and E. Mohacsi, *Tetrahedron Lett.*, No. 25, 1207 (1962).

[58] G. B. Butler and M. A. Raymond, *J. Polym. Sci.*, *A3*, 3413 (1965).

[59] G. B. Butler, A. Crawshaw, and W. L. Miller, *J. Am. Chem. Soc.*, *80*, 3615 (1958).

[60] J. F. Jones, *J. Polym. Sci.*, *33*, 15 (1958).

[61] D. J. Carlsson and K. U. Ingold, *J. Am. Chem. Soc.*, *90*, 7074 (1968).

[62] A. A. Frost and R. G. Pearson, *Kinetics and Mechanism*, Wiley, New York, 1961, p. 297.

[63] J. C. H. Hwa, W. A. Fleming, and L. Miller, *J. Polym. Sci.*, *A2*, 2385 (1964).

[64] J. C. Bevington and B. W. Malpass, *J. Polym. Sci.*, *A2*, 1893 (1964).

[65] M. Azori, N. A. Plate, G. D. Rudkovskaya, T. A. Solsolova, and V. A. Kargin, *Vysokomol. Soedin.*, *8*, 759 (1966).

[66] G. B. Butler and G. R. Myers, *J. Macromol. Sci. Chem.*, *A5*, 1135 (1971).

[67] P. J. Flory, *Principles of Polymer Chemistry*, Cornell Univ., Ithaca, 1953, p. 158.

[68] J. Brandrup and E. H. Immergut, *Polymer Handbook*, Interscience, New York, 1965, p. II-59.

[69] C. Aso, *Pure Appl. Chem.*, *23*(2-3), 287 (1970).

[70] M. S. Kharasch, E. T. Margolis and F. R. Mayo, *J. Org. Chem.*, *1*, 393 (1937).

[71] F. S. Pilar, *J. Chem. Phys.*, *29*, 1119 (1958).

[72] (a) G. B. Butler, A. H. Gropp, R. J. Angelo, W. J. Huck and E. P. Jorolan, *Fifth Quarterly Report*, U.S. Atomic Energy Commission Contract AT-(40-1)-1353, Sept. 15, 1953; (b) G. B. Butler, Gordon Research Conference on Ion Exchange, June 15, 1955 [*Science, 121*, 574 (1955)]; (c) G. B. Butler and R. J. Angelo, *J. Am. Chem. Soc.*, *79*, 3128 (1957); (d) G. B. Butler, A. Crawshaw, and W. L. Miller, *J. Am. Chem. Soc.*, *80*, 3615 (1958).

[73] W. A. Pryor, *Free Radicals*, McGraw-Hill, New York, 1966, p. 221ff.

[74] R. W. Lenz, *Organic Chemistry of Synthetic High Polymers*, Interscience, New York, 1967, p. 342ff.

[75] C. S. Marvel and R. D. Vest, *J. Am. Chem. Soc.*, *79*, 5771 (1957).

[76] T. Miyake, *Kogyo Kagaku Zasshi*, *64*, 1272 (1961).

[77] K. Sultanov and I. A. Arbuzova, *Uzb. Khim. Zh.*, *9*(6), 38 (1965); *C.A. 64*, 12799h (1966).

[78] (a) D. H. Solomon and D. G. Hawthorne, *J. Macromol. Sci. Rev.*, *C15*(1), 143 (1976); (b) J. E. Lancaster, L. Baccei, and H. P. Panzer, *Polymer Lett.*, *14*, 549 (1976).

[79] T. A. Sokolova and G. C. Rudkovskaya, *J. Polym. Sci.*, *C16*, 1157 (1967).

[80] F. Gotzen and G. Schroder, *Makromol. Chem.*, *88*, 133 (1965).

[81] Y. M. Boyarchuk, *Vysokomol. Soedin.*, *A11*, 2161 (1969).

[82] T. Kodaira and F. Aoyama, *J. Polym. Sci.*, *12*, 897 (1974).

[83] W. S. Friedlander, Abstr. of Papers, 133rd Nat. Meet. of ACS, San Francisco, Calif., April, 1958, Paper 18N.

[84] (a) N. O. Brace, *J. Am. Chem. Soc.*, *86*, 523 (1964); (b) *J. Org. Chem.*, *31*, 2879 (1966); (c) *ibid.*, *32*, 2711 (1967); (d) *ibid.*, *34*, 2441 (1969); (e) *J. Polym. Sci.*, *A1*, *8*, 2091 (1970).

[85] J. I. G. Cadogan, D. H. Hey, and A. O. S. Hock, *Chem. Ind.* (*London*), 753 (1964).

[86] R. C. Lamb, P. W. Ayers, and M. K. Toney, *J. Am. Chem. Soc.*, *85*, 3483 (1963).

[87] A. L. Glasebrook and W. G. Tonell, *J. Am. Chem. Soc.*, *61*, 1717 (1939).

[88] H. Hart and P. D. Wyman, *J. Am. Chem. Soc.*, *81*, 4891 (1959).

[89] C. Walling and M. S. Pearson, *J. Am. Chem. Soc.*, *86*, 2262 (1964).

[90] J. K. Kochi and P. J. Krusic, *J. Am. Chem. Soc.*, *91*, 3940 (1969).

[91] G. B. Butler, *J. Polym. Sci.*, *48*, 279 (1960).

[92] M. Julia, *Accounts Chem. Res.*, *4*, 386 (1971); *Rec. Chem. Prog.*, *25*, 1 (1964).

[93] S. Arai, S. Sato, and S. Shida, *J. Chem. Phys.*, *33*, 1277 (1960).

[94] A. S. Gordon and S. R. Smith, *J. Phys. Chem.*, *66*, 521 (1962).

[95] E. L. Eliel, *Stereochemistry of Carbon Compounds*, McGraw-Hill, New York, 1962, pp. 265–269.

[96] T. W. Smith and G. B. Butler, *J. Org. Chem.*, *43*, 6 (1978).

[97] C. Walling and A. Cioffari, *J. Am. Chem. Soc.*, *94*, 6059, 6064 (1972).

[98] H. G. Kuivila and O. F. Beumel, Jr., *J. Am. Chem. Soc.*, *83*, 1246 (1961).

[99] C. Walling, J. H. Cooley, A. A. Pouaras and E. J. Racak, *J. Am. Chem. Soc.*, *88*, 5361 (1960).

[100] J. W. Wilt, S. N. Massie and R. B. Dabek, *J. Org. Chem.*, *35*, 2803 (1970).

[101] A. L. J. Beckwith and W. B. Gara, *J. Am. Chem. Soc.*, *91*, 5691 (1969).

SOME PROBLEMS CONCERNING THE DEGRADATION OF LINEAR VINYL POLYMERS AND COPOLYMERS IN NONISOTHERMAL CONDITIONS

I. A. SCHNEIDER

Department of Physical Chemistry,
Polytechnical Institute, R-6600 Iassy, Romania

SYNOPSIS

A critical review of the experimental results concerning the thermal degradation of linear vinylic polymers in dynamic conditions shows that generally the use of a single thermogravimetric curve may lead to ambiguous conclusions concerning both the steps and the mechanism of the thermal degradation, as well as the corresponding evaluated kinetic parameters. The ultimate values depend not only on the transformation degree, but also on the heating rate used, so that extrapolations may be recommended. However, the significance of the extrapolated kinetic parameters is still arguable, so that to compare thermogravimetric results, standard working conditions must be used. In such conditions the relative thermal stability of the polymers and the influence of comonomer units on the thermal behavior of copolymers may be appreciated.

As it is well known, the thermal degradation of linear vinylic polymers $+CH_2—CRX+_n$ occurs depending on their structure by two basic different reaction mechanisms. Either the polar character of the substituent X causes a homolytic scission of the then less strong C—X bond, or it splits randomly a C—C bond of the main chain, if the macromolecule does not contain weak end bonds (especially, but not only double bonds).

In the first case the scission of the C—X bond is followed by the elimination of HX molecules, according to an "unzip" mechanism (which may be a molecular or a chain reaction), the C—C skeleton of the macromolecule remaining unaffected. It degrades only in a second step at higher temperatures. If in the absence of a weak terminal link a C—C bond of the main chain splits randomly, then again at least two different ways are possible, which seem to depend particularly on the presence and mobility of tertiary H-atoms (R$=$H) (see Scheme I). The random initiation is followed either by the so-called "depolymerization," i.e., an "unzip"-like expulsion of monomer molecules (generally accepted to be a chain reaction), or it is succeeded by an accidental inter- or/and intramolecular transposition of mobile tertiary hydrogen atoms, accompanied by a β-scission of the macromolecular chain. In this latter case may arise chain fragments of different lengths (which may be or not volatile), the yield of monomer being practically equal to zero.

However, these situations are only idealized limit reaction mechanisms, and

Journal of Polymer Science: Polymer Symposium 64, 95–116 (1978)
© 1978 John Wiley & Sons, Inc.
0360-8905/78/0064-0095$01.00

Scheme I

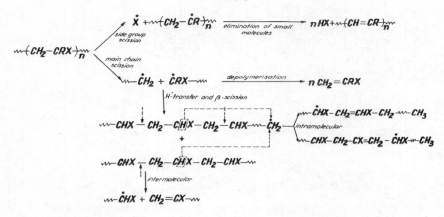

depending on the structure of the polymer, the nature of the terminal links, and the presence of irregularities in the structure or/and the suprastructure of the polymer, the real degradation mechanism comprises a more or less pronounced superposition of these limit mechanisms. The thermal degradation of the co-polymers may then be analyzed in the light of a more or less regulated structure and the introduction by copolymerization of weak links in the macromole-cules.

In the present work we tend to establish whether the method of dynamic thermogravimetry, which showed in the sixties such an impetuous development, offers any possibilities to differentiate and to estimate the kinetics and the mechanism of the thermal degradation of linear vinylic polymers and copoly-mers. The mentioned uncommon enlargement of the dynamic thermogravimetry was determined not only by the sophisticated improvements of the available instruments, but also by the development of many methods for computing kinetic parameters. Classifications of these methods are well known and an excellent review was presented by Šesták et al. [1].

Concerning the degradation of those polymers accompanied by the elimination of small HX molecules, this process occurs in dynamic thermal conditions always in at least two independent thermogravimetric steps, while the scission of the main C—C skeleton, followed by the elimination of volatile products (monomer or higher fragments of the macromolecular chain) happens usually in a single, more or less complex thermogravimetric step.

Therefore a distinction between these two kinds of degradation processes seems to be evident in dynamic thermogravimetry. In Figure 1 we show the corresponding thermogravimetric curves of a suspension-PVC and an anionic-PMMA, taken in the same working conditions.

However, this statement is unfortunately not always positive as we can see in Figure 2, which compares the thermogravimetric curves for the thermal de-struction in nitrogen at a heating rate of $10°C/min$ of 2.5 mg samples of a radical, $M_v = 84,500$, and an anionic PMMA, $M_v = 93,800$.

The kinetic analysis of the thermogravimetric curves obtained at different

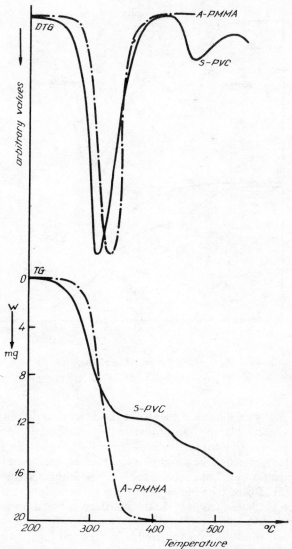

FIG. 1. Thermogravimetric curves for the degradation in air of PVC and PMMA. Initial weight, $w_0 = 20$ mg; heating rate, h.r. $= 12.4°C/min$. Full line, suspension PVC "Geon" $M_w = 63,000$; dotted line, A-PMMA, by anionic polymerization, $M_n = 47,000$.

heating rates by the method of Flynn et al. [2] suggests a serious dependence of the apparent activation energy of the degradation in air, on the transformation degree as seen in Figure 3.

At the same time the kinetic parameters also depend on the heating rate, which may be explained by the extension of the temperature interval corresponding to the same transformation interval, indicating thus an apparent decrease of the temperature coefficient of the reaction. These two phenomena are interdependent as we have demonstrated elsewhere [3].

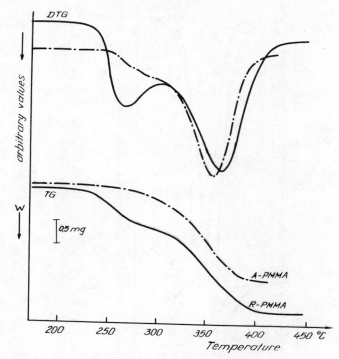

FIG. 2. Thermogravimetric curves for the degradation of PMMA in nitrogen, w_0 = 2.5 mg, h.r. = 10°C/min. Full line, R-PMMA, by radical polym., M_v = 84,500; dotted line, A-PMMA, M_v = 93,800.

There are many reasons for a possible dependence of the kinetic parameters of a solid–gas reaction on the transformation degree, especially in nonisothermal conditions and the different factors are largely discussed by Garn [4].

First we thought about the usual physical factors, such as sample size, diffusion, and heat transfer, but a minute study of the dehydrochlorination reaction of the PVC [5] showed that the explanation must be sought in the complexity of the reaction mechanism itself of the thermal degradation processes of polymers in nonisothermal conditions.

Some of these results are summarized in Table I and Figure 4 and we can see that in the working conditions used the apparent activation energy, as computed by the method of Freeman and Carroll [6], depends neither on the sample size nor on the surrounding atmosphere, but only on the heating rate used. It seems therefore that in the working conditions utilized, the potential of the evolved volatiles is high enough to ensure a degradation in their atmosphere, regardless of the surrounding atmosphere.

The same conclusion seems to be valid for the thermogravimetric destruction of the anionic PMMA, and other results of a minute study concerning the thermogravimetric behavior of PMMA confirm this overall statement [7].

There may be some pure kinetic reasons for a modification of the reaction mechanism of the thermal degradation of polymers in nonisothermal conditions.

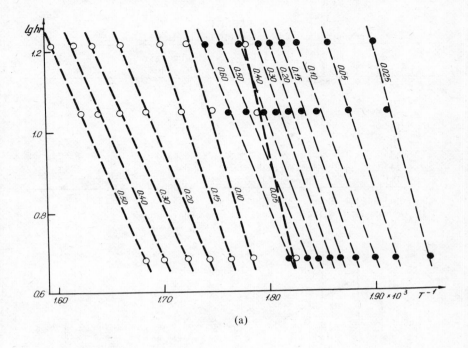

(a)

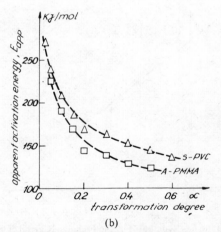

(b)

FIG. 3. (a) Flynn–Wall–Ozawa (FWO) plot of the degradation in air of PVC and PMMA. ●, s-PVC; ○, A-PMMA. The figures on the lines indicate the constant transformation degree. (b) Dependence of the apparent activation energy, calculated by the FWO method on the transformation degree.

First, the initiation process may change due to the presence of chemical bonds of different strength, causing a more or less marked modification of the whole reaction mechanism. Second, a temperature rise during the degradation process will influence preferentially one by one the possible parallel reactions.

In both cases the modifications of the reaction mechanism may be so impor-

TABLE I
Thermogravimetric Behavior of PVC; First Degradation Step[a]

Sample Size (mg)	Atmosphere[b]	Heating Rate (°C/min)	E_{app} (kJ mole^{-1})
5	air (D)	12.8	236.6
10	air (D)	12.8	222.8
20	air (D)	12.8	229. 1
30	air (D)	12.8	233.7
20	air (D)	4.8	288.4
59	air (L)	4.8	294.7
74	vacuum (L)	4.8	285.1
78	nitrogen (L)	4.8	286.7

[a] M_w = 143,000, polydispersity 3.5.
[b] (D), Derivatograph, MOM Budapest; (L), thermobalance "Linseis."

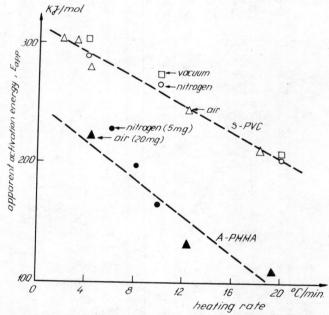

FIG. 4. Dependence of the E_{app}, computed by the Freeman–Carroll (FC) method, on the heating rate. Open points, first degradation step of s-PVC; full points, A-PMMA.

tant that kinetic computations are at most possible for very low transformation degrees as seen in Figure 5 for radical PMMA, which degrades, as is well known, at lower temperatures by end group initiation, while at higher temperatures by random chain scission [8].

Last but not least, because of the temperature rise, even in a simple chain

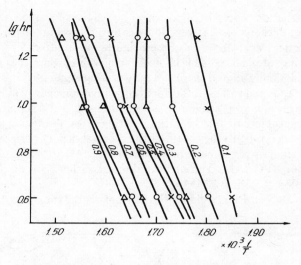

FIG. 5. FWO plot of the degradation in air of R-PMMA.

depolymerization, a stationary state will never occur, so that the kinetic parameters must modify with the transformation degree [9].

Consequently it is not commendable to compute kinetic parameters of the

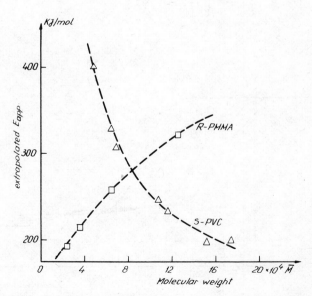

FIG. 6. Dependence of the E_{app} extrapolated to zero h.r. (or to zero transformation degree) on the molecular weight. $\triangle$, s-PVC, M_w = 143,000; $\square$, R-PMMA, M_v = 84,500.

thermal destruction processes of polymers by the use of a single thermogravimetric curve. Perhaps only at very low heating rates and/or transformation
degrees, or better by extrapolation of these figures for zero transformation degree
(or zero heating rate), it is possible to obtain some values which may characterize
the initial reaction rates and therefore may present some physical meaning.

These extrapolated values of the apparent activation energy may depend on
the molecular weight of the polymer as seen in Figure 6 and supposing that the
figures are characteristic for the initial reaction rates, the shown dependences
will indicate according to the theory of the thermal depolymerization of Boyd
[10] an initiation by random chain scission and long "zip"-length for the degradation of the PVC and end group initiation and rather short "zip"-length for
the degradation of radical PMMA.

If the above statement is valid for the thermal degradation of radical PMMA

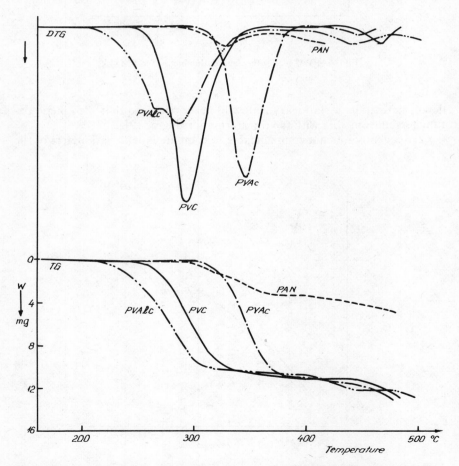

FIG. 7. Thermogravimetric curves for the first thermogravimetric step in air of some vinyl polymers.
– · · – PVAlc, Hoechst, $K_w = 70$, 98% saponification; — s-PVC, $M_w = 143,000$; – · – PVAc, BDH,
$M_v = 170,000$; - - - PAN, $M_v = 121,000$.

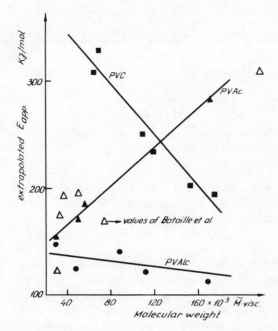

FIG. 8. Dependence of the E_{app} extrapolated to zero h.r. (or zero t.g.) of the first degradation step on the molecular weight of the polymers.

at very low transformation degrees [8], the conclusion concerning the reaction mechanism of PVC needs some discussion.

First it is necessary to show that Boyd's theory of thermal depolymerization applies to the dehydrochlorination of PVC and similar processes only if we suppose that these reactions can be assumed as "depolymerization" processes of model compounds formed by fixed $(HX)_n$ polymers to stable polyenic macromolecular chains [11].

$$\left(\begin{matrix} H\text{---}X \\ | \\ C\!\!=\!\!\!=\!\!C \\ | \quad | \\ H \quad H \end{matrix}\right)_n \longrightarrow {}_nHX + \left(\begin{matrix} C\!\!=\!\!C \\ | \quad | \\ H \quad H \end{matrix}\right)_n$$

Second, if an initiation of the PVC-dehydrochlorination by random chain scission seems to be acceptable, a long "zip"-length may be explained only by the admission of transfer reactions, because Braun and Thallmaier [12] have shown that generally the length of polyene sequences is not too high in partially dehydrochlorinated PVC.

Thermogravimetric curves for the destruction in air of vinylic polymers which degrade by elimination of small HX molecules are shown in Figure 7.

As seen from the data given in Table II, the percentage weight loss in the first degradation step, calculated as HX molecules, decreases from poly(vinyl alcohol)

TABLE II
Percentage Weight Loss of Functional Groups of the First Thermogravimetric Step in Air[a]

Polymer	Weight Loss Calculated as	Weight Loss (%)
PVAlc	H_2O	112.5
PVC	HCl	98.1
PVAc	CH_3COOH	81.3
PAN	HCN	32.4

[a] Heating rate 12.4°C/min, w_0 = 20 mg.

(PVAlc), to poly(vinyl chloride) (PVC) to poly(vinyl acetate) (PVAc), and to poly(acrylonitril) (PAN). The higher value for PVAlc may be explained by a lower saponification degree as that indicated by the supplier, while the much lower loss in the case of PAN usually is accounted by internal cyclization [13].

The dependences of the extrapolated values of the apparent activation energy (Fig. 8) suggest different reaction mechanisms. If PVC degrades, as shown before, by random chain initiation and long "zip," in PVAc end group initiation and shorter "zips" seem to be the characteristic kinetic steps. PVAlc behaves intermediate, i.e., the extrapolated values of the apparent activation energy (E_{app}) depend not essentially on the molecular weight. However, this behavior enables no conclusions concerning the reaction mechanism.

Our data for PVAc are also in accordance with those of Bataille and Van [14], who studied the degradation in an inert dynamic atmosphere of helium at a heating rate of 10°C/min. These authors attributed the remarkable spread of their results to an influence of the polydispersity of the used probes.

However, we wish to note supplementarily, that the endothermic character of the degradation in air decreases in the order PVC > PVAc > PVAlc, the latter showing even a slight exothermic DTA curve. At the same time, in opposition to the behavior of PVC, the corresponding Flynn–Wall–Ozawa plots exhibit linear dependence only for PVAlc probes with lower molecular weight [15] and are more or less curved for PVAc [16], suggesting modifications in the reaction

TABLE III
Thermogravimetric Behavior of VC–VAc Copolymers[a]

Polymer	$\overline{M}_{visc}$	Loss of Functional Groups (%)	$E_{app}/$ kJ mole^{-1}
PVC 56% Cl	40,370	97.6	212.5
PVAc 85% Acetate	485,000	87.4	250
VC/VAc 80/20 44.1 Cl	28,300	76.6	189
VC/VAc 85/15[b]	M_n = 28,000		163.4

[a] Degradation in air, heating rate = 12.4°C/min, w_0 = 20 mg.
[b] Studied by Malhorta et al., degradation in nitrogen, heating rate = 20°C/min.

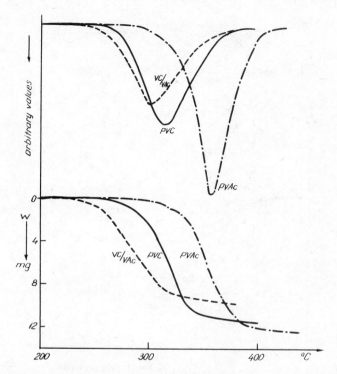

FIG. 9. Thermogravimetric curves for the first degradation step in air of random vinyl chloride/vinyl acetate copolymers, $w_0 = 20$ mg, h.r. $= 12.4°C/min$. For the characteristics of the polymer probes see Table III.

mechanism during the degradation. The thermogravimetric analysis of a statistic vinyl chloride–vinyl acetate copolymer with a vinyl chloride content of 80 wt % [17], showed that the copolymer is less thermostable then the corresponding homopolymers.

This result is in accordance with the data of Malhorta et al. [18] (which are included in Table III) concerning the degradation of a copolymer with 85% vinyl chloride in a nitrogen atmosphere at different heating rates and with those of Grassie et al. [19] who used a pyrotechnic analysis method. These authors affirmed that copolymers having a vinyl acetate content of about 50% are truly the least stable.

On the contrary, in the case of acrylonitrile–vinyl acetate copolymers it seems that statistic copolymers with 7–8 wt % content of vinyl acetate are more thermostable than the homopolymers (Fig. 10). It may be of interest to show that these copolymers have also the best spinning properties. The thermogravimetric behavior of vinyl polymers which degrade by main chain scission is shown in Figure 11.

It seems that those polymers which depolymerize preponderantly (i.e., poly(α-methylstyrene) (P-α-MS) and poly(methylmethacrylate) (PMMA) are less thermostable than the polymers which degrade essentially by tertiary

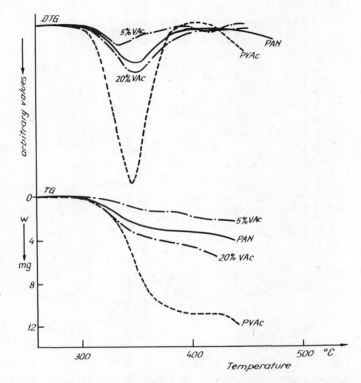

FIG. 10. Thermogravimetric curves of random vinyl acetate/acrylonitrile copolymers. Degradation in air, $w_0 = 20$ mg, h.r. = 12.4°C/min. The figures indicate the wt % content in vinyl acetate of the copolymers.

H transfer and random β-scission, i.e., linear poly(ethylene) (l-PE). However, this statement may be arguable because of the fact that in dynamic thermal conditions the weight loss, especially in the last case, may be controlled by the elimination of the volatiles. The higher the temperature, the more probable will be volatilization of molecules of larger chains, so that apparently the weight loss will be more intensive.

This assumption is supported by the aspect of the thermogravimetric curves of l-PE obtained at different heating rates (Fig. 12), taking into account that different oligomers, but not ethylene, occur in the pyrolysis of PE [20].

As seen, the evolution of the weight loss with temperature is much faster the higher the heating rate. Against it at lower heating rates, after an initial period of slow weight loss, the transformation gets impetuous at a certain temperature and the only explanation is that these then become also volatile certain larger chain fragments.

In such a situation kinetic computation based on thermogravimetric results are not only difficult as demonstrated by the Flynn–Wall–Ozawa and Free-man–Carroll plots in Figure 13, but perhaps they have also no sense. This may explain why Anderson and Freeman [21] obtained no linear plots for the thermal degradation of PE in vacuum.

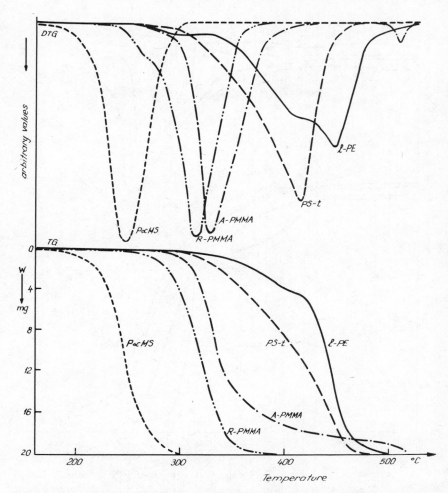

FIG. 11. Thermogravimetric curves for the degradation in air of some vinyl polymers, w_0 = 20 mg, h.r. = 12.4°C/min. - - - P-alphaMS, M_v = 25,200; - · · - R-PMMA, M_v = 84,500; - · - A-PMMA, M_v = 93,800, M_n = 56,000; - - t-PS, by thermal polymerization, M_v = 132,000; and — 1-PE, SRM 1475 M_n, 18,310, 0.15 CH_3 groups/100 C atoms.

If the results concerning the thermal degradation of PMMA, as shown before, are in accordance as well with the statements of the isothermal studies [8] and with the observations of the few nonisothermal studies [22], the data concerning the thermal behavior of polystyrene (PS) need some supplementary explanation.

If isothermal studies [23] reached not to final conclusions concerning the initiation, the influence of the molecular weight and the complexity of the degradation process, all nonisothermal studies agree about the occurrence of at least two parallel reactions. This is confirmed by the rise of E_{app} during the degradation process. Figure 14 shows some of our results [24] and those of Malhorta et al. [25] concerning the thermogravimetric behavior of anionic PS.

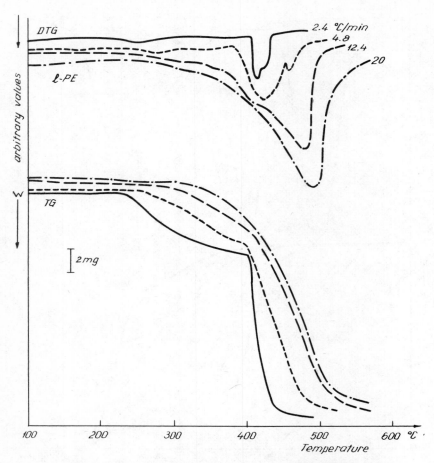

FIG. 12. Thermogravimetric curves for the degradation in air of 1-PE at different heating rates.

However, in an earlier work on thermally polymerized PS [26] we observed an opposite dependence, but that time the calculations of the E_{app} were performed by another method [27]. The dependence of the E_{app} extrapolated of the degradation in air of the thermal polymerized PS for zero heating rate suggests random chain initiation and long "zip" (Fig 15). The same mechanism seems to be also valid for the degradation of anionic PS according to the data of Malhorta et al.

However, Kokta et al. [28] observed an opposite influence of the molecular weight on the activation energy, while Funt and Magill [29] found no influence of this factor. But in both works the heating rates used were unusually high.

The study of the thermogravimetric behavior of P-α-MS was not extended. Although it has no *semnification,* we mention that the E_{app} of the degradation in air, as computed by Freeman–Carroll's method, was of about 203 kJ mole^{-1}. This value is somewhat lower than those reported for the isothermal depoly-

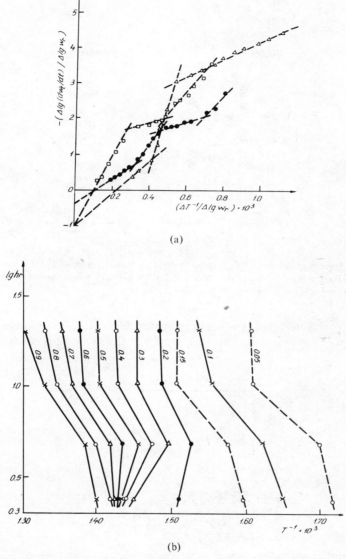

FIG. 13. (a) FC and (b) FWO plots for the thermogravimetric degradation of 1-PE in air. (a) h.r. = □, 4.8; ●, 12.4, and △, 20°C/min; (b) the figures on the curves indicate the constant t.g.

merization in absence of air [30], but if we take into account that generally E_{app} decreases with increasing heating rate and we used 12.4°C/min, the discrepancy is not too high. The analysis of the thermogravimetric behavior of methyl-methacrylate–styrene copolymers (PMS) offered some interesting results. Using different copolymers with a styrene content of about 50 mole %, we observed in constant working conditions that a ternary block copolymer behaves thermogravimetrically between the two homopolymers, while the alternate and

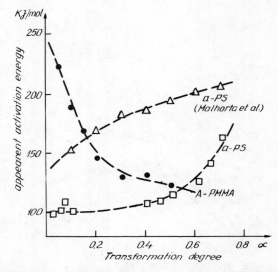

FIG. 14. The dependence of the E_{app} of the thermal degradation in air of PMMA and PS on the transformation degree; ●, A-PMMA, M_n = 47,000; □, a-PS, by anionic polymerization M_n = 52,000; △, a-PS, M_w = 1,800,000, data of Malhorta et al.

random copolymers show a different character (Fig. 16). This was also confirmed by a study of pyrolysis gas chromatography of the same copolymers [31].

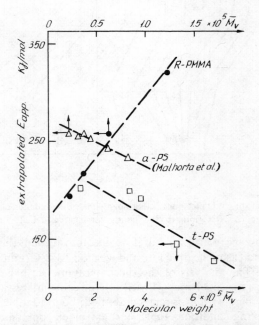

FIG. 15. Dependence of the extrapolated E_{app} on the molecular weight of the polymer.

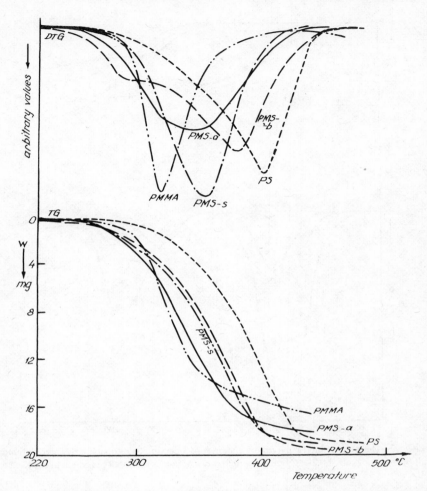

FIG. 16. Thermogravimetric curves of methyl methacrylate/styrene copolymers. Degradation in air, w_0 = 20 mg, h.r. = 12.4°C/min. For characteristics of the polymers see Table IV.

At the same time the figures of the E_{app} (Table IV) show that both the two last copolymers, but especially the random, are more thermostable. A possible explanation is that in random copolymers the normal depolymerization "zip" is interrupted. Therefore, the longer the "zip" in a homopolymer, the more thermostable will be a random homopolymer. The only condition, however, is, that weak links which favor the initiation are not introduced by copolymerization. On the other hand in random copolymers longer chain fragments may occur, which volatilize with more difficulty. However, our statement is in contrast to the conclusions of Grassie and Farish [32], who used an isothermal method and of McNeill [33], who used the TVA method in the study of PMS copolymers. These authors concluded that random copolymers behave *thermic* between the corresponding homopolymers.

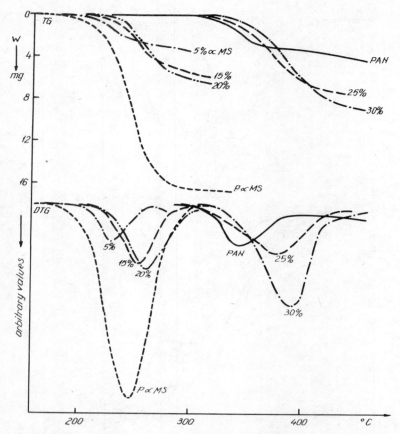

FIG. 17. Thermogravimetric curves of random α-MS-acrylonitrile copolymers. Degradation in air, $w_0 = 20$ mg, h.r. = 12.4°C/min. The figures on the curves indicate the content of α-MS in wt %.

Very interesting is the thermogravimetric behavior of copolymers obtained from monomers which give polymers degrading by polar bond scission and by main chain C—C bond scission, respectively.

So in Figure 17 we show the thermogravimetric curves of random α-methylstyrene–acrylonitrile copolymers. Unexpectedly, the copolymers with an α-MS content higher than 20 wt % are more thermostable than PAN. Also, pyrolysis gas chromatography data confirmed that these copolymers behave peculiarly [34]. If in copolymers with an α-MS content less than 20% exists a proportionality between the α-MS/AN ratio of the pyrolysis products and the α-MS content of the copolymer, at higher contents the ratio not only does not rise further in the same extent, but even diminishes slightly.

Also, unexpected results show styrene–vinyl chloride copolymers obtained by Friedel–Crafts reaction of the PVC with benzene [34] (Fig. 18). Regardless of the substitution degree, even at figures as high as 85%, the copolymer behaves PVC-like. Only the initial degradation temperature increases somewhat as seen

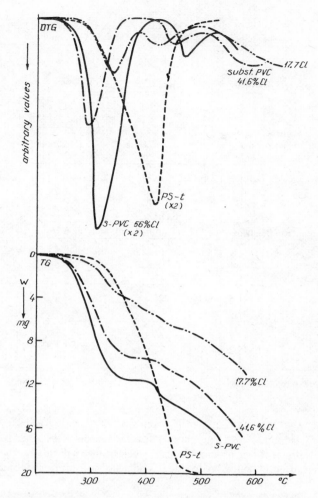

FIG. 18. Thermogravimetric curves of probes of PVC substituted with benzene by Friedel–Crafts reaction. The figures on the curves indicate the chloride content of the substituted PVC.

in Table V. Pyrolysis gas chromatography data indicated that these copolymers show always beside styrene structures due to the simple Friedel–Crafts substitution, also double substitution in the benzene ring in the *ortho*-position against the same macromolecular chain, or in *para*-position against a second polymer molecule [36].

In conclusion, dynamic thermogravimetry offers some useful results concerning the degradation of polymers. However, kinetic studies require a great volume of thermogravimetric curves taken necessarily at different heating rates and the computation of the dependence of the apparent activation energy on the transformation degree. Even in this case, kinetic calculations are not always possible, especially when parallel reactions are probable. However, kinetic parameters will be never computed with a single thermogravimetric curve. Sup-

TABLE IV
Thermogravimetric Behavior of MMA-S Copolymers[a]

Polymer	Content of Styrene (mol %)	$\overline{M}_n$	E_{app} (kJ mole^{-1})	Nature of the Process
PMMA—anionic		52,000	138.0	exo
PS—anionic		47,000	148.0	exo
PMSM—three sequenced block cop.	56	90,000	136.0	exo
PMS—a alternate copolymer	50	—	208.0	endo/exo
PMS—s random copolymer	~55	80,000	246.5	endo

[a] Degradation in air, heating rate = 12.4°C/min., w_0 = 20 mg.

TABLE V
Thermogravimetric Behavior of s-PVC Substituted with Benzene by Friedel–Crafts Reaction

Polymer	Cl Content (%)	Substitution Degree (%)	T_i (°C)[a]	Weight Loss (%) Theor.	Weight Loss (%) Exp.
PVC 41.6	41.6	26.6	222	43.3	46.6
PVC 33	33.0	41.8	234	34.0	36.5
PVC 23.7	23.7	58.2	240	24.4	26.1
PVC 17.7	17.7	68.8	252	18.2	22.3
PVC 12.7	12.7	77.6	257	13.1	14.2
PVC 8.3	8.3	85.4	260	8.5	12.3

[a] T_i temperature of initial weight loss in air at a heating rate of 12.4°C/min.

plementary thermogravimetry fails when the polymer degrades in larger fragments by β-scission. These fragments will volatilize only at higher temperatures, so that initially at lower temperatures no weight loss will be observed, although the polymer already begins to degrade. The possible situations which may occur

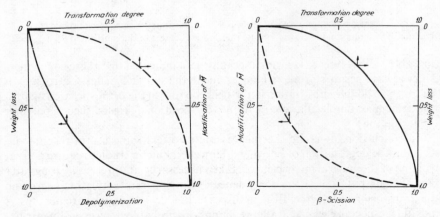

FIG. 19. Thermal transformation curves of polymers for typical degradation processes. Full line, weight loss; dotted line, decrease of the mean molecular weight.

during the thermal degradation processes of polymers are illustrated in Figure 19.

As seen, there exist always discrepancies between the weight loss and the decrease of the main molecular weight of the polymer during the degradation processes. However, this depends on the reaction mechanism.

In a polymerization the weight loss will be more rapid than the decrease of the mean molecular weight, the difference being the greater, the longer the "unzip" (the kinetic chain length) of the degradation reaction. A similar behavior shows the degradation processes by elimination of small molecules, HX, which can be considered as a "depolymerization" of an unstable "$(HX)_n$ polymer" fixed on a stable polyenic backbone.

In contrast, in the degradation processes by tertiary H transfer followed by β-scission of the main chain, the decrease in molecular weight will always be faster than the weight loss, this time the difference being smaller the higher the reaction temperature. In this situation, dynamic thermogravimetry fails.

Therefore, it is fortunate that thermogravimetry will be connected with other methods of analysis such as pyrolysis gas chromatography, gas evolution analysis, but especially with a method of determination of the molecular weight modifications. Copolymers may behave similarly or intermediate to the corresponding homopolymers. This depends on whether the structure is essentially maintained by copolymerization, as for example in the case of block copolymers. However, block copolymers will always be as stable or even less stable than the less thermostable homopolymer.

If the structure is perturbed, then the copolymer will show a different thermal behavior. This will be observed mainly in random copolymers. Either it will be less thermostable, if weak links, which may initiate the degradation, are introduced by copolymerization, or more thermostable, if the normally depolymerization "zip"-length of the homopolymer is perturbed by the insertion of the copolymer sequences.

In a limiting case, alternate copolymers may be considered as real homopolymers, showing therefore their own thermal behavior.

REFERENCES

[1] J. Šesták, V. Šavata, and W. W. Wendlandt, *Thermochim. Acta, 7,* 333–556 (1973).

[2] J. H. Flynn and L. A. Wall, *J. Polym. Sci., B4,* 323 (1966); T. Ozawa, *Bull. Chem. Soc. Jpn., 38,* 1881 (1965).

[3] I. A. Schneider, *Makromol. Chem., 125,* 201 (1969); *Rev. Roumaine Chim., 17,* 291 (1972).

[4] P. D. Garn, *CRC Critical Rev. Anal. Chem.,* 66–111 (1972).

[5] D. Furnică and I. A. Schneider, *Makromol. Chem., 108,* 182 (1967); I. A. Schneider, C. Vasile, D. Furnică, and A. Onu, *Makromol. Chem., 117,* 41 (1968).

[6] E. S. Freeman and B. Carroll, *J. Phys. Chem., 62,* 394 (1958); *J. Polym. Sci., 54,* 299 (1961).

[7] Natalia Hurduc and I. A. Schneider, *Rev. Roumaine Chim.,* in press.

[8] S. Madorsky, *Thermal Degradation of Organic Polymers,* Interscience, New York, 1964, pp. 176–188; J. R. MacCallum, *Makromol. Chem., 83,* 137 (1965); A. Barlow, R. S. Lehrle, J. C. Robb, and D. Sunderland, *Polymer, 8,* 557 (1967).

[9] I. A. Schneider, *Wiss. Z. Techn. Hochsch. Chem. Leuna-Merseburg, 15,* 61 (1972); *Rev. Roumaine Chim., 17,* 291 (1972).

[10] R. H. Boyd, private communication.

[11] I. A. Schneider and A. Cs. Biró, *J. Therm. Anal., 5,* 293 (1973).

[12] D. Braun and M. Thallmaier, *Makromol. Chem., 99,* 59 (1966).

[13] N. Grassie and I. C. McNeill, *J. Polym. Sci., 27,* 207 (1958); *39,* 211 (1959); J. Schurz, *J. Polym. Sci., 28,* 438 (1958).

[14] P. Bataille and B. T. Van., *J. Therm. Anal., 8,* 141 (1975).

[15] M. Popa, C. Vasile, and I. A. Schneider, *J. Polym. Sci., Polym. Chem. Ed., 10,* 3679 (1972).

[16] E. Mihai, A. Cs. Biró, A. Onu, and I. A. Schneider, *Makromol. Chem., 175,* 3437 (1974).

[17] I. A. Schneider and C. Vasile, *Eur. Polym. J., 6,* 687 (1970).

[18] S. L. Malhorta, J. Hesse, and L.-P. Blanchard, *Polymer, 16,* 269 (1975).

[19] N. Grassie, I. F. McLaren, and I. C. McNeill, *Eur. Polym. J., 6,* 679, 865 (1970).

[20] H. Seeger and H.-J. Cantow, *Makromol. Chem., 176,* 1411 (1975).

[21] D. A. Anderson and E. S. Freeman, *J. Polym. Sci., 54,* 253 (1961).

[22] Yu. N. Sazanov, E. P. Skvortsevich, and E. B. Milovskaia, *J. Therm. Anal., 6,* 53 (1974); I. C. McNeill, *Eur. Polym. J., 4,* 21 (1968).

[23] H. H. G. Jellinek, *J. Polym. Sci., 3,* 850 (1948); N. Grassie and W. W. Kerr, *Trans. Faraday Soc., 53,* 234 (1957); *55,* 1050 (1959); S. L. Madorsky, *J. Polym. Sci., 9,* 133 (1953); G. G. Cameron and J. R. MacCallum, *J. Macromol. Sci., Rev. Macromol. Chem., C1,* 327 (1967).

[24] I. A. Schneider and N. Hurduc, *Makromol. Chem.,* in press.

[25] S. L. Malhorta, J. Hesse, and L.-P. Blanchard, *Polymer, 16,* 81 (1975).

[26] C. N. Caşcaval, C. Vasile, and I. A. Schneider, *Makromol. Chem., 131,* 55 (1970).

[27] L. Reich, *Polymer Lett., 2,* 621 (1964); L. Reich and D. W. Levi, *Makromol. Chem., 66,* 102 (1963).

[28] B. V. Kokta, J. L. Valada, and W. N. Martin, *J. Appl. Polym. Sci., 17,* 1 (1973).

[29] J. M. Funt and J. H. Magill, *J. Polym. Sci., Polym. Phys. Ed., 12,* 267 (1974).

[30] H. H. G. Jellinek and H. Kachi, *J. Polym. Sci., C23,* 97 (1968).

[31] N. Hurduc, C. N. Caşcaval, I. A. Schneider, and G. Riess, *Eur. Polym. J., 11,* 429 (1975).

[32] N. Grassie and E. Farish, *Eur. Polym. J., 3,* 267 (1967).

[33] I. C. McNeill, *Eur. Polym. J., 4,* 21 (1968).

[34] C. N. Caşcaval and I. A. Schneider, *Rev. Roumaine Chim., 20,* 575 (1975).

[35] P. Teyssié and G. Smets, *J. Polym. Sci., 20,* 351 (1956).

[36] C. N. Caşcaval, I. A. Schneider, and I. C. Poinescu, *J. Polym. Sci., Polym. Chem. Ed., 13,* 2259 (1975).

FREE RADICAL AND IONIC GRAFTING

JOSEPH P. KENNEDY

Institute of Polymer Science, The University of Akron, Akron, Ohio 44325

SYNOPSIS

Advantages and disadvantages of graft copolymer syntheses by free radical and ionic (i.e., anionic and cationic) techniques are compared. Recent significant advances in the field of cationic graft copolymerizations are summarized and discussed. Particular emphasis is placed on cationic syntheses which employ alkylaluminum coinitiators in conjunction with macromolecular initiators which produce unadulterated, pure graft copolymers, e.g., EPM-*g*-polystyrene, SBR-*g*-polyisobutylene, chlorobutyl-*g*-polystyrene. In these systems grafting efficiencies are close to 100%, which demonstrates by direct gravimetric analysis that chain transfer can be completely eliminated in carbenium ion polymerizations. The synthesis of PVC-*g*-polytetrahydrofuran which involves the grafting of THF from PVC using silver salts also leads to pure graft. Finally, cationic systems which either directly or after selective extraction produce pure grafts and thus allow the examination of physical properties of pure materials will be presented.

INTRODUCTION. BRIEF COMMENTS ON GRAFT SYNTHESES BY FREE RADICAL AND ANIONIC METHODS

While this Seminar is mainly devoted to discussion of problems in radical copolymerizations including graft copolymerization, in view of recent significant advances in ionic, particularly cationic, graft copolymerizations, the Organizing Committee felt that a limited expansion of the scope of this Seminar was desirable and that a first-hand report on this subject should be made.

Among the techniques that have been successfully employed for the synthesis of graft copolymers, the free radical method is certainly the oldest and the most investigated one. Graft copolymers may be synthesized by free radical techniques utilizing (a) chemical methods, (b) photolytic methods, (c) high energy irradiation techniques, and (d) mechanochemical techniques.

Chemical methods involve the generation of free radicals by suitable chemical reactions and the utilization of these radicals by "grafting from" or "grafting onto" techniques [1]. In grafting from, the free radical (or other active site) is generated on the backbone and subsequently it initiates the polymerization of monomers to produce branches; in grafting onto, a growing free radical (or other active species) attacks another preformed polymer preferentially carrying suitable substituents and thereby produces a branch on the preformed backbone.

Journal of Polymer Science: Polymer Symposium 64, 117–124 (1978)
© 1978 John Wiley & Sons, Inc.   0360-8905/78/0064-0117$01.00

Well-investigated free radical chemical methods include grafting processes involving unsaturation in the chain, peroxide or diazo groups in the chain, macromolecular redox initiators, and trapped radicals. Photolytic methods can be either direct (illumination of photolyzable groups on backbones which induce branch formation) or indirect (illumination of photosensitizers which in turn lead to grafting by a radical mechanism). Similarly, high energy radiation techniques may involve direct or mutual irradiations, and preirradiation techniques. Finally, a variety of mechanochemical methods have been described that produce radicals which, among other reactions, may also lead to grafts [1].

Except for a few notable exceptions, i.e., the ceric ion redox technique developed by Mino and Kaizerman [2] that results in essentially homopolymer-free grafts, free radical induced grafting processes usually not only lead to the desired graft copolymer but also to homopolymer(s) and other disturbing side reactions. Consequently, the exploration and detailed characterization of grafts produced by free radical methods is often cumbersome or sometimes impossible by present day analytical techniques. Indeed, in spite of numerous misleading titles of scientific publications and patents which purport to describe the synthesis of graft copolymers, very few unadulterated, pure grafts have been produced by free radical techniques. Published physicochemical properties of the products are usually those of mixtures comprising a bewildering combination of grafted and ungrafted polymer sequences and fragments, and do not reflect the properties of pure materials.

Only since very recently, i.e., since the discovery of the excellent physical properties of well-defined sequential copolymers, particularly those of thermoplastic elastomers, have scientists started to pay attention to the inherent properties of pure block and graft copolymers.

Thus it was inevitable that the search for new useful sequential polymers was rapidly extended beyond the boundaries of the more familiar, less expensive, free radical polymerizations, to the largely uncharted territory of ionic polymerizations.

Historically, the significant breakthrough in this field was the clear recognition of the concept of "living" anionic polymerizations including the potential for the synthesis of sequential (block) copolymers by M. Szwarc et al. in ~1956 [3, 4].

Looking back on this field from a perspective of about 20 years, phenomenal progress both in the scientific and technological–commercial sector is quite evident: hundreds of scientific publications including many books on blocks and grafts, and literally hundreds of patents bespeak scientific growth; and several commercial products, even "second generation" thermoplastic elastomers, made by this technique, indicate sustained commercial interest.

In contrast to the great advances made in anionic block copolymerization, development of anionic graft systems remains much less impressive. Both anionic grafting "from" and "onto" [1] methods have been employed; however, neither of these techniques produced satisfactory products and ill controlled side reactions often obscured the desired reaction paths. Anionic grafting from was used by several workers who metallated (lithiated) various preformed backbones, e.g., polybutadiene and PVC and subsequently contacted these reactive intermediates with anionically initiable monomers, e.g., styrene [5–10]. Anionic grafting onto was used by others [11–13] who prepared living macromolecules, e.g., living polystyryl anions, and contacted these with macromolecules carrying suitable reactive substituents, e.g., ester groups and halides. Readers interested in deepening their general understanding in regard to problems peculiar to free radical and anionic graft synthesis methods may want to consult several excellent books available [14–20].

RECENT DEVELOPMENTS IN CATIONIC GRAFTING

Recently, in the course of our fundamental studies on the mechanism of cationic polymerizations, we have discovered simple and efficient methods for the synthesis of some unique graft copolymers. In many instances these products can be directly synthesized in essentially pure form and they exhibit interesting physical–mechanical properties.

Historically, cationic methods employing conventional Lewis acids such as $AlBr_3$, $TiCl_4$, have been examined on several occasions by most able investigators as possible routes to grafts; however, they soon found that these techniques produced but small amounts of grafts together with large quantities of undesirable homopolymers and degradation fragments. The source of homopolymers may have been protogenic impurities in the system which in conjunction with the conventional Lewis acids used rapidly and efficiently initiated the homopolymerization of monomers in the system, and led to unavoidable chain transfer reactions. The grafting of styrene from PVC by Friedel–Crafts halides serves as an example. Plesch proposed [21] that polymeric halides, e.g., PVC, could be used to generate polymer-cations in the presence of Lewis acids such as $AlCl_3$, $TiCl_4$ and that the PVC carbocation could initiate the polymerization of styrene. Experimentally, Plesch dissolved PVC in chlorobenzene or nitrobenzene, added $AlCl_3$ or $TiCl_4$ and finally styrene. The presence of graft copolymer in the product was evidenced by its solubility behavior. In this grafting process, however, large quantities of homopolystyrene must have also formed because $AlCl_3$ or $TiCl_4$ under the particular conditions employed rapidly polymerizes styrene. Also, upon treatment with $AlCl_3$ and $TiCl_4$, PVC must have decomposed and this was indicated by discoloration of the initially colorless product. Moreover, dehydrochlorination which occurs upon the addition of $AlCl_3$ or $TiCl_4$ to PVC produces HCl even in an initially pure system, which in turn may lead to homopolymerization–initiation and consequently to reduced grafting efficiencies. While changing the reagent addition sequence to PVC–styrene–Lewis acid might have alleviated some of these problems, it would not have eliminated them. A detailed discussion of early cationic grafting research is given in a book soon to be published [22].

Dormant research in the field of cationic grafting was revitalized by the discovery [23] that dialkyl- and trialkyl-aluminum compounds are unable to coinitiate the polymerization of even very reactive olefins like isobutylene or

α-methylstyrene, in the absence of a suitable protogenic or cationogenic initiator. It was recognized [23] that, for example, Et_2AlCl in conjunction with cation sources, be these small molecules (t-BuCl) or polymeric (PVC, polychloroprene, chlorobutyl), efficiently initiates olefin polymerizations. The events leading to initiation may be schematized as follows:

$$(Et_2AlCl)_2 \rightleftharpoons 2Et_2AlCl \tag{1}$$

$$Et_2AlCl + MeCl \rightleftharpoons Et_2AlCl{\leftarrow}MeCl \tag{2}$$

$$Et_2AlCl{\leftarrow}MeCl + t\text{-}BuCl \rightleftharpoons Et_2AlCl{\leftarrow}t\text{-}BuCl + MeCl \tag{3}$$

$$Et_2AlCl{\leftarrow}t\text{-}BuCl \rightleftharpoons t\text{-}Bu^{\oplus} Et_2AlCl_2^{\ominus} \tag{4}$$

$$t\text{-}Bu^{\oplus} Et_2ACl_2^{\ominus} \rightleftharpoons t\text{-}Bu^{\oplus} \cdot Et_2AlCl_2^{\ominus} \rightleftharpoons \cdots t\text{-}Bu^{\oplus} + Et_2AlCl_2^{\ominus} \tag{5}$$

$$t\text{-}Bu^{\oplus} + C{=}C \rightarrow t\text{-}Bu\text{-}C\text{-}C^{\oplus} \tag{6}$$

First, the dimeric Et_2AlCl coinitiator dissociates to monomeric species [eq. (1)]. The rate of this process is largely determined by the nature (mainly polarity) of the medium [24]. In the presence of the tertiary chloride t-BuCl and MeCl diluent, the initially formed $Et_2AlCl{\leftarrow}MeCl$ complex [eq. (2)] is converted to the $Et_2AlCl{\leftarrow}t$-BuCl complex [eq. (3)] which ionizes [eq. (4)] and, depending on the reaction parameters, proceeds to form contact, solvent separated, etc. ion pairs and, ultimately free ions [eq. (5)]. The position of this series of equilibria is again a function of conditions. Finally, initiation is completed by the electrophilic attack of one of the carbocations on the olefin in the system.

According to eqs. (1)–(6) t-BuCl will produce homopolymer with a t-Bu head-group, and, in a similar vein suitable polymer chlorides, for example those carrying tertiary, benzylic or allylic chlorines, will lead to polymer head groups, i.e., graft copolymers [22].

While this development greatly improved the controlled synthesis of cationic graft copolymers and removed the most important source of homopolymer contamination, that due to initiation by protogenic impurities, it did not completely eliminate the problem: Homopolymer contamination due to chain transfer could still occur:

$$\text{\small\raise1pt\hbox{$\sim$}}C\text{-}C^{\oplus} + C{=}C \rightarrow \text{\small\raise1pt\hbox{$\sim$}}C{=}C + C\text{-}C^{\oplus} \tag{7}$$

$$C\text{-}C^{\oplus} + nC{=}C \rightarrow \text{homopolymer} \tag{8}$$

Grafting efficiency is a quantitative measure of this problem, i.e., homopolymer formation due to chain transfer,

$$GE(\%) = \frac{P_b}{P_b + P_h} \times 100$$

where P_b = the weight of the newly formed grafted branch and P_h = weight of homopolymer. Extensive research with a variety of systems led to the definition of reaction conditions under which the problem of homopolymer formation could be eliminated and graft copolymers with 100% GE could be produced.

TABLE I
Effect of the Nature of Alkylaluminum Coinitiator on Grafting Efficiency [25][a]

Temp ^0C	Solvent nC_6H_{14}/EtCl v/v	Grafting Efficiency (%)		
		Et_2AlCl	Me_3Al	Et_3Al
-10	50/50	24	49	78
-35	70/30	35	69	87
-35	50/50	61	67	87
-50	50/50	69	78	88

[a] 20 g/liter chlorinated EPM (with 3.0 wt % Cl); [Styrene] = $1.5M$; conversion range 10–35%; $[Et_2ACl] = 2 \times 10^{-3}M$, $[Et_3Al] = 4 \times 10^{-2}M$, $[Me_3Al] = 1 \times 10^{-2}$ to 4×10^{-2} M; total volume = 300 ml.

One of the systems which was found to produce essentially homopolymer-free grafts was that of EPM-g-PSt, i.e., a graft comprising an EPM backbone (a random ethylene–propylene copolymer) and polystyrene branches [25].

We found that Cl-EPM (i.e., chlorinated EPM) in conjunction with Et_3Al coinitiator using a 50/50 v/v mixture of n-hexane/ethyl chloride at −50°C produced essentially pure graft copolymer, i.e., GE up to 88%. Table I shows some representative data.

Similarly, pure SBR-g-PIB grafts can be prepared by initiating the grafting of isobutylene by chlorinated SBR (styrene–butadiene rubber) in conjunction with Et_2AlCl in the temperature range from −30 to −70°C. Table II shows some data [26].

Another cationic synthesis which leads to practically pure graft copolymers is the chlorobutyl-g-PSt system [27]. Thus polymerization of styrene initiated by the chlorobutyl/Et_3Al initiator/coinitiator system using n-pentane/meth-

TABLE II
Grafting Isobutylene from SBR [26][a]

Temp ^0C	EtCl vol. %	Time min.	Yield g	Conv. %	GE %	SBR content(%)	$\overline{Mn}$ x10^{-5}
-30	35	3	11.0	22	83	45.5	1.10
-50	35	8	11.9	25	89.7	46	1.52
-70	20	8	11.8	25	100	45.5	1.73
-70	35	15	11.9	25	100	44.9	2.06
-70	50	8	12.3	27	97.6	43.3	2.13

[a] Total vol: 300 ml = n-heptane + EtCl + isobutylene; 5 g chlorinated SBR per charge; Cl content of SBR in solution: $3.9 \times 10^{-3}M$; $[Et_2AlCl] = 2 \times 10^{-2}M$.

TABLE III
Influence of Temperature on Grafting Sytrene From Chlorobutyl[a]

Temp ^{0}C	Conv. %	GE %	PSt in graft %	$\overline{Mn}$ x 10^{-3}
Et$_2$AlCl				
-30	15	71	31	200 ± 10
-40	20	65	30	228 ± 8
-50	16	72	43	260 ± 15
-65	9	78	26	–
Me$_3$Al				
-30	17	50	19	–
-40	15	65	32	–
-50	13	61	24	–
Et$_3$Al				
-30	12	86	39	–
-40	8	87	40	–
-50	10	92	40	–
-60	6	90	40	–

[a] [St] = 1.01M, [Cl-IIR] = 20 g/liter of $\overline{Mn}$ = 145 × 10^3, n-C$_5$H$_{14}$/CH$_2$Cl$_2$ = 70/30 v/v, [Et$_2$AlCl] = 0.024M, [Me$_3$Al] = 0.08M, [Et$_3$Al] = 0.13M.

TABLE IV
Polymer Sequences Utilized in Cationic Graft Syntheses by "Grafting From" Technique

Backbones		Branches	
Rubbers	Glasses	Rubbers	Glasses
poly(ethylene-co-propylene) (EPM)	poly(vinyl chloride)	polyisobutylene	polystyrene
poly(ethylene-co-propylene-co-diene) (EPDM)	polystyrene	butyl rubber (IIR)	poly(α-methylstyrene)
chlorobutyl rubber		polychloroprene	polyindene
chlorosulfonated polyethylene		polytetrahydrofuran	poly(indene-co-α-methylstyrene)
styrene-butadiene rubber (SBR)			polyacenaphthylene
polybutadiene			polydioxolane
polychloroprene			poly(1,4-dichloro-2,3 epoxybutane)
poly(p-chloromethyl-styrene-co-butadiene)			polyoxetane

ylene chloride (70/30 v/v) diluent gave virtually pure graft copolymer. Grafting efficiencies up to ~75% were also readily obtained using Et$_2$AlCl or Me$_3$Al. Table III shows some representative data.

The fact that such essentially pure graft copolymers can be obtained by direct synthesis is significant not only from the point of view of materials science but also for the kinetic theory of cationic olefin polymerizations in general. Thus, 100% grafting efficiencies demonstrate by direct gravimetric analysis and without resorting to sometimes quite cumbersome kinetic experimentation (e.g., Mayo plots) that chain transfer can be completely suppressed and eliminated even in such notoriously chain transfer prone systems as reactive olefins.

Another advantage of the use of higher alkylated organoaluminum compounds resides in their relatively mild Lewis acidities. In contrast to conventional strong Lewis acids, e.g., $AlCl_3$, higher alkylaluminums have been found to be much less aggressive toward preformed polymer chains, thus reducing the possibility for disturbing degradative side-reactions.

Another grafting method of great promise recently developed in our laboratories involves the grafting of heterocycles, e.g., tetrahydrofuran, frrom various backbones, e.g., PVC and polychloroprene [28, 29]. In this method we induce the polymerization of tetrahydrofuran by a macromolecular halide and a silver salt, for example, silver triflate. The synthesis exploits the fact that triflate esters induce the polymerization of cyclic esters and that initiating triflates can be produced by mixing polymers containing active (e.g., allylic) chlorines with silver triflate [28]:

$$P\text{—}Cl + AgOSO_2CF_3 \longrightarrow P\text{—}OSO_2CF_3 + AgCl$$

$$P\text{—}OSO_2CF_3 + THF \rightarrow P\text{—}O(CH_2)_4\text{—}OSO_2CF_3 \rightleftharpoons P\text{—}O^+ \rbrack \ ^-OSO_2CF_3$$

$$\downarrow THF$$

$$P\text{—}PTHF \text{ graft}$$

where P = preformed polymer backbone, e.g., PVC, polychloroprene.

Table IV summarizes polymer sequences that have been recently utilized in preparing graft copolymers by cationic grafting from techniques by our group and other researchers around the world. The sequences are divided into backbones and branches which are further subdivided into rubbers and glasses. While the synthesis of many backbone/branch combinations have already been demonstrated, obviously, not all the possibilities offered by this compilation could be explored and many backbone/branch pairs are yet to be examined and evaluated. Research along these lines is intensively explored in our laboratories.

Financial support by the Division of Materials Research of the National Science Foundation is gratefully acknowledged.

REFERENCES

[1] J. P. Kennedy, in *Recent Advances in Polymer Blends, Grafts and Blocks,* L. H. Sperling, Ed., Plenum, New York, 1974. p. 3.

[2] G. Mino and S. Kaizerman, *J. Polym. Sci., 31,* 242 (1958).

[3] M. Szwarc, M. Levy, and R. Milkovich, *J. Am. Chem. Soc., 78,* 2656 (1956).

[4] M. Szwarc, *Nature, 178,* 1168 (1956).

[5] G. Greber and G. Egle, *Makromol. Chem., 53,* 208 (1962).

[6] G. Greber and G. Egle, *Makromol. Chem., 71,* 47 (1964).

[7] M. Y. Yampolskaya, O. Y. Okhlobitsin, S. L. Davydova, and N. A. Plate, *Vysokomol. Soedin., 8,* 771 (1966).

[8] H. Harada, K. Shiima, and Y. Minoura, *J. Polym. Sci., A1, 6,* 559 (1968).

[9] J. C. Falk and R. J. Schlott, *J. Macromol. Sci., Chem., A7,* 1663 (1973).

[10] K. Schiima and Y. Minoura, *J. Polym. Sci., A1, 4,* 1069 (1966).

[11] A. F. Halasa, G. B. Bitchell, M. Stayer, D. P. Tate, A. E. Oberster, and R. W. Koch, *J. Polym. Sci., Chem. Ed., 14*, 497 (1976).

[12] J. Roth, P. Rempp, and J. Parrod, *Comp. Rend., Paris, 251*, 3356 (1960).

[13] P. Rempp, J. Parrod, G. Laurent, and Y. Gallot, *Comp. Rend., Paris, 260*, 903 (1965).

[14] H. A. J. Battaerd and G. W. Tregear, *Graft Copolymers,* Interscience, New York, 1967.

[15] W. J. Burlant, *Block and Graft Polymers,* Reinhold, New York, 1960.

[16] R. J. Ceresa, *Block and Graft Copolymers, Butterworths,* London, 1962.

[17] *Block and Graft Copolymerization,* R. J. Ceresa, Ed., Wiley, New York, Vol. I (1973); Vol. 2 (1976).

[18] D. C. Allport and W. H. James, *Block Copolymers,* Wiley, New York, 1973.

[19] J. A. Manson and L. H. Sperling, *Polymer Blends and Composites,* Plenum, New York, 1976.

[20] A. Noshay and J. E. McGrath, *Block Copolymers, Overview and Critical Survey,* Academic, . New York, 1977.

[21] P. H. Plesch, *Chem. Ind., 1958,* 954.

[22] *Cationic Graft Copolymerization,* J. P. Kennedy, Ed., *J. Appl. Polym. Sci., Appl. Polym. Symp., 30* (1977).

[23] J. P. Kennedy and F. P. Baldwin, Belgian Patent 701,850 (Jan. 26, 1968) and U.S. Patent 3,904,708 (Sept. 9, 1975).

[24] J. P. Kennedy and G. E. Milliman, *Adv. Chem. Series, 91,* 287 (1969).

[25] J. P. Kennedy and R. R. Smith *Recent Advances in Polymer Blends, Grafts and Blocks,* L. H. Sperling, Ed., Plenum, New York, 1974, p. 303.

[26] J. Oziomek and J. P. Kennedy in Ref. [22].

[27] J. P. Kennedy and J. J. Charles in Ref. [22].

[28] P. Dreyfuss and J. P. Kennedy, *Polym. Lett., 14,* 135 (1976).

[29] P. Dreyfuss and J. P. Kennedy, *Polym. Lett., 14,* 139 (1976).

GRAFTING PROCESS IN VINYL ACETATE POLYMERIZATION IN THE PRESENCE OF NONIONIC EMULSIFIERS

I. GAVĂT, VICTORIA DIMONIE, D. DONESCU, C. HAGIOPOL, MARIANA MUNTEANU, KRISTIANA GOŞA, and TH. DELEANU

Institute of Macromolecular Chemistry "P. Poni," 6600 Jassy, Romania

SYNOPSIS

The influence of vinyl acetate grafting reactions or nonionic type emulsifiers on the kinetics and on properties of emulsions has been studied. The emulsifiers used were polyvinyl alcohol and block copolymers ethylene oxide–propylene oxide. Due to the grafting reactions the polymerization rate decreased or increased with the increase in polyvinyl alcohol concentration, depending on the emulsifier/monomer ratio. During polymerization, graft copolymers of the monomer on the emulsifier chain are obtained. These copolymers are either water-soluble, benzene-soluble, or insoluble in water and benzene. The number of grafting reactions increases with the rise in initiator concentration. This leads to a decrease of benzene-soluble fractions and to an increase of the water-soluble and water-and-benzene-insoluble fractions. The concentration of emulsifier in water-and-benzene-soluble fractions was established by NMR analysis. The IR spectra of products proved the formation of graft copolymers.

INTRODUCTION

Emulsion polymerization is one of the most widespread industrial procedures for polymer synthesis, due to certain technological advantages. Apart from its technological importance, emulsion polymerization presents also a scientific interest because of the complexity of the process resulting from the overlapping of the polymerization reaction with colloidal phenomena occurring at the interface.

Several theories have been advanced for emulsion polymerization and it is difficult to make a generalization in this sense due to the specificity and diversity of reaction mechanisms as a function of the reagent types participating in polymerization. For the same monomer the emulsion properties, the size and number of polymer particles, depend on several factors, the most important being the emulsifier and initiator nature and concentration and the monomer water-solubility.

In regard to the nature of the emulsifier, emulsion polymerization may be divided into three main groups, each of them exhibiting characteristic features:

Journal of Polymer Science: Polymer Symposium 64, 125–140 (1978)
© 1978 John Wiley & Sons, Inc.   0360-8905/78/0064-0125$01.00

1. Emulsion polymerization in the presence of ionic emulsifiers has been extensively studied and Smith and Ewart [1] have advanced a quantitative theory, subsequently developed and completed by Medvedev [2, 3].

For partially water-soluble monomers the Smith–Ewart micellar theory of particle formation has no longer been verified.

In the case of such monomers [4, 5] and even of styrene [6] it has been admitted that radical formation and polymerization initiation occur in aqueous phase and the resulting oligomer radicals precipitate from solution, followed either by homogeneous self-nucleation and formation of initial polymer particles, or by capture to the already existing polymer particles.

Quantitative aspects regarding the number of formed particles based on the model of initiation and nucleation in the homogeneous phase have been presented by Fitch [7]. The experimental data on emulsion polymerization of acrylates and methacrylates verify the suggested mechanism [8].

All these mechanisms excluded the possibility of polymer initiation in monomer droplets because of the relatively small global surface compared to that of micelles or of primary polymer particles. However, recently, Ugelstad and Vanderhoff [9] have shown that when monomer dispersion occurs in droplets of very reduced dimensions producing a drastic growth of their surface, the initiation takes place in monomer droplets protected by emulsifiers.

2. Emulsion polymerization in the presence of nonionic emulsifiers has been far less studied. The systematic investigation carried on lately by Medvedev et al. [3] has revealed essential differences between emulsion polymerization mechanisms in the presence of ionic and nonionic emulsifiers.

Although these emulsifiers also form micelles in solution, Medvedev et al. have shown in a number of papers [10–12] that in this case the whole monomer amount is distributed in discrete monomer–polymer particles whose surface is entirely covered by the emulsifier. The number and size of particles is dictated by the emulsifier/monomer ratio, the determining parameter of the process being the ratio between the surface and the volume of droplets.

3. The third group includes polymerizations carried out in presence of nonionic macromolecular emulsifiers, among which ethylene oxide–propylene oxide block copolymers and polyvinyl alcohol (PVA) are most largely used.

The use of macromolecular emulsifiers complicates even more the picture of phenomena occurring in emulsion polymerization for the following reasons: (a) Depending on the conditions existing in the medium, the conformation of emulsifier macromolecules in solution will radically change and, hence, accordingly also their emulsifying properties; (b) the radical transfer processes entail the formation of graft polymers which also greatly affect the emulsifying properties, hence, in general, all the emulsion characteristics.

One of the most important problems of emulsion polymerization, as it also results from the above, is the locus in which the elementary processes take place during polymerization, i.e., monomer droplets, monomer–water interface, monomer-saturated emulsifier micelles, or the monomer aqueous solution. Depending on the existing conditions (emulsifier, monomer and initiator concentrations) the polymerization reaction will preponderantly develop in one of these possibilities.

The literature reports a relatively large number of papers on vinyl acetate polymerization in the presence of PVA, but yet the problems regarding the mechanism of this process are far from being totally elucidated [13, 14].

The present paper undertakes to clarify some aspects of vinyl acetate (VAc) emulsion polymerization in the presence of PVA in relation to the emulsifier influence on the formation of polymer particles and also on the characteristics of the resulting emulsion. Likewise, comparative data are presented with regard to the VAc emulsion polymerization in the presence of other nonionic emulsifiers, EO–PO block copolymers.

EXPERIMENTAL

Vinyl acetate, commercial reagent, was purified on a high efficiency column. No traces of aldehydes could be identified after distillation. Polyvinyl alcohol, commercial product, was used as an aqueous solution of 7–14% concentration, and has a molecular mass of about 150,000. Block copolymers EO–PO were synthetized by Petroleum Research Institute–Cîmpina. Their composition was established by NMR and IR analysis.

Polymerizations were performed at $63 \pm 0.5°C$, in a four-necked flask equipped with thermometer, dropping funnel, reflux cooler, and mechanical stirrer.

The polyvinyl alcohol solution was introduced into the polymerization flask, and distilled water then added to adjust to the desired concentration. After thermostating, formic acid was added up to pH = 3, then the iron (II) sulfate solution and, subsequently, the vinyl acetate drop by drop for 10 min. The moment when the initiator was introduced was considered the zero moment of the reaction. In all the polymerizations, the overall reaction volume, the stirring speed, and the rate of addition of reactants were maintained constant.

During polymerization, samples were withdrawn to determine the gravimetric conversion and to analyze the polymer obtained.

From the samples taken in the course of polymerization, films were obtained by drying at room temperature and were subjected to successive extractions in water and in benzene at boiling for 5–6 hr each, obtaining finally three fractions. The amounts of polymer soluble in water, soluble in benzene, and insoluble in benzene were determined gravimetrically.

The water-soluble fraction was analyzed from the point of view of chemical composition by determination of the acetate groups.

IR spectra were performed by means of a Hilger and Watts spectrophotometer.

The NMR spectra were performed by means of a Jeol JNM 60 HL instrument on the polymer solution of 5% concentration in deuterium oxide or benzene at 40 and 100°C, respectively.

The UV spectra were measured on a UV-vis Speroid Carl Ziess Jena spectrometer at room temperature.

RESULTS AND DISCUSSION

As seen in Figure 1, VAc polymerization in the presence of PVA, initiated by the H_2O_2–$FeSO_4$ redox system at high monomer concentration has a classical three-stage development. We remark on the existence of an acceleration period in the polymerization rate at the beginning of the process, followed by a constant rate interval (R_{max}) over a large conversion range, 15–90%.

The variation of the polymerization rate (R_{max}) with the initial monomer concentration represents a direct proportionality between the rate and the monomer concentration, at a constant PVA and initiator concentration (Fig. 2) $R_{max} \simeq (M_0)$ [1]. Taking into account the fact that the polymerization rate is of a zero order relative to the instant monomer concentration and of a first order relative to the initial monomer concentration, it can be admitted that the number of monomer–polymer particles formed in the first stage of the process is determined by the initial monomer concentration. In the constant rate stage, as in the case of other macromolecular emulsifiers [15, 16], the particle number remains constant.

For initial monomer concentrations below the limit of VAc water-solubility, the polymerization rate is smaller by about one order of magnitude and comparable with that found in the case of polymerization in aqueous solution [17].

The above-mentioned experimental data have suggested to us the hypothesis that the initiation takes place in discrete monomer–polymer particles whose

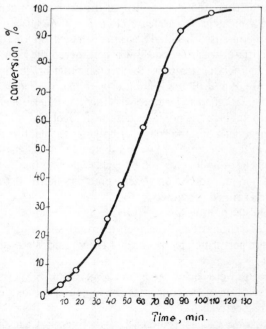

FIG. 1. Time conversion curve for the polymerization of VAc at 63°C [23] [VAc]$_0$, 3.44 mole/liter; [PVA]$_0$, 35.4 g/liter; [H$_2$O$_2$]$_0$, 3 × 10^{-3} mole/liter; [FeSO$_4$]$_0$, 5.85 × 10^{-5} mole/liter.

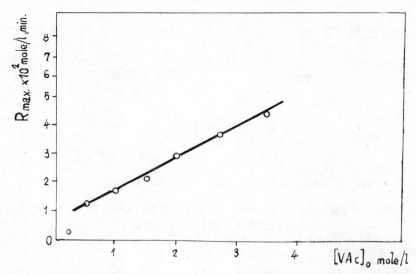

FIG. 2. Dependence of the maximum polymerization rate on the monomer initial concentration at 63°C [23]. $[PVA]_0$, 35.4 g/liter; $[H_2O_2]_0$, 3×10^{-3} mole/liter; $[FeSO_4]_0$, 5.85×10^{-5} mole/liter.

number/volume unit for a constant initiator concentration is proportional to the initial monomer concentration. This mechanism implies that the initiation is in monomer microdroplets which are superficially protected by a PVA layer. Studies of emulsifying with PVA solutions, performed by Abramzon [18], have demonstrated that these stabilize the emulsions, forming a structural–mechanical barrier at the surface of droplets under the form of a protective network.

The emulsifying process is well known as being a dynamic process and the transition between large monomer drops and those with the smallest dimensions takes place at equilibrium and continuously by a large range of intermediates with different sizes. This fact has been clearly evidenced for PVA by Biehn and Ernsberger [19] who have photographed various dibutylphthalate emulsions obtained with aqueous PVA solutions of various concentrations. According to the data reported by these authors, the higher the PVA concentration, the smaller is the size of droplets and the narrower their distribution by dimensions. If our hypothesis claiming that VAc polymerization in the presence of PVA occurs in microdroplets protected by PVA aggregates is real, then through polymerization, the phenomena observed on emulsifying VAc should be reflected in the dimension and distribution of the resulting polymer particles.

For this purpose we have performed a series of polymerizations at variable PVA/VAc ratios, aiming at monitoring the variation of polymer particle dimensions and their size distribution. It has been unanimously admitted that initiation takes place preferentially in particles having the smallest diameter and the largest surface related to the volume.

Therefore, the reaction conditions chosen by us involved a very high polymerization rate such that the number of free radicals per volume unit at a certain

TABLE I

Vinyl Acetate Polymerization with PVA and the Redox System H_2O_2–$FeSO_4$ at 65°C[a]

PVA/VAc (% Grav)	Solid Content of Emulsion	Viscosity (20°C) (cP)
8.2	50.7	3,150
6.5	50.2	850
5.0	50.3	642
3.0	50.3	170
1.5	49.9	37
0.25	50.2	suspension
0	traces	—

[a] PVA, 99% hydrolyzed; time, 120 min; H_2O_2, 1.18 g/liter; $FeSO_4 \cdot 7\ H_2O_2$, 0.054 g/liter.

moment would largely increase; this considerably enlarges the probability for a radical to penetrate into a droplet with large dimensions. We have deliberately worked with a PVA having a high hydrolysis degree, and hence weaker emulsifying abilities in the formation of polymer particles. The polymerization conditions are given in Table I.

The results obtained confirm the formulated hypothesis. Thus in the absence of PVA, polymerization does not take place in this condition. The introduction of very small PVA amounts produces the initiation of the polymerization and its continuation until complete monomer consumption. Very large particle dimensions are obtained in this case, resulting practically in a suspension.

The sedimentometric analysis of the dimensions and distribution of polymer particles of these emulsions shows that in parallel with the decrease of the PVA/VAc ratio, the maximum of distribution curves shifts towards higher values of the particles dimension and the dimensional polydispersity of particles rises (Fig. 3).

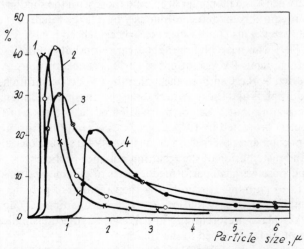

FIG. 3. Particle size distribution curves. Polymerization conditions are given in Table I. PVA/VAc (%): 1, 8.2; 2, 6.5; 3, 5; 4, 1.5.

Our data concerning the initiation of the polymerization disagree with the hypothesis reported by the Japanese authors [20], which admits that the initiation occurs in aqueous solution, followed by the oligomer precipitation and the formation of initial polymer particles. We base this statement on the following considerations: In the absence of PVA, polymerization has not been initiated in the conditions given; polymerization begins readily with the introduction of very small emulsifier amounts, which demonstrates the role of microdroplets formation for the initiation of the polymerization; polymerization in VAc aqueous solution at the limit of its solubility even in the PVA presence takes place at very low rates. At the moment when the monomer concentration exceeds the solubility limit, the polymerization rate presents a jump (Fig. 2).

Based on entirely different experiments, Nikitina et al. [16] have reached similar conclusions. They have shown by electron microscopy that for the VAc emulsion polymerization in the presence of EO–PO copolymers, the dimensions of polymer particles are identical with those of the initial microemulsion.

It is well known that in parallel with emulsion polymerization, vinyl acetate grafting reactions on polyvinyl alcohol will also occur [13, 21, 22, 23].

Studying the VAc grafting on PVA in the presence of $K_2S_2O_8$, Hartley [21] has reported the formation of two categories of graft copolymers: one water-soluble with a high PVA content, and another water-insoluble with a high PVAc content.

Quantitative studies on the grafted copolymer formation are not yet reported. The grafting process has important consequences both on the kinetics of reaction and the final properties of the emulsion. The dependence of the rate of polymerization on the emulsifier concentration has been reported by many authors but the kinetic data are very spread.

Thus we observed a decrease of the polymerization rate with the increase in the emulsifier concentration, a very unusual fact for emulsion polymerization

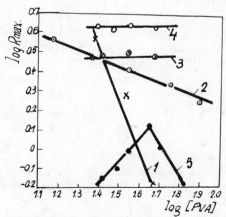

FIG. 4.   Dependence of polymerization rate of VAc on PVA concentration [23]. $[H_2O_2]_0$, 3×10^{-3} mole/liter; $[FeSO_4]_0$, 5.85×10^{-5} mole/liter; $[VAc]_0$: 1, 1.5; 2, 2.0; 3, 2.7; 4, 3.5 mole/liter; 5, $[VAc]_0$, 2.0 mole/liter, $[diallylether]_0 = 6 \times 10^{-3}$ mole/liter.

(Fig. 4). This decrease is more pronounced as the initial monomer concentration is smaller.

Since in the case of PVA the number of emulsion particles increases with the emulsifier concentration, as was shown above, we consider that the only explanation referring to the decrease of the polymerization rate may be correlated with the grafting process.

As a consequence of the transfer reactions with PVA, macroradicals result whose activity in polymerization is more reduced than that of the polyvinyl acetate radical. This fact was revealed by a number of experiments in which the polymerizations were performed in the presence of diallyl ether, a strong transfer agent (Fig. 4, curve 5).

When the diallyl ether transfer reactions are predominant, a normal dependence of the polymerization rate on the emulsifier concentration is obtained. This condition is fulfilled only at low PVA concentrations. Over a certain PVA concentration, the cumulative effect of the two transfer reactions with PVA and diallyl ether is obvious.

It is evident that the ratio between radicals formed on the PVA chain and the other radicals formed in polymerization is a function of the relative concentrations of reagents.

In order to determine the intensity of the grafting processes in emulsion polymerization, the polymer in the form of a film was subjected to successive extractions with water and benzene [23]. In all the cases, three polymer fractions have been obtained: water-soluble, benzene-soluble, and insoluble in water and benzene. The formation of an insoluble polymer fraction in VAc emulsion polymerization was also noted by Traaen [24].

The increase of the number of grafting reactions with the increase of the concentration of PVA is evident: the difference between the calculated and the experimental amounts of water-soluble polymer rises with the increase of the PVA concentration (Fig. 5). The PVA content in the aqueous fraction determined by chemical analysis has always been smaller than the calculated amount from the PVA initial content.

Concomitant with the decrease in the polymer amount soluble in benzene, an increase of the insoluble fraction is observed (Fig. 6).

As expected, when the PVA concentrations are maintained constant, the increase of the VAc concentration in the system also favors the grafting processes. The polymer amount soluble in benzene will always be smaller than the calculated one corresponding to the formed polyvinyl acetate (Fig. 7).

Important data on the development of the grafting process have been obtained by studying the variation in time of the water-soluble polymer amount. As can be seen (Fig. 8) the variation of the aqueous extract as a function of conversion is represented in all the cases by a curve with a maximum. The value of this maximum represents an increase of the water-soluble fraction by about 50% of the initially PVA amount.

We think that the ascending branch of the curve corresponds to the formation of graft copolymers with a high content of —OH groups and hence water-solubility. As the grafting process goes on and hence also the accumulation of a

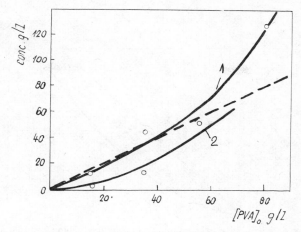

FIG. 5. Variation of the water-soluble fraction (1) and of PVA concentration existing in this fraction (2) with the PVA initial concentration. $[VAc]_0$, 172 g/liter; $[H_2O_2]_0$, 0.1 g/liter; $[FeSO_4]_0$, 8.7 × 10^{-3} g/liter; - - - calculated.

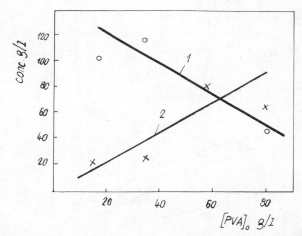

FIG. 6. Variation of the benzene-soluble fraction (1) and of the insoluble fraction (2) with PVA initial concentration. Polymerization conditions are the same as in Fig. 5.

larger number of polyacetate branching on the PVA chain, the water-solubility of the graft copolymer decreases, which characterizes the descending branch of the curves.

It is significant that the maximum of the curves is always reached at the same values of the ratio between the formed PVAc and PVA, i.e., a mean value of 1.6. At the same time the existing PVA amount in the aqueous extract decreases monotonously throughout the polymerization process.

The decrease of the water-soluble polymer fraction to even below the value of the initial PVA amount is due to the substantial changes in the PVA composition.

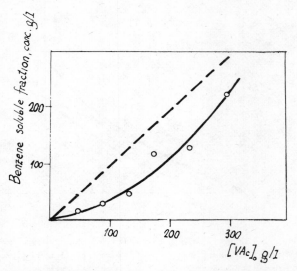

FIG. 7. Dependence of the benzene-soluble fraction on the initial monomer concentration. $[PVA]_0$ = 35.4 g/liter; reaction conditions are the same as in Fig. 2.

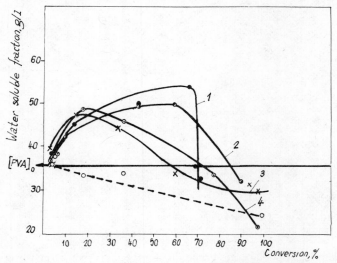

FIG. 8. Variation of the water-soluble fraction with conversion [23]. Reaction conditions are the same as in Fig. 2. $[VAc]_0$: 1, 1.0; 2, 1.5; 3, 2.72; 4, 3.44 mole/liter; - - -: the calculated weight of PVA in water-soluble fraction.

The existence of grafting reactions largely complicates the whole emulsion polymerization process since their direct consequence is the continuous change during polymerization of both: first, the emulsifying qualities of PVA by the modification in the balance of hydrophilic–hydrophobic parts, and, second, of the available emulsifier amount.

The decrease of the emulsifier amount in the aqueous phase below certain limits causes the agglomeration of emulsion particles, due to the absence of the protecting colloid. This phenomenon generally has been observed in the last stages of the polymerization, at high conversions.

The above data lead to the conclusion that an important parameter of the grafting reactions is the PVA/VAc ratio; but this ratio also determines a significant variation of the emulsion's viscosity (see Table I).

Since PVA undergoes strong transfer reactions and grafting, we have undertaken to evidence the influence of the PVA transformation degree on the emulsion viscosity, one of the important properties of the emulsions.

In a series of experiments, the PVA content, the initiator concentration, temperature, and reaction time have been kept constant, varying only the monomer amount, hence implicitly the PVA/VAc ratio (Table II). To intensify the grafting reactions we used PVA with a low hydrolysis degree capable of readily undergoing grafting reactions. The initiator was potassium persulfate, known to be a strong grafting agent [13].

The amount of water-soluble polymer and the viscosity of the resulting emulsion was determined (Fig. 9).

A comparison of the percentage of water-soluble polymer obtained experimentally with that calculated based on the PVA amount used in each sample shows two domains: in the first part of the curve at low VAc/PVA ratios, up to about 1.5, the water-soluble polymer amount is larger than that corresponding to the PVA amount. Throughout this domain of ratios we think that the grafting of polyvinyl acetate on PVA chains fails to entail its insolubilization.

Further, as the ratio increases, the water-soluble polymer fraction decreases below the PVA corresponding value, which proves an advanced grafting degree of PVA and produces its insolubilization in water. The fact must be mentioned that PVA insolubilization due to grafting starts in this case too at a VAc/PVA ratio very close to that previously found (a 1.6 value in Fig. 8), when the PVA grafting was studied as a function of the conversion.

Interesting results have been obtained by viscosity measurements of the corresponding emulsions. The results obtained with a Rheotest viscometer at three different shear gradients are presented in the Figure 10.

The variation of viscosity with the increase of the VAc/PVA ratio is illustrated

TABLE II
Vinyl Acetate Polymerization with PVA and Potassium Persulfate at 65 °C[a]

PVA/VAc (% grav.)	$K_2S_2O_8$/VAc (% grav.)	Solid content of emulsion (%)
181.0	1.20	9.8
105.0	0.77	14.1
75.5	0.55	16.3
62.1	0.44	17.2
52.7	0.37	18.5
43.9	0.31	21.9
35.2	0.25	24.1

[a] PVA, 84% hydrolyzed; time, 120 min.

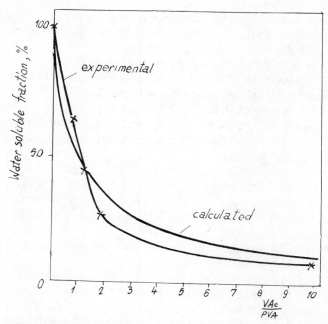

FIG. 9. Dependence of water-soluble fraction on VAc/PVA ratio. Polymerization conditions are given in Table II.

by an unusual curve aspect which however fully agrees with the results concerning the variation of the water-soluble polymer fraction.

A maximum viscosity is noted at a VAc/PVA ratio about the same at which the PVA insolubilization starts.

The viscosity increase may be explained if account is taken of the gradual modification in the PVA characteristics as a result of the more intense grafting with the increase in the VAc/PVA ratio. At the moment when a part of PVA leaves the aqueous phase because of a strong grafting, a decrease of the emulsion viscosity is observed. The following viscosity increase is probably due to the increased content of solid substance in the system (particularly polyvinyl acetate).

These findings present also a practical importance, as they emphasize the influence of the polymerization conditions on the emulsion viscosity, permiting the control of the reaction parameters and in particular the VAc/PVA ratio to obtain emulsions with expected properties.

In order to verify whether the characteristics observed in VAc emulsion polymerization using PVA as an emulsifier are in general specific of the macromolecular emulsifiers, a comparative study has been carried out regarding the possibilities of the PVAc grafting also on other macromolecular emulsifiers, at various initiator concentrations.

For this purpose we used an EO–PO block copolymer with an 80% gravimetric content of EO units, known to be an efficient emulsifier for VAc emulsion polymerization [15, 16, 25, 26].

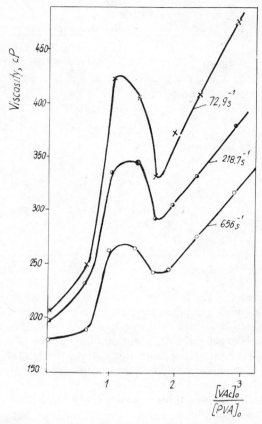

FIG. 10. Dependence of emulsion viscosities on VAc/PVA ratio. Polymerization conditions are given in Table II.

Although there are many studies of the VAc emulsion polymerization in the presence of such block copolymers, none of them has considered the possibility of VAc grafting on the emulsifier molecules.

Under similar polymerization conditions as in the case of polyvinyl alcohol, studies were made on the influence of the increasing initiator concentration ($K_2S_2O_8$); similar behaviors were found with regard to the composition of the polymer obtained from the emulsion.

When EO–PO block copolymers were used, three fractions were again obtained: water-soluble, benzene-soluble, and insoluble in water and benzene. From the values presented in Table III, it can be seen that also the EO–PO copolymer undergoes grafting reactions, proving to be a stronger transfer agent than polyvinyl alcohol, since the water-soluble polymer amount is much smaller than that corresponding to the introduced emulsifier amount and in general below the values obtained for polyvinyl alcohol. Correspondingly, the benzene-soluble fraction assumes higher values than those obtained for polyvinyl alcohol and that insoluble in benzene is situated below the values found for PVA.

TABLE III
Effect of the Initiator Concentration on the Polymer Composition[a]

Emulsifier	$K_2S_2O_8$ (g/l)	Water-Soluble Fraction (% grav.)	Benzene-Soluble Fraction (% grav.)	K	Insoluble Fraction (% grav.)
Polyvinyl alcohol (99% hydrolyzed)	0.2	3.8	85.5	78	10.7
	0.6	6.3	76.0	57	17.7
	1.2	15.4	77.9	80	6.7
Polyvinyl aacohol (84% hydrolyzed)	0.48	8.7	81.7	89	9.6
	0.9	6.4	60.0	76	33.6
	1.2	6.0	69.0	66	25.0
	1.58	16.9	57.8	76	25.3
Block copolymer EO–PO (80% EO)	0.64	6.2	81.0	52	12.8
	1.11	3.4	83.0	50	13.6
	1.59	5.1	83.0	60	11.9
	3.17	6.8	77.3	61	15.9

[a] VAc, 317 g/liter; emulsifier, 49.2 g/liter; temperature of polymerization, 65°C.

With the increase in the initiator concentration the water-soluble polymer fraction increases while the benzene-soluble polymer fraction decreases (Table III).

This behavior is explained by the fact that the number of graftings increase with the rise in the initiator concentration and this leads to the increase of the polymer amount insoluble in benzene, due to the increased cross-linking possibilities.

The analysis of fractions soluble in water and benzene by NMR spectroscopy confirms the above results and proves that the grafting reactions occur also on the EO–PO block copolymer chains. The amount of emulsifier found in the fraction soluble in water rises with the increase of the initiator concentration and the benzene-soluble fraction decreases (Fig. 11). The existence of the grafted polymers was also evidenced by IR spectroscopy.

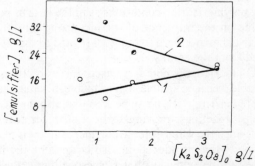

FIG. 11. Effect of the initiator concentration on the concentration of block copolymer EO–PO existing in water-soluble fraction (1) and benzene-soluble fraction (2). [VAc]$_0$, 317 g/liter; [emulsifier]$_0$ = 49.2 g/liter, 80% EO in emulsifier.

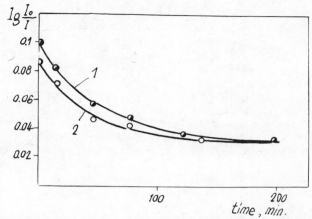

FIG. 12. Variation of lg I_0/I in time for aqueous solution of PVA (1, 84% hydrolyzed; 2, 99% hydrolyzed). [PVA], 30 g/liter; [$K_2S_2O_8$] = 1.6 g/liter; temperature, 65°C.

Thus, when the PVA was used as emulsifier the IR spectrum of the water-soluble polymer exhibits both the bands of the PVA and of the PVAc. Additionally two new bands appear at 1136 and 1046 cm^{-1} and a shoulder at 810 cm^{-1}. According to literature data [20] these bands prove the existence of grafted copolymers. Also in the case of benzene-soluble polymer fractions, besides the bands characteristic of PVAc, some bands specific to PVA have been observed. A new band was observed at 807 cm^{-1} and its intensity changed from sample to sample.

Therefore we consider that this band, which appears in the domain of 805–810 cm^{-1} in all the spectra, may be characteristic of grafted copolymers.

In addition to hydrogen atoms (—CH—; CH$_3$—COO) capable of reacting with the free radicals mentioned in the literature [27], the PVA chains also contain double bonds, capable of adding free radicals. The existence of carbonyl groups conjugated with double bonds was evidenced by UV spectroscopy [28, 29].

Using the same method, we tried to determine the modification of double bonds in the presence of radical initiators. The characteristic domains of absorption are 225, 280, and 325 mμ [28]. Since the $K_2S_2O_8$ initiator and the residual acetate groups in PVA give absorption changes in the domain 200–220 mμ, we studied the absorption change only in the domain 280 mμ.

According to the Lambert–Beer law, representing the log I_0/I variation as a function of the reaction time with $K_2S_2O_8$ for two types of PVA (99% and 84% hydrolyzed, respectively), the curves are exhibited in Figure 12.

The decrease of log I_0/I in time demonstrates the consumption of the conjugated double bonds adjacent to the carbonyl groups. Due to their high reactivity, these double bonds may constitute sites where free radicals attack the PVA chains. They may therefore constitute one of the causes for the formation of cross-linked polymers, whose concentration depends, as previously remarked, on concentration of PVA and initiator.

The authors wish to thank Dr. V. Bărboi and E. Popa for NMR and UV analysis.

REFERENCES

[1] W. V. Smith and R. W. Ewart, *J. Chem. Phys., 16,* 592 (1948).

[2] S. S. Medvedev, *Int. Symp. on Macromol. Chem.,* Pergamon, New York, 1959, p. 174.

[3] S. S. Medvedev, *Kinetics and Mechanism of Polyreactions,* Akademiai Kiado, Budapest, 1971, p. 39.

[4] R. M. Fitch, *Off. Dig. J. Paint. Technol. Eng., 37,* 32 (1965).

[5] R. M. Fitch and C. H. Tsai, *Polymer Colloids,* Plenum, New York, 1971, p. 73.

[6] C. P. Roe, *Ind. Eng. Chem., 60* (9), 20 (1968).

[7] R. M. Fitch, *Brit. Polym. J., 5* (6), 467 (1973).

[8] N. Süterlin, H. J. Kurth, and G. Markert, *Makromol. Chem., 177,* 1549 (1976).

[9] J. Ugelstad, N. S. El-Aasser, and J. W. Vanderhoff, *J. Polym. Sci., Polym. Lett., 11,* (8), 503, (1973).

[10] I. A. Gritskova, S. S. Medvedev, and M. F. Margaritova, *Koll. Zh., 26,* 168 (1964).

[11] V. V. Dudukin, I. A. Gritskova, I. N. Medvedeva, S. S. Medvedev, Z. M. Utsinova, and N. M. Fodiman, *Visokomol. Soedin., 10-A,* 456 (1968).

[12] S. S. Medvedev, I. A. Gritskova, A. V. Zuikov, L. I. Sedakova, and G. D. Berejnoi, *J. Macromol. Sci. Chem., A-7* (3), 715 (1973).

[13] M. K. Lindeman, *Vinyl Polymerization,* G. E. Ham, Ed., Vol. 1, Pt. I, Marcel Dekker, New York, 1967, p. 288.

[14] C. Hagiopol, D. Donescu, and V. Dimonie, *Stud. Cercet. Chim., 21* (8), 929 (1973).

[15] D. N. French, *J. Polym. Sci., 32,* 395 (1958).

[16] S. A. Nikitina, V. A. Spiridinova, and A. B. Taubmann, *J. Polym. Sci., A-1, 8,* 3045 (1970).

[17] J. W. Vanderhoff, *Vinyl Polymerization,* G. E. Ham, Ed., Vol. 1, Pt. II, Marcel Dekker, New York, 1969, p. 89.

[18] A. A. Abramzon, E. V. Gromov, and N. N. Macagonova, *Koll. Zh., 35* (1), 122 (1973); A. A. Abramzon and E. V. Gromov, *Koll. Zh., 31* (6), 795 (1969).

[19] G. F. Biehu and M. L. Ernsberger, *Ind. Eng. Chem., 40* (8), 1449 (1948).

[20] M. Furuta, *J. Polym. Sci., Polym. Lett., 11,* 113 (1973); *ibid., 12,* 459 (1974).

[21] F. D. Hartley, *J. Polym. Sci., 34,* 397 (1959).

[22] E. V. Gulbekian, "Emulsion Polymerization," in *Polyvinyl Alcohol,* C. A. Finch, Ed., Wiley, New York, p. 427, 1973.

[23] V. Dimonie, D. Donescu, M. Munteanu, C. Hagiopol, and I. Gavăt, *Rev. Roum. Chim., 19,* 931 (1974).

[24] A. H. Traaen, *J. Appl. Polym. Sci., 7,* 581 (1963).

[25] E. Y. Gulbekian, *J. Polym. Sci., A-1, 6,* 2265 (1968).

[26] S. G. Rogova, *J. Prikladnoi Himii, 44,* 1103 (1971).

[27] Y. Ikada, Y. Nishizaki, and I. Sakurada, *J. Polym. Sci., Polym. Chem., 12,* 1829 (1974).

[28] J. G. Pritchard, *Polyvinyl Alcohol,* MacDonald, London, 1970.

[29] V. T. Shirinian, S. S. Mnatzakanov, G. A. Shirikova, F. O. Pozdniakova, G. S. Popova, and T. V. Hvotintzeva, *Plast. Massî, 8,* 15 (1974).

TOPOLOGICAL MODEL OF THE CRYSTALLINE MORPHOLOGY IN POLYETHYLENE WITH GRAFTED DEFECTS

GH. DRĂGAN

Institute of Chemical Research, ICECHIM, Spl. Independentei 202, Bucharest, Romania

SYNOPSIS

The most important results obtained on polyethylene morphology as a function of thermal, mechanical and chemical treatments are reviewed. Amorphous–crystalline coupling in high density polyethylene with grafted defects by photochlorination is quantitatively studied by differential thermal analysis using six parameters. The model of this coupling is discussed in terms of the topological thermodynamics.

This work is a short presentation of the principal results obtained during approximately 6 years of study of polyethylene (PE) morphology in correlation with thermal, mechanical, and chemical treatments. The starting point of this study series was the adjustment of a calorimetric method for making evident phase transitions connected to the processing conditions of the polymers [1, 2]. This method, known as "flash calorimetry," differs from the classical adiabatic calorimetry (or drop calorimetry), by the fact that the heat fluxes have great amplitude and short periods of time, thus inducing, during unidirectional heat transfer, secondary processes of relaxation. Temporary dependence of response function C_p (heat capacity) is revealed in this calorimetric system as a function of the characteristics of the perturbation heat flux, and thus this method becomes a dynamic method.

Figure 1 shows two typical thermograms for a branched and linear species of PE, respectively [1, 2]. It can be observed that for the two samples processed by injection molding in similar conditions, two melting effects are revealed, namely the melting of the lamellar areas T_m^0 and a group of endothermal effects centered on 110°C, denoted as T_m^1 effect. By studying the appearance of this last effect, especially for injection molding specimens, the following characteristics are revealed [2]: (i) for linear species, T_m^1 transition is less dispersed and tends to a transition with lambda shape, which is characteristic of an order–disorder transition, and (ii) with the increase in the degree of branching, the T_m^1 transition group becomes more dispersed and extended over a large temperature domain. This phenomenon is similar to that previously observed and reported in the literature on a highly branched PE and in parallel on a copolymer of ethylene and butene, both gradually annealed in steps of 5° below

Journal of Polymer Science: Polymer Symposium 64, 141–148 (1978)
0360-8905/78/0064-0141$01.00

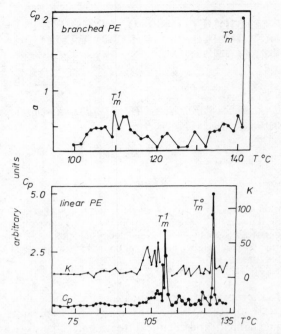

FIG. 1. Dependence of C_p versus temperature resulted in the flash calorimetric system for two species of polyethylenes. K is the coefficient of thermoconductivity [1].

T_m^0 melting point [3]. Both samples show in DTA scans a group of T_m^1 effects in addition to T_m^0 effect, which reproduce exactly the annealing temperatures.

Secondary melting processes are reported by dynamic methods also for other crystalline polymers, in connection to the annealing, mechanical, and chemical treatments. By analyzing these facts comparatively to the proper results, it has been concluded that T_m^1 effect corresponds to an interlamellar order initiated in these samples by row crystallization along shearing lines during injection molding, and perfected by gradual annealing during calorimetric scan [2]. This endothermal effect is frequently reported also for amorphous morphology where a nodular order forms as a result of annealing [4]. In crystalline morphology it has been observed that the lamellar areas separate from the amorphous domains as a result of the applied treatments, and the kinetics of this process strongly depends on molecular conformation and configuration; more exactly it depends on conformational and/or conformational defects in lamellar areas. These defects are responsible for these morphological modifications, by their migration towards an amorphous domain. These amorphous–crystalline transformation processes imply transport processes and it can be supposed that they are basically thermally activated relaxation processes. Indeed, the data obtained by Hoseman and co-workers [5] on a linear species of PE by plastic deformation allow the measurement of an activation energy of 35 kcal/mole [6]. A similar value of 31.4 kcal/mole was subsequently found in creep experiments [7]. This process is somewhat new, because annealing studies, especially for linear species, have revealed the thickening process on the $|001|$ direction only. Because this process implies two phases in PE morphology, via molecular conformation and/or configuration, it has been considered that the T_m^1 effect really defines an amorphous–crystalline coupling [2].

To perform this study, it has been necessary to chose an appropriate material and an analytic method adequate of visualizing this coupling. Three sorts of PE with essentially linear molecular species and with exclusively conformational lamellar defects were chosen. Their different defect concentrations can be spectroscopically established because the chain sequences containing G species as GTG, GG, TGT, and GTTG (G and T are gauche and trans isomers, respectively) are active in the infrared band of 1350 cm^{-1} [8]. The mobility of these defects in the unprocessed samples is greatly reduced [6], so that the thickening process is exclusively revealed after their annealing [9]. In this way we have "decorated" these defects by photochlorination, knowing the fact that the solid reactivity is proportional with defect concentration. This technique was already reported in literature even for PE as gamma irradiation, thermo-oxidative attack, halogenation, styrene grafting, and other chemical treatments, which separate crystalline domains from the amorphous ones in a so-called shish-kebab structure.

For a rapid and efficient comparison of the numerous samples resulting in different annealing conditions, differential thermal analysis was chosen. This method permits quantitative assessment of the T_m^1 effect by means of some parameters measured from the thermogram. Figure 2 reproduces a DTA thermogram of a chlorinated PE (CPE) annealed at 125°C for an hour, and where the T_m^1 effect exceeds the amplitude of the T_m^0 effect. Six parameters are used to define the amorphous–crystalline coupling, namely: h_1, h_2, the heights corresponding to the endothermal effects T_m^0 and T_m^1, respectively; $\alpha = h_2/(h_1 + h_2)$ the splitting coefficient; ΔH_m the melting enthalpy of the all melting process; and $\alpha \Delta H_m$ and $H_m^1 = (1 - \alpha)\Delta H_m$ the melting enthalpy associated with T_m^1 and T_m^0 effects, respectively. This system of characterization is called an "in vitro" system, because it does not directly characterize the formation of the amorphous–crystalline coupling during annealing. Three CPE samples with an equal chlorine content of 40% by weight, have been compared, corresponding to the three linear originating PE. The kinetics of the amorphous–crystalline coupling as a result of annealing between 110–130°C, were studied by this technique [9].

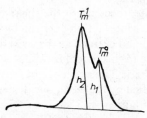

FIG. 2. Typical DTA thermogram obtained on a CPE annealed sample. $\alpha = h_2/(h_1 + h_2)$; h_1, h_2; $\Delta H_m, \alpha \Delta H_m$; $\Delta H_m^1 = (1 - \alpha)\Delta H_m$.

In the 1960s and 1970s, some relaxation properties of the response functions especially the calorimetric response functions as heat capacity, enthalpy, entropy, and encraty, were found for a category of low molecular weight compounds, defined as "glassy crystals" [10]. Multimelting effects appear in C_p versus temperature spectrum as a result of the thermal treatments, even for high purity samples. The mobility of the conformational and/or configurational defects induces in $C_p(T)$ dependence the addition of an excess heat capacity (enthalpy) ΔC_p, which is correlated for annealing at an equilibrium temperature, T below T_m^0 point, on the classical concept of thermodynamics, according to the following kinetic equation given by Baughman and Turnbull [11]:

$$\ln(T^2 \Delta C_p) \sim -\frac{E}{RT}$$

and which denotes its thermally activated aspect.

This reasoning has been applied for CPE samples, considering the values h_2, $\alpha \Delta H_m$ and perhaps α, as being proportional to the heat capacity in excess. The splitting coefficient α does not obey this dependence because it was observed that after annealing the process of defect precipitation progressively destroys the lamellae by their fragmentation, so that h_1 decreases while h_2 increases. Annealed samples as such, by dry annealing, and annealed in mixture with silicone oil are compared, respectively, and equal values of activation energy have been found for both annealing conditions and according to both parameters h_2 and $\alpha \Delta H_m$. For the three CPE samples the activation energies are between 24 and 53 kcal/mole [9]. It is important to note, however, that the silicone oil as annealing medium affects the T_m^1 process, also the amorphous–crystalline coupling, so that DTA thermograms of dry annealed samples, which are then mixed with silicone oil, show a modified T_m^1 effect compared to the dry samples.

A model of the amorphous–crystalline coupling that resulted after defect precipitation during annealing was proposed, considering the previous observations of Nagasawa and Kobayashi [12] on migration of gamma-reticles in PE. Precipitation process of decorated defects by grafting in the crystalline domains is also observed for other materials along a glide direction which defines the minimum elasticity modulus. This glide direction is found to be for PE a combination between $|100|$ and $|001|$ directions, according to Nagasawa and Kobayashi's results [12].

Between lamellar fragments and the newly formed amorphous domains an elastic energy is accumulated because of the shear stresses in their interfaces [Fig. 3(a)]. This situation of interaction between two elastically differing materials was recently studied and defined as the twist disclination state [13]. Interlamellar order according to this model is similar with a nematic structure in liquid–crystalline morphology. The existence of twist disclination state in annealed CPE samples is verified by the technique of shear lines development [14] and can be sketched as in Figure 4.

Compounds of annealed or unannealed CPE samples with polyvinyl chloride were successively melt rolled, row crystallized by air quenching, melted and

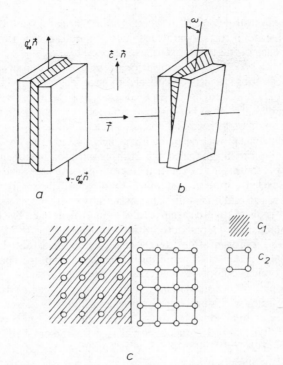

FIG. 3. The model of the disclination state (a,c) performed after defect precipitation, and its relaxation during T_m^1 process (b).

recrystallized in static conditions (namely in the absence of an external shear flow) by compression molding of the superposed sheets, and ultimately by the mechanical splintering of a cross section of the resulted plate. For compounds containing annealed CPE samples, even for the annealing of the originating PE previous to chlorination, shear lines induced in melt rolling and frozen by row crystallization are developed in the cross section of the plates. This effect is explained by two important assumptions: (i) the fact that the interlamellar order induced by PE and/or CPE annealing and reproduced by row crystallization is kept also in molten state, and (ii) the property of the disclination state which it relaxes on melting by reciprocal rotating and simultaneous approaching of the lamellar fragments [Fig. 3(b)]. The first fact was experimentally verified

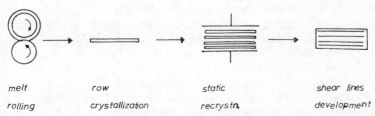

| melt | row | static | shear lines |
| rolling | crystallization | recrystn. | development |

FIG. 4. Schematically drawn technique used for development of the shear lines in CPE/PVC compounds.

for many paracrystalline orders induced in solid state and which are conserved also in molten state in static conditions. The second fact was theoretically deduced for the case of the twist disclination state [13].

The reordering process by precipitation of the grafted defects by photochlorination has been subsequently established to be an endothermal effect by using the isothermal DTA system as an "in vivo" system [9]. In this way a paradoxic situation is reached because both transformation processes, "in vitro" and "in vivo" systems, are of endothermal direction, and can not be explained by classical thermodynamic concepts of energy balance-sheet considering the material being unitary. An adequate language to explain this paradox has been imposed. So, topological representations of the transformation processes are considered to be the most adequate manner. In short, topological thermodynamics has an object to replace the real nonequilibrium system by components with elementary behavior, such as dissipative, capacitive, and inductive, defined by the constitutive equations. The basic concepts were initially introduced in 1971 [15]. The adaptation of these representations to real cases led to the establishment of basic principles as the fact that all nonequilibrium systems are of the composite nature [16]. PE is a typical composite material made of amorphous and crystalline domains as topological components. A kinetic equation for transformation processes in systems with purely dissipative coupling between components was deduced [17] starting from the constitutive equations for thermally driven processes [16]. This equation has been verified for the crystallization process in the three linear PE [17], and for the crosslinking process in mixtures of epoxy resins with curing agents [18].

Amorphous–crystalline coupling in CPE samples was experimentally established not to obey the approximation of purely dissipative coupling. In topological terms, the formation of the disclination state is a viscous process expressed by the transfer and accumulation of kinetic momentum [19]. This may be represented in a so-called bond diagram as follows: the two components [Fig. 3(c)] which capacitively accumulate in differing reference systems are connected by a dissipative medium and an inductive element [Fig. 5(a)]. This inductive element is explained by the fact that the kinetic momentum of translation which drives this process, suffers a modification of its direction by passing from C_1 reference system to C_2 reference system [19]. Considering the energy equivalence of the transformation modes, we may rewrite the process as thermally driven, for its study in a calorimetric system [Fig. 5(b)]. In this way the inductive flux w_L represents the accumulation of an elastic energy, which tends to annihilate by definition the flux w_c of capacitive accumulation by the formation of the interlamellar amorphous domain C_2, so that w_2 (transferred in the laboratory system of C_1) $= w_L + w_c \rightarrow 0$. This phenomenon results experimentally in a conduction calorimeter system so that the transformation flux is near zero, however the amplitude of w_L and w_c exceeds the apparatus sensitivity. In this representation the energetic effect observed in the "in vivo" system (in the isothermal DTA system) represents a w_c effect, by elimination of the inductive effect, and on the other hand the energetic effect associated with the T_m^1 process corresponds to the accumulation of the elastic energy equal to w_L, but of an opposite direction [20].

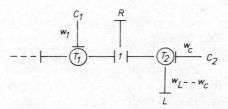

FIG. 5. Internal energetic circuit represented as viscous (a) and thermal (b) process, respectively, during annealing.

The more appropriate procedure to verify the existence of the inductive element in the internal circuit of the CPE samples during annealing, is their inductive coupling by simultaneous annealing of two identical samples [21]. This is experimentally performed in a cylindrical block made of brass which has four orifices, also cylindrical and vertical having the same distance between them and in relation to the center of the block, and two other horizontal ones with the centers at half the height of the metallic block whose axes form an angle of 90 degrees (Fig. 6). Two of the vertical orifices and one of the horizontal ones have the inner diameter equal to the inner diameter of the used DTA containers (almost 6 mm). Here the samples are directly annealed by using the brass as thermal dissipative medium. The other orifices have a 10 mm diameter so that the Pyrex cylindrical tubes, whose inner diameter equals that of the DTA container, are placed therein. The samples annealed in these "in vivo" conditions of symmetry are then cut lengthwise for thermal analysis without changing the thermal radial symmetry of "in vitro" conditions. Three reciprocal orientations of the samples during annealing were selected, and denoted as following: (I)

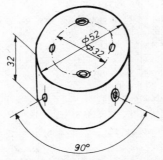

FIG. 6. Metallic block used for inductive coupling experiments.

individual annealing of one sample in the vertical position, as reference position; (II) simultaneous annealing of two samples in vertical positions; (III) simultaneous annealing of two samples, one in vertical position and other in horizontal position (we refer to the position of the axis of the orifice). All experimental results, both by annealing in brass and annealing in Pyrex glass, reveal that the splitting coefficient α shows a systematic coupling during annealing, so that

$$\alpha_{II} > \alpha_I > \alpha_{III}$$

In this way it is observed that the gravity field of the earth does not affect the annealing "in vivo" process because here is no difference between horizontal and vertical values.

The inductive element has been theoretically defined by its temporary relativistic effect between the reference system of time proper to the topologic components in the system [19]. To use this idea I take again the previous experiments but without keeping the spatial symmetry of the "in vivo" system in the "in vitro" system. In this way only a temporary dependence is revealed (spatial dependence is removed), and the obtained results concerning the six parameters suggest that the inductive coupling of two CPE samples behaves like the scattering process for a two particle system.

REFERENCES

[1] Gh. Drăgan, *Lucr. III-a Conf. Nat. Chim. Anal.*, IV, pp. 351, Sept., 22–26, 1971, Braşov.
[2] Gh. Drăgan, *Mater. Plst., 10*, 87, (1973).
[3] A. P. Gray and K. J. Casey, *J. Polym. Sci., B 2*, 381 (1964).
[4] K. Neki, and P. H. Geil, *J. Macromol. Sci.-Phys., B 8*, 295 (1973).
[5] J. Loboda-Cackovic, R. Hoseman, and H. Cackovic, *Koll.-Z.Z. Polym., 247*, 824 (1971).
[6] Gh. Drăgan, *Rev. Roum. Chim., 20*, 687 (1975).
[7] N. G. Mc Crum and D. L. Pearce, *Nature* (*Phys. Sci.*), *243*, 52 (1973).
[8] G. Zerbi, L. Piseri and F. Cabassi, *Mol. Phys., 22*, 241 (1971).
[9] Gh. Drăgan, *Rev. Roum. Chim., 21*, 1381 (1976).
[10] See for example: K. Adachi, H. Suga and S. Seki, *Bull. Chem. Soc. Jpn. 45*, 1960 (1972).
[11] R. H. Baughman and D. Turnbull, *J. Phys. Chem. Solids, 32*, 1375 (1971).
[12] T. Nagasawa and K. Kobayashi, *J. Appl. Phys., 44*, 4332 (1973).
[13] T. W. Chou, *J. Appl. Phys., 42*, 4931 (1971).
[14] Gh. Drăgan, *Rev. Roum. Chim., 22* (1977).
[15] G. F. Oster and D. M. Auslander, *J. Franklin Inst., 292*, 1 (1971).
[16] Gh. Drăgan, *J. Thermal Anal., 9*, 405 (1976).
[17] Gh. Drăgan, *Rev. Roum. Chim., 22* (1977).
[18] Gh. Drăgan and F. Stoenescu, to be published.
[19] Gh. Drăgan, *Rev. Roum. Phys.*, in press.
[20] Gh. Drăgan, *Rev. Roum. Chim.*, in press.
[21] Gh. Drăgan, *Rev. Roum. Chim., 21*, 1537 (1976).

MECHANOCHEMICAL SYNTHESES

CRISTOFOR SIMIONESCU, CLEOPATRA VASILIU
OPREA, and CLAUDIA NEGULIANU
*Department of Macromolecular and Organic Chemistry,
Polytechnic Institute of Jassy, R-6600 Jassy, Romania*

SYNOPSIS

In the present paper some main directions in mechanochemical synthesis are developed, based on the mechano-destructive process induced under different conditions in homo- and heterochain solid polymers (polyethylene, cellulose and derivatives, polyesters and polyamides). The fact is known that the elastic energy conferred to crystalline amorphous polymers focuses on structural and composition defects distributed statistically, causing molecular ruptures of the homolytic type which generate nascent microcracks propagating and increasing within the solid leading finally to a fracture. The nascent microcrack represents in fact the existence within small volumes of a number of active macroradicals (or of a number of reactive functional groups after the stabilization of the destruction fragments) capable of reacting with appropriate chemical reagents existing in the working medium. On this basis became possible the initiation of graft polymerization, block copolymerization, polycondensation, and complexing reactions. The concentration of the elastic energy not on molecular underlayers but on metallic surfaces, even when these are the reaction vessel walls, leads to the formation of active centers of a particular type, capable of initiating homo- or copolymerization reactions.

INTRODUCTION

The starting point of the investigations in the polymer mechanochemistry field was the finding of negative, destructive effects induced by the mechanical stress in these materials, either by working them in industrial equipment or during the slow process of using different finished articles. Thus the molecular mechanochemical mechanism of fracturing, chemical relaxation, and corrosive cracking has been discovered.

The progress of these investigations resulted from the requirement of finding the proper ways for hindering or limiting such effects in their early stages as well as of using them for converting and synthesizing new macromolecular compounds.

Thus two main directions of study developed in the mechanochemistry field: (1) the mechanochemical destruction based on investigations on the macromolecular chain splitting under the action of mechanical energy and (2) the mechanochemical synthesis by use of the active fragments formed by destruction as promoters of the chemical reactions.

Journal of Polymer Science: Polymer Symposium 64, 149–180 (1978)
0360-8905/78/0064-0149$01.00

Nowadays the investigations in the polymer mechanochemistry field are more and more numerous and complex. There are all over the world specialized groups of investigators devoting their activity to detecting, limiting, hindering, or using the effects of the mechanochemical energy on polymers. The outstanding contributions are well known of the English [1–28], German [29–39], Soviet [40–71], American [72–89], and Japanese [90–97] schools. For more than a decade such studies have been carried out in Romania, some obscure aspects of the polymer mechanochemistry being approached. Some contributions deserve mention, namely those elucidating the destruction mechanism of some heterochain polymers [98–104], the role of supramolecular structure in the development of the mechano-destructive process and the achievement of some syntheses such as grafting and block copolymerization [105–109], polycondensation [110–116], complexing [117–123] and recently homo- and copolymerization using for activation different forms of mechanical energy (vibratory milling, ultrasonics, and cryolisis).

Nascent Cracks Generated Mechanodestructively as Initiating Centers of Chemical Reactions

Mechanochemical reactions occur and develop within the solid macromolecular compounds as a result of the elastic energy concentrations in small volumes or thin superficial layers according to specific mechanisms. However, the mechanical impact is not the only energy form existing in the system since under its action high temperatures and pressures develop as well as strong *electrizations,* electron streams and magnetic fields associated with these, whose action and succession are extremely rapid since the regions of maximum concentrations of the mechanical energy have small size (radii of 10^{-3}–10^{-7}) and short life duration (10^{-3}–10^{-6} sec).

The energy developed in this way is high enough to promote homolytic breaks which in small volumes generate nascent active cracks (Fig. 1) which propagate and grow within the solid, leading to a fracture.

In fact, the nascent microcrack represents the existence in small volumes of particularly active radicals capable of reacting with appropriate chemical reagents existing in the working medium. For this reason the mechanochemical

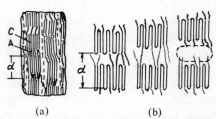

(a)   (b)

FIG. 1. Schematic representation of the mechanism of a submicroscopic crack formation within a crystalline–amorphous polymer [127]. (a) The internal amorphous–crystalline structure of the fibrils (c, crystallites, A, amorphous range, d, long period); (b) the behavior of an amorphous–crystalline element of the fibril type under the action of a tensile force. Formation of a nascent submicroscopic crack.

reactions cannot be considered as simple interactions between kinetic units and the structural ones with molecular character, their promoters being in fact the active, elementary volumes, the nascent cracks which concentrate a variable number of active centers. Regarded from this point of view they belong rather to the supermolecular level.

The study of the destructive effect caused by the action of the mechanical forces on the polymers should generally clarify the following two important problems: (1) which bond type breaks with the highest probability in a macromolecular compound of a given structure under given deformation condition and (2) what is the mechanism of this process?

Elucidation of these aspects would determine the nature of the reagent capable of reacting with either active center in the stage of existence of the nascent cracks or with the functional groups concentrating along the fracture obtained at the process end after the stabilization of the destructive fragments.

If the mechanical activation of the polymers is supposed to consist mainly in the modification of the valence angles and interatomic distances within macromolecules causing the increase of the potential energy of the system which passes into the "mechano-excited" or "strained" state, certain reactions are expected to be promoted even by the deformation states before breaking of the chemical bonds.

The mechanochemical transformation of the macromolecular compounds proceeds generally by stages, passing through the following principal ones:

(1) occurrence of destruction microcenters on the strongly strained interatomic bonds, due to the uniform charge distribution;

(2) breaking of the overstrained bonds under the action of the thermal fluctuations and formation of nascent microcracks;

(3) fusion of the nascent microcracks into larger cracks and finally into magistral cracks which leads to the macroscopic rupture of the solid.

The first two stages represent elementary acts of a mechanochemical nature and supply active centers usually of a radical nature; these have been intensively studied by means of ESR spectroscopy [127-132], X-ray diffraction under small angles [133-136] or laser beams [136]. In the third stage, named "defracturation," new surfaces are formed which promote different syntheses when moistened with suitable reagents. In fact, along the fracture (magistral crack) the active centers formed in the previous stage are stabilized, hence an increasing number of functional groups are concentrated, identifiable by IR spectral measurements [137-141].

From the above considerations the initiation of certain chemical reactions appears possible in each of the three stages.

The stimulation of one stage or another is possible by selecting the polymer types, the mechanical energy type which should be in agreement with the physical state of the worked polymer, the temperature, and the medium (liquid, gaseous, inert or active).

Thus the polymers with a higher initial molecular weight will be destroyed more rapidly than those with shorter chains. For instance, different types of cellulose reach the destruction limits $\overline{M}_\infty$ after 10 hours of mechanical work by

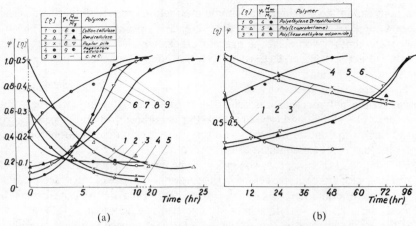

FIG. 2. The intrinsic viscosity ($[\eta]$) and destruction degree (φ) versus duration: (a) for a series of cellulose materials; (b) for polyethylene terephthalate and polyamides.

vibratory milling, while for polyethylene terephthalate, polyhexamethylene adipamide, and poly-ϵ-caprolactam, much longer durations are required for the complete degradation (Table I, Fig. 2(a), curves 1–5 and Fig. 2(b), curves 1–3). In this last case the destruction limits remain at much higher values.

TABLE I

Influence of the Cellulose Type on the Mechanochemical Destruction by Vibratory Milling

Cellulose Type	Milling Duration (hr)	Molecular Weight	Crystallinity Index (%)
Cotton cellulose	Unmilled	391,716	0.233
	1.5	221,778	—
	4	105,786	0.165
	8	25,596	0.133
	10	24,462	—
Reed cellulose	Unmilled	125,712	0.222
	1.5	111,132	—
	4	64,152	0.220
	8	45,684	0.204
	20	23,814	—
	24	22,032	0.145
Poplar pubes cellulose	Unmilled	83,636	0.223
	1.5	50,544	—
	4	31,752	0.155
	8	12,636	0.147
	10	10,106	—
Regenerated cellulose of viscose type	Unmilled	62,856	0.143
	1.5	42,606	—
	4	33,372	0.125
	8	28,026	0.113
	10	25,272	—

By comparing the efficiency of the molecular factor $\overline{M}_0$ with that of the supramolecular one, crystallinity, the latter seems more significant. For instance, cellulose in poplar pubes having an incomparably smaller polymerization degree than that of cotton cellulose is destroyed in a quite similar manner which can be explained by the smaller resistance of the less crystalline polymer. These materials although essentially different as regards their molecular ($\overline{M}_0$) and supermolecular (I_{cr}) characteristics are destroyed almost equally rapidly, attaining approximately the same destruction degree, φ (Fig. 2(a), curves 6, 8, Table I). The maximum number of split bonds in time is attained by the cellulose in poplar pubes (Fig. 3(a), curve 7). However, a more rapid decrease of this quantity, measured by the parameter $\overline{A} = (\overline{M}_\xi - \overline{M}_\infty)/\overline{M}_\infty$ is noticed for the more rigid material, with higher crystallinity, namely the cotton cellulose (Fig. 3, curve 1). Other macromolecular materials, with lower molecular weights and crystallinities exhibit an increased stability towards degradation under mechanical action, the increase in number of the split bonds, Z, as well as of the parameter A being slower during their working (Fig. 3(b), curves 4–6 and 1–3 for polyesters and polyamides, respectively).

The fact that the efficiency of the mechanochemical reactions expressed by the number of formed macroradicals increases with decreasing temperature is illustrated in Figure 4, on natural, artificial, and synthetic polymers as models. This result might be explained by reduction of the chain flexibility.

The medium in which the mechanochemical reaction proceeds is decisive for its mechanism. Under the inert gaseous atmosphere (nitrogen, argon) the molecular weight and destruction degree change more slowly than in the presence of radical acceptors existing in the same state of aggregation, nitrogen oxides or air oxygen, which spontaneously intercept the radical fragments at the very moment of their formation, thus preventing them from recombining into stable destruction products (Fig. 5 for reed cellulose and carboxymethylcellulose, and Fig. 6 for polyethylene terephthalate).

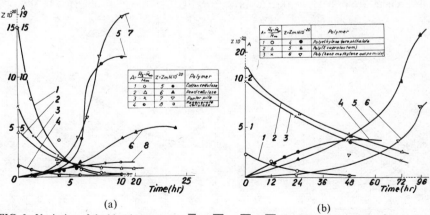

(a) (b)

FIG. 3. Variation of the kinetic parameter $\overline{A} = (\overline{M}_\xi - \overline{M}_\infty)/\overline{M}_\infty$ and of number of the split bonds (Z) with duration: (a) for a series of cellulose materials; (b) for polyethylene terephthalate and polyamides.

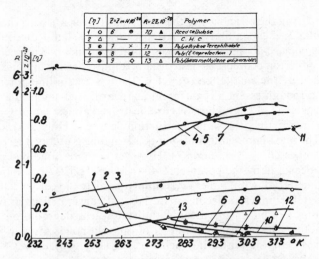

FIG. 4. Influence of the temperature on the mechano-degradation process of some natural, artificial, and synthetic polymers.

The number of split chemical bonds is always greater in the presence of radical acceptors, the nitrogen oxides being the most efficient among those studied by us (Fig. 6, curves 6(a) reed cellulose and (b) polyethylene terephthalate). However, their existence in the working medium along with other reagents (monomers, for instance) active with respect to free radicals often limits the reaction effectiveness of the latter.

When using liquid inert media, a restriction occurs of the mechano-destructive process due to the partial retention of mechanical energy by the micromolecular component. In all cases investigated (different polymers and liquids) the variation of the characteristics under study indicates destruction processes more intensive for the mechanical working in dry inert media [Fig. 7(a), polyethylene terephthalate in hexane and heptane; Fig. 7(b), poly-ε-caprolactam in acetone

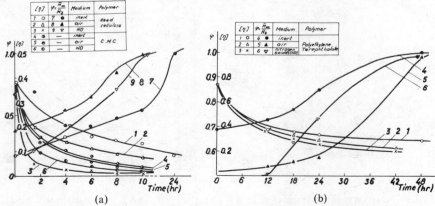

FIG. 5. Influence of the nature of the medium on mechano-degradation: (a) polyethylene tereph-thalate; (b) cellulose and CMC.

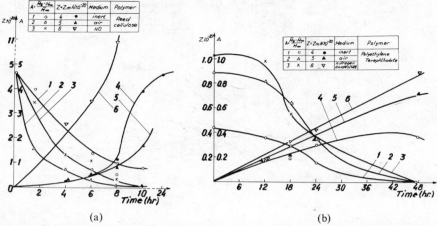

FIG. 6. Variation of the split bonds (Z) and of the parameter A with mechano-degradation duration in the presence of some radical acceptors, compared to that in inert medium: (a) reed cellulose; (b) polyethylene terephthalate.

and dioxane]. In these cases the number of split bonds and forming radicals is diminished [Fig. 8(a) and (b) for the same polymers].

To promote different chemical reactions the mechanical degradation seems to be favored by dry inert media; the liquid addition proves useful to protect the polymer from degrading and to promote certain transformations at the supramolecular level bringing about useful changes of the superficial and volume properties.

In the following, some synthesis directions approached by us using the active centers formed by splitting the macromolecular chain by mechanical action are presented.

Mechanochemical Grafting and Block Copolymerization

One of the first directions approached in mechanochemical synthesis based on the formation of radical centers (initiators of the vinyl monomer polymerization) during the mechano-destructive process was graft–block copolymerization.

For this purpose both polymer–monomer and polymer–polymer systems have been used in experiments. In the first version polyethylene terephthalate and polyamides, mainly poly-ε-caprolactam, have been used as grafting supports, while gaseous vinyl chloride, acrylonitrile and vinyl acetate in the first case as well as vinyl chloride, acrylonitrile and styrene, in the second case, were chosen as vinyl components.

As shown in Figures 9 and 10 the amount of new polymer synthesized mechanochemically increases gradually with increasing duration of vibratory milling (Fig. 9, curves 5–7 and Fig. 10, curves 1–3) while the content of polymer support diminishes (curves 8–10 and 4–6, respectively).

It is worth noticing that in all cases the homopolymerization remains quite

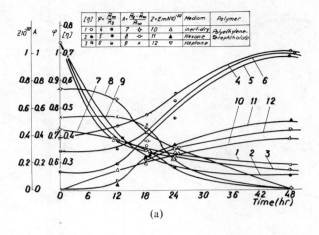

(a)

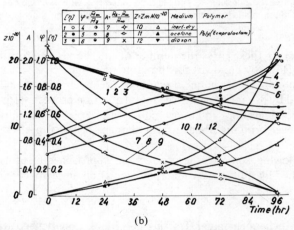

(b)

FIG. 7. Variation of the destruction degree, φ, viscosity, and A parameter with mechano-degradation duration in different liquid media: (a) poly-ϵ-caprolactam; (b) polyethylene terephthalate.

insignificant (Fig. 9, curves 11, 12 and Fig. 10, curves 7, 8) and sometimes does not take place at all (for instance, poly-ϵ-caprolactam grafting with acrylonitrile).

By grafting these polymers at low temperatures ($-75°C$) with acrylic and methacrylic acids by vibratory milling the introduction becomes possible of a significant number of polar acidic groups which increase markedly the wetability of these macromolecular compounds.

The graft–block copolymerization reactions are also possible by simultaneous mechanical working of two or several macromolecular components. The mechanical energy produced by either vibratory milling, mastication in a Brabender masticator, or during rolling was used as an activation source and polyethylene–poly-ϵ-caprolactam, polyethylene–ethylcellulose, and polyvinyl chloride–synthetic rubber as reaction couples.

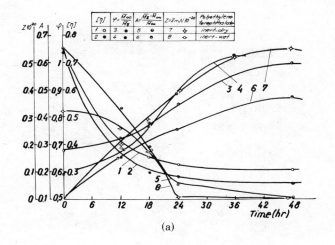

(a)

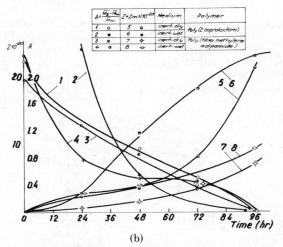

(b)

FIG. 8. Variation of number of the split bonds and of the A parameter versus duration of mechano-degradation in inert and humid media (water traces): (a) polyamide (poly-ε-caprolactam
and polyhexamethylene-adipamide); (b) polyethylene terephthalate.

The vibratory milling was accomplished in inert medium at low temperatures
(−40, −70°C) for short durations. In the other cases the working conditions
were those usually specified.

An important factor affecting the characteristics of the block copolymer
obtained by working the first couple of polymers was the component ratio.

The content of the polymer soluble in the solvents of the support was found
to increase continuously with increasing nitrogen percentage (Fig. 11, curve 3,
fraction extracted with specific solvents for polyamide; curve 4, unextracted
fraction) and the degree of conversion into copolymer (curve 1) to decrease. The
result is that the fraction depicted by curve 2 is not polyethylene entirely but
a copolymer with poly-ε-caprolactam. The forming of block copolymer with

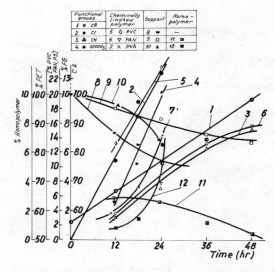

FIG. 9. Mechanochemical grafting of the polyethylene terephthalate with different vinyl monomers by vibratory milling: (1) standard curve of polyethylene terephthalate degradation under the same conditions in terms of the acidity index variation (CA); (2)–(4) variation of the functional group number (-Cl, -CN, -OCOCH$_3$); (5)–(7) variation of the percentage of the polymer chemically bound by grafting (5-PVC, 6-PAN; 7-PAC); (8)–(10) variation of polymer support content; (11, 12) variation of the homopolymer percentage.

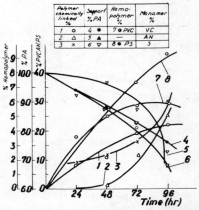

FIG. 10. Mechanochemical grafting by vibratory milling of the poly-ε-caprolactam with different vinyl monomers: (1)–(3) variation of the percentage of the chemically bound polymer.

increasing duration along with the change of the characteristics of the support polymer is illustrated in Figure 12 ($\overline{M}_\eta$, curve 1; Z, curve 2; R, curve 3). In the same manner the influence is shown of the temperature at constant duration. The lowest temperatures are noticed to be the most efficient for obtaining block copolymers.

The analysis of some features shows the product to become partially soluble under optimum working conditions (low temperatures, e.g., 40°C) its flow index

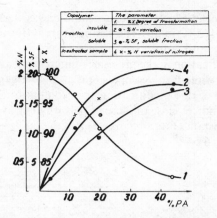

FIG. 11. Influence of the component ratio in case of the polyethylene block copolymerization with poly-ε-caprolactam on the following parameters: (1) X%, conversion degree in polymer; (2) %SF, soluble fraction; (3) %N, variation of the nitrogen content in insoluble fraction; (4) %N, variation of the nitrogen content in unextracted sample.

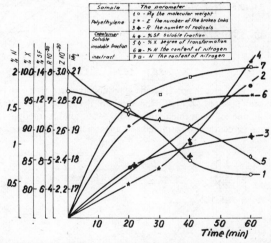

FIG. 12. Influence of the milling duration for polyethylene block copolymerization with poly-ε-caprolactam on the following parameters: (1) molecular weight of polyethylene during vibratory milling under the same conditions; (2) Z, number of the split bonds; (3) R, number of the formed radicals; (4) %SF, soluble fraction; (5) %X, conversion degree; (6, 7) nitrogen content in extracted and unextracted samples, respectively.

diminishing significantly from about 1.24 to 0.35 which suggests a cross-linking competitive process along with the main graft copolymerization reaction.

As a matter of fact, the vibratory milling of the support polymer alone (polyethylene) under the same conditions indicates the same effects.

Hence, the modifications of this compound by vibratory milling at low temperatures are induced not only by destructive processes but also by cross-linking reactions.

Another polymer couple, polyethylene–cellulose, was worked in a Brabender masticator at about 165°C in air. The temperature influence on the same parameters is illustrated in Figure 13.

The support modification in the presence of increasing amounts of cellulose derivatives as well as without this component is given in Figure 14. The assertion can be made that it is this modification that confers to the new compound physicomechanical indices close to those of polyethylene excepting for the improved dielectric constant (Fig. 14, curve 5) and the tangent of the angle of dielectric losses (curve 6).

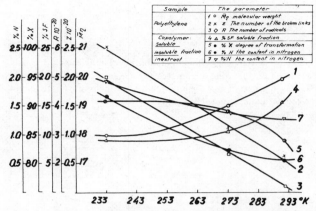

FIG. 13. Temperature influence in polyethylene block copolymerization with poly-ε-caprolactam on the following parameters: (1) $\overline{M}$, molecular weight of the support polymer; (2) Z, number of the split bonds; (3) number of the formed radicals; (4) %SF, soluble fraction; (5) %X, conversion degree; (6, 7) nitrogen percentage in extracted and unextracted samples, respectively.

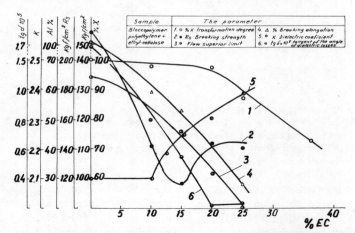

FIG. 14. Variation of the physicomechanical indices by mastication of polyethylene with ethyl-cellulose versus the component ratio; (1), %X, conversion degree; (2) Rr, tensile strength; (3) upper yield value; (4) elongation on break, %; (5) dielectric constant, K; (6) tangent of the loss angle tan $\delta \times 10^5$.

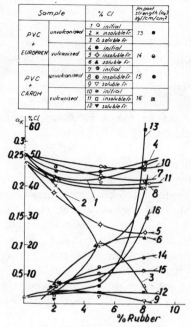

Sample		% Cl	Impact Strength (αₖ) kg/cm/cm²	
PVC + EUROPREN	unvulcanised	1 ○ initial 2 × insoluble fr. 3 △ soluble fr.	13	■
	vulcanised	4 ● initial 5 ◇ insoluble fr. 6 ▲ soluble fr.	14	◉
PVC + CAROM	unvulcanised	7 ● initial 8 ◐ insoluble fr. 9 ▽ soluble fr.	15	●
	vulcanised	10 ● initial 11 □ insoluble fr. 12 ▼ soluble fr.	16	◩

FIG. 15. Variation of the chlorine percentage and of shock resistance as a function of the component ratio for block copolymerization by comastication of polyvinyl chloride and rubber (CAROM and EUROPREN).

Finally, the last couple under study consisted of polyvinylchloride rubber (CAROM and EUROPREN). The most important result obtained by block copolymerization of these two components was the improvement of shock resistance. In Figure 15 the correlation is depicted between this quantity (a_k) and the variation of chlorine percentage obtained by block copolymerization of polyvinyl chloride with variable amounts of rubber (EUROPREN and CAROM, vulcanized and unvulcanized).

In all cases the comastication on rollers at 170°C of the two components taken in variable ratios results in two fractions, one soluble and the other insoluble in benzene (a solvent used for removing the eventual traces of unreacted rubber) whose chlorine percentage variation is compared to that of some physical mixtures obtained under the same conditions but without rolling (Fig. 15, curves 1, 4, 7, 10).

Thus, for the copolymer obtained by simultaneous comastication of polyvinyl chloride with variable amounts (up to 5%) of unvulcanized EUROPREN, a significant decrease of the chlorine percentage with the growing amount of the elastomer is noticed, a minimum being attained (Fig. 15, curve 2).

When the elastic component percentage in the reaction mixture is increased, a significant modification in the copolymer composition is noticed, the chlorine bound chemically greatly diminishing (portion of curve 2 after the minimum). For the benzene-soluble fraction, the variation of this parameter is opposite, a maximum of chlorine content being noticed at about the same position.

The runs carried out with the CAROM Romanian rubber under quite similar conditions indicated the same influence of the elastic component excepting for the maxima and minima evidenced less clearly (Fig. 15, curves 8, 9).

Based on these results the insoluble fractions whose chemical compositions are described by the first part of the above mentioned curves might be assumed as representing block copolymers in which the elastic component predominates quantitatively. On the contrary, those described by the ascendant portion of Figure 15, curve 2 might be regarded as block copolymers with a dominant rigid component (polyvinyl chloride). For the soluble fractions, opposite compositions may be assumed.

The conclusions derived from this interpretation are in good agreement with the results obtained when determining the shock resistance, a quantity increasing significantly for the benzene-insoluble copolymers (Fig. 15, curves 13, 15).

By using in the reaction the same elastomers (but vulcanized) and following the same parameters (variation of the chlorine content and shock resistance) the same results are obtained qualitatively but not quantitatively. Generally, the variations obtained for the quantities taken as a criterion are less than those characteristic of the block copolymers obtained under identical conditions, but in case of the unvulcanized elastomer maxima and minima clearly evidenced for the former, are attained for the latter for compositions probably beyond those examined by us (Fig. 15, curves 5, 6 and 11, 12 compared to 3, 4 and 8, 9). The shock resistance also increases in this case for the benzene-insoluble block copolymers (Fig. 15, curves 14, 15).

The data thus obtained indicate the possibility of changing favorably some physicomechanical indices, especially the shock resistance, by means of comastication under conditions specific for the industrial technology of working these two components.

Mechanochemical Polycondensation

Starting from the idea that by mechanical destruction of the heterochain polymers, an increased number of functional groups accumulate as the process proceeds, which means at the supermolecular level an accumulation of such groups along the magistral crack, we turned our attention toward their use in reactions termed by us "mechanochemical polycondensations." Polyethylene terephthalate, polyamides, cellulose, and polyvinyl alcohol have been chosen as supports.

On the polyethylene terephthalate model, which on destruction by vibratory milling at the room temperature (or lower) generates an increasing number of carboxylic groups, the reactions with a series of aromatic and aliphatic diamines have been achieved.

From Figure 16 the correlation may be inferred between the chemical bonding of diamine (ethylene diamine) and the quantities characterizing the mechano-destructive act promoting this synthesis, namely the increase of the percentage of nitrogen chemically bound (curve 1) is in agreement with the increase of number of split bonds, hence the destruction fragments formed under the same

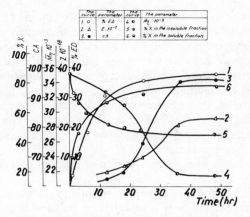

FIG. 16. Variation of the conversion degree and of the amount of ethylenediamine chemically bound versus milling duration compared to the number of split bonds (Z), viscosimetric molecular weight ($\overline{M}_\eta$) and acid value (CA) (determined on a standard sample submitted to milling under the same conditions but in the absence of diamine): (1) amount of ethylenediamine chemically bound; (2) number of split bonds; (3) acid value; (4) molecular weight; (5) conversion degree into insoluble product; (6) conversion degree into soluble product.

conditions but in the absence of diamine (curve 2), of the acidity index characterizing the increasing number of carboxylic groups in the system (curve 3) which are correlated with the decrease in molecular weight (curve 4).

The mechanochemical reaction between the destruction fragments of the support polymer and the diamine used leads to two fractions, one soluble and the other insoluble, whose variation versus duration is described by Figure 16, curves 5 and 6. Figure 17 shows the former to decrease quantitatively with increasing duration (curves 1–3, 7–9) and the latter to increase (curves 4–6, 10–12) but in both cases a limit of the conversion degree (X) is attained. Like any other mechanochemical transformation, this type of reaction is promoted by tem-

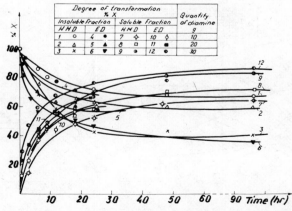

FIG. 17. Variation of the conversion degree ($X\%$) into insoluble and soluble products obtained from polyethylene terephthalate and hexamethylenediamine and ethylenediamine, respectively, with milling duration.

perature decreasing and may be started under the action of mechanical energy of different types: vibratory milling, ultrasonics, or cryolisis as shown in Figure 18(a) and (b). Furthermore, the reaction may be carried out even under destruction conditions other than those obtained by mechanical energy. We have also tested the efficiency of thermal energy and γ irradiation of ^{60}Co [123].

In elucidating the reaction mechanism, the following questions had to be answered: (1) Are the active centers obtained by the scission of chemical bonds important? Could they be free radicals in the presence of condensing reagents? (2) Are the functional groups appearing at the end of the stabilization of destruction fragments more important?

To elucidate the above questions, different types of polymers have been investigated:

(1) polymers capable of providing terminal functional groups by the destruction process (polyethylene terephthalate, polyamides);

(2) polymers carrying a great number of preexistent functional groups on the chain being able at the same time to increase their number by mechano-destruction (cellulose);

(3) polymers without functional groups on the main chain but with a great number of side groups (polyvinyl alcohol);

(4) polymers without any functional groups (polyethylene).

The polyethylene terephtalate, polyamides, and cellulose were found to bind significant percentages of diamine while the polyvinyl alcohol and polyethylene to a small extent only.

This result led to the conclusion that the end functional groups are not obligatorily responsible for the reaction with the condensing agent, but the intermediate active centers. In order to test the existence of the radical centers in the presence of diamines, different acceptors have been added which, as can be seen in Figure 19 and Table II significantly disturb the attachment of the condensing agent, the nitrogen percentage taken as a criterion varying within large limits.

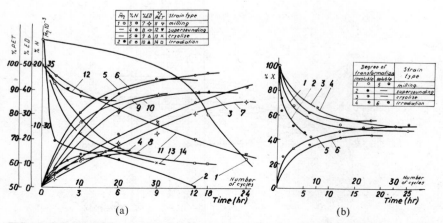

(a)   (b)

FIG. 18. Influence of the mechanical energy nature on the conversion degree ($X\%$) in the case of polyethylene terephthalate condensation with ethylenediamine.

TABLE II

Influence of the Radical Acceptor on the Mechanical Polycondensation Reaction between
Poly(ethylene–Terephthalate) and Ethylene Diamine[a]

Component Ratio	Acceptor Nature	Acceptor Formula	Synthesis Time (hr)	Content of Chemically Bound Nitrogen (%)
Poly(ethylene-terephthalate): Ethylenediamine 1:1 (gravimetric)		Cl–⬡–Cl with SH $(C_{10}H_7OH)$	9	6.7
		HO–⬡–OH $C_{12}H_{25}SH$	24	16.0
	Dichlorothiophenol		9	5.0
	β-naphtol	C_6H_4 ring C–SH	9	5.54
	Hydroquinone		9	4.95
	Dodecyl mercaptan		9	6.6
	2-Mercaptobenzo-thiazole	$(C_6H_5)_2 N–\dot{N}$–⬡ NO_2, NO_2, NO_2	9	4.1
	Diphenyl picryl hydrazyl	O_2	24	14.0
	Molecular oxygen	NO	9	8.6
	Nitrogen oxide	⬡–OH	9	8.9
	Phenol		9	2.25

[a] Temperature: 18°C; filling ratio: 0.4%.

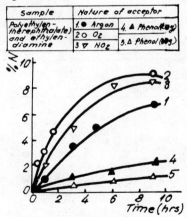

FIG. 19. Variation of the nitrogen percentage with milling duration for the products of mechanochemical polycondensation of polyethylene terephthalate with ethylenediamine: (1) argon; (2) oxygen; (3) nitrogen oxide (NO); (4) phenol (20 g); (5) phenol (30 g). Polyethylene terephthalate to ethylenediamine ratio = 1.

Figure 19 shows some of the acceptors added to behave as accelerators of the reaction with diamines (oxygen and nitrogen oxide, curves 2, 3, compared to curve 1 in argon) and others as retarders (curves 4, 5 obtained for different amounts of phenol). Table II lists the products capable of modifying in one sense or another the reaction of the destruction fragments with diamines.

The obtained results certify the presence and significance of the radical centers in the reaction systems of this type. On this basis and taking the fact into account that polyethylene which is not able to get active functional groups by destruction reacts, however, with diamines, the conclusion was drawn that the last may act as radical acceptors, this way contributing at least partially to the development of "mechanochemical polycondensation."

More detailed studies have been accomplished on polyethylene terephthalate for which the conditions required to react either directly with the aliphatic and aromatic diamines or by means of acyl chlorides, e.g., sebacyl chloride, have been settled.

The polyamides may be converted similarly by means not only of diamines but also of ethylene glycol, phenol, diacids, and diamines.

The polymers obtained in this way (the insoluble fractions) are thermostable, resistant to chemical agents, exhibit features of organic semiconductors, being recommended for either direct use when substances with such properties are required or as intermediates in other reactions.

For instance, the polymers of the polyester–polyamide type obtained from polyethylene terephthalate with different diamines, especially aromatic, due to their end amino groups, can be converted by diazotization and coupling into colored polymers (Table III); if the coupling components contain in their turn diazotable primary amino groups the adjustment of the material nuance according to one's desire becomes possible [112].

TABLE III

Coupling Components Used for Synthesis of Colored Polymers

Coupling Components		pH of the Coupling	Color[a]
Name	Formula	Solution	Obtained
β-naphthol		pH > 7 (coupling in Δ)	red
Resorcin		pH > 7 (coupling either in ortho or para position with respect to -OH group)	yellow
Naphtol AS-g (diacetyl acetotoluidine)		pH > 7	yellow
Acid H		pH ≥ 4 the first coupling at pH = 4; the second coupling at pH = 8	violet[b] (after two succesive couplings)

Δ, coupling position.

[a] The polymer was used in powder form.

[b] According to literature data [23] the sulfur atom confers to the polymer ion-exchanging properties.

On the other hand, starting from the fact that these polymers contain along the molecular chain electronegative atoms such as nitrogen and oxygen which possess unshared electrons at the chemical bond, they might be used as ligands in complexing reactions.

Mechanochemical Complexations

In order to synthesize polychelates consisting of macromolecular ligands which represent products of mechanochemical polycondensation of the destruction fragments produced from different macromolecular supports (polyethylene terephthalate, polyamides, cellulose and proteins) with different diamines, Fe^{3+}, Mn^{2+}, V^{3+} from the corresponding metallic salts were used as complexing centers.

The mechanical activation has been achieved by vibratory milling, ultrasonics, and cryolisis. The complexing reaction has been carried out simultaneously with ligand synthesis [113–123].

The chemical binding of the metal was followed by the change of nitrogen percentage which was smaller in complex compared to that in polycondensate obtained under the same conditions, as well as by the amount of thermostable residue obtained by combustion at $1000°C$.

The variation of these characteristics in complexes of polyethylene terephthalate and poly-ϵ-caprolactam as a function of duration or amount of condensing agent is shown in Figure 20(a)–(c). It should be mentioned that the content of thermostable residue increases with increasing duration and amount of the condensing agent; by IR spectral measurements absorption bands common to those of the ligand (polycondensate, respectively) are evidenced.

The introduction of metal into the structural unit of the macromolecular complexes of this type results in an essential change of certain properties in comparison with the support polymer. For instance, the polyethylene terephthalate converted by means of metallic salts and diamines into complex loses the isolating character, a fact proved by ESR spectroscopy. The obtained spectra show clear signals which are indicative of a permanent magnetic moment of the polychelates synthesized mechanochemically (Fig. 21). The values of activation energy and electrical conductivity are shown in Figure 22. The σ values (in ohm cm^{-1}) of these polymers vary with temperature laying within 10^{-12}–10^{-6} ohm cm^{-1}. The variation of weight losses versus temperature, as determined thermogravimetrically, is given in Figure 23 and characterizes these polymers as thermostable materials.

Mechano-Chemical Homo- and Copolymerization [124–126]

Finally, another type of synthesis with which we have been concerned recently is based on the mechanical activation of the compounds with small molecules. The investigations in the field of polymer mechanochemistry have shown that to take the mechanical energy for activating, the substances should be either macromolecular or in a condensing (crystalline) state. For this reason the

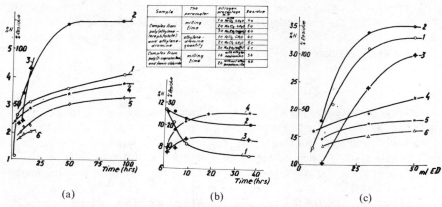

FIG. 20. Variation of the nitrogen and thermostable residue percentages with milling time for the complex polyethylene terephthalate–ethylenediamine and different metallic salts; (a) (1) variation of nitrogen in the complex with ferric chloride; (2) *idem* with manganese chloride; (3) *idem* with manganese acetate; (4) variation of the thermostable residue for the complex with ferric chloride. (b) Variation of nitrogen percentage and of thermostable residue with milling time for the complex on polyamide support; (1) variation of nitrogen percentage for the complex of poly-ε-caprolactam and ferric chloride; (2) variation of nitrogen percentage for the complex without ethylenediamine; (3) variation of thermostable residue for the complex of poly-ε-caprolactam and ferric chloride; (4) variation of thermostable residue for the complex without ethylenediamine. (c) Variation of nitrogen percentage and thermostable residue with ethylenediamine quantity in complex of poly-(ethylene terephthalate), ethylenediamine, and different metallic salts: (1) variation of nitrogen in the complex with ferric chloride; (2) *idem* with manganese acetate; (3) *idem* with manganese chloride; (4)–(6) *idem,* variation of thermostable residue.

mechanochemical homo- and copolymerizations have given rise to particular problems.

In our laboratory the polymerizations were achieved of acrylonitrile and styrene as well as their copolymerizations by vibratory milling in a V_2A apparatus without adding any special additive, by prolonging the working duration to about 200 hours and keeping the filling ratio constant ($\eta = 1\%$). In Figure 24 some results are given, comparing homo- and copolymers.

A main feature that should be emphasized is the appearance of the curves plotting the conversion versus duration (Fig. 25, curves 1, 2 for copolymerization and 3, 4 for homopolymerization) which in both cases is similar, namely the conversion attains a maximum at durations of about 120 hours for homopolymer (representing 82.5%, curves 3, 4) and 168 hours for copolymer (curves 1, 2).

In the same figure the influence is shown of another mechanical parameter, namely the serviceability of the milling bodies, V_2A spherical balls, $\phi = 9$ mm, some of which are used for about 10 years in mechanochemical reactions. It may be noticed that the use of old balls (Fig. 24, curves 2, 3) results in the best conversions in comparison with the balls used for the first time in this synthesis (Fig. 24, curves 3, 4) which suggests a correlation between the polymerization process thus activated and that of corrosion produced by the monomers on the apparatus walls. The determination of the nitrogen percentage in mechanochemically synthesized polymer shows the obtained values to be always smaller than those

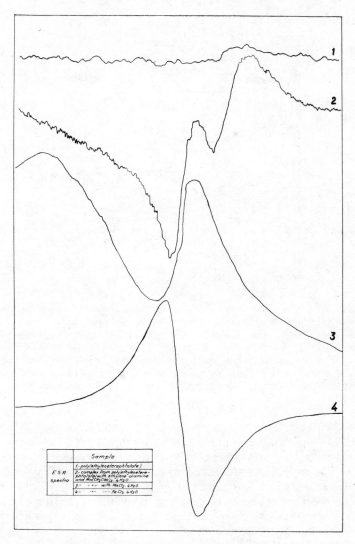

ESR spectra	Sample
	1 - poly(ethyleneterephtalate)
	2 - complex from poly(ethyleneterephtalate) with ethylene diamine and Mn(CH₃COO)₂ · 4H₂O
	3 - " - " with MnCl₂ · 4H₂O
	4 - " " --- FeCl₃ · 4H₂O

FIG. 21. ESR spectra of insoluble compounds synthesized from polyethylene terephthalate (1); (2) with ethylenediamine and manganese acetate; (3) with ethylenediamine and manganese chloride; (4) with ferric chloride.

obtained for the common polyacrylonitrile (Fig. 25 curve 7). This characteristic may be considered constant for both homo- and copolymerization until about 120 hours, a slight diminution being then noticed in all cases (Fig. 25, curves 5–7).

This type of synthesis results also in two sorts of products, soluble and insoluble in the solvents specific for the corresponding polymers obtained by common industrial methods.

The elemental analysis of the soluble homo- and copolymers synthesized

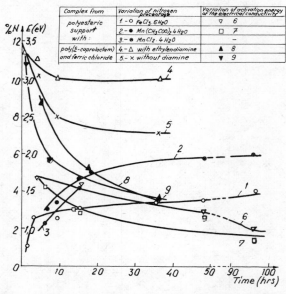

FIG. 22. Dependence of the activation energy of the electrical conductivity on the nitrogen percentage introduced at different synthesis durations in complexes obtained on polyester supports; (1) variation of nitrogen percentage with duration for complexes with Fe^{3+}; (2) *idem* with Mn^{2+} (from acetate); (3) *idem* with Mn^{2+} (from chloride); (4, 5) variation of the nitrogen percentage with duration for complexes of poly-ϵ-caprolactam support with ferric chloride in the presence of diamine and its absence, respectively; (6, 7) variation of the energy of activation of the electrical conductivity in polyethylene terephthalate complexes with ferric chloride and manganese acetate, respectively.

mechanochemically indicates the presence of important amounts of thermostable residue, which within the studied range varies in time (Fig. 25, curves 7–10) proving the chemical attachment of the metal to the structural units of these polymers. The crystallinity of these compounds evidenced by X-ray analysis was shown to increase to a maximum and then to decrease.

Another mechanical parameter directing the synthesis, the filling ratio, could be varied in two ways, namely by varying either the ball [Fig. 25] or the monomer [Fig. 24(b)] amounts. As can be seen the higher conversions are obtained at the lowest filling ratios (Fig. 24(a), curves 1, 4).

The curves 3 in the same plotting depict the variation of the nitrogen percentage against this parameter while curves 4 that of the thermostable residue.

Figure 26 refers exclusively to the copolymer synthesis, the dependence being plotted of the conversion degree (X) and nitrogen and thermostable residue percentages versus the ratio of the two components (comonomers). The conversion degree diminishes with increasing concentration of styrene (curve 1).

It is worth noticing that by choosing the optimum conditions (long duration, small filling ratios, temperature of $18°C \pm 2°$), conversion degrees of 82.5% and even 100% have been obtained for homo- and copolymer, respectively.

To obtain information regarding the nature of the active centers leading to

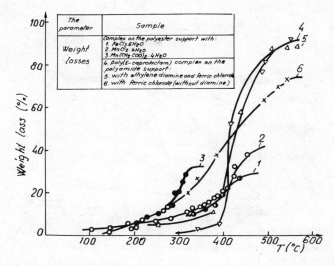

FIG. 23. Variation of the weight losses with decomposition temperature for different complexes obtained on polyester support: (1) with ferric chloride; (2) with manganese chloride; (3) with manganese acetate on polyamide support; (4) poly-ε-caprolactam (standard sample); (5) complex with ethylenediamine and ferric chloride; (6) complex with ferric chloride, in diamine absence.

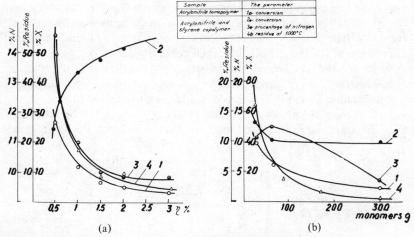

FIG. 24. Influence of the filling ratio on some characteristics of homopolymer of acrylonitrile and the copolymer with styrene.

such reactions, experiments have been performed in the presence of air, with oxygen considered as radical acceptor (Fig. 27). It was noticed that this reagent really disturbed the reaction, the conversion attained in the air presence being lower than that under inert medium.

The first information regarding the structure and the properties of these macromolecular compounds has been obtained by elemental analysis, IR, ESR and Mössbauer spectral measurements, thermal differential analysis as well as by determinations of the solubility and of the softening range.

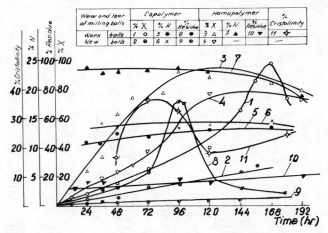

FIG. 25. Influence of the milling duration on some characteristics of acrylonitrile homopolymer and its copolymer with styrene.

The fact was already mentioned that, both by homo- and copolymerizations, soluble and insoluble fractions have been obtained, and that by increasing the duration the conversion of the former is higher. The recorded IR spectra are given in Figure 28 for homo- and copolymers (soluble fractions) together with those of the corresponding monomers and of a copolymer industrially manufactured. As can be seen in the spectrum of the mechanochemically synthesized polyacrylonitrile (curve 4) more absorption bands are to be found compared to the spectrum of the industrially obtained polymer (curve 3).

A series of bands can be regained in both products, the most important being that at 2250 cm^{-1} corresponding to the —C≡N group; the polymer prepared mechanochemically exhibits numerous additional bands with respect to the standard, namely the peaks at 850 and 1150 cm^{-1} which can be attributed to the C—N bond, clear absorptions at 1380 cm^{-1} and especially at 1650 cm^{-1},

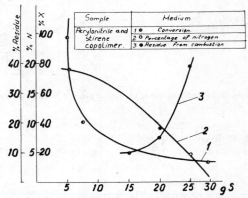

FIG. 26. Influence of the comonomer ratio on the conversion degree and some characteristics of the copolymer.

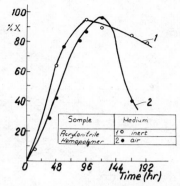

FIG. 27. Medium influence on the mechanochemical homopolymerization of acrylonitrile; (1) inert; (2) oxygen in air (air oxygen).

which are not found in the standard sample and correspond to alternate $C=C$ and $C\equiv N$ bonds. Finally, the bands at 2060 and 2180 cm^{-1}, which are absent in the standard sample spectrum and usually attributed to inorganic azides, can be assigned in the present case to iron chemically bound in these products.

The analysis of the copolymer IR spectrum (curve 5) evidences the existence of both polyacrylonitrile (a weak absorption of the —C$\equiv$N group at 2250 cm^{-1}) and polystyrene sequences (within 700–900 and 2800–3100 cm^{-1} ranges).

The chemical attachment of the metal to the structural units of the new products was tested not only by elemental and IR spectral analyses but also by Mössbauer spectroscopy (Fig. 29).

These recordings have been made at room temperature at a source rate of 3 mm/sec. The spectrum of sodium nitroprussiate (Fig. 29, curve 1) taken as standard, obtained at a source rate of 10 mm/sec, served for detecting the traces of iron included (unbound chemically) in the samples synthesized mechanochemically. The spectra illustrated by the curves 2 and 3 show the iron to be chemically bound, only traces of powder iron being noticed as a distinct phase. These recordings indicate hyperfine electrical models, hence only an electrical field gradient at the nucleus. The iron appears to be situated in two neighborhoods as indicated by the two quadrapolar doublets in the spectrum. From the data on the isomeric shifting it may be inferred that the iron bonding is strongly covalent, especially in neighborhood II. From the quadrapolar splittings it may be seen that the electric field gradient at nucleus (q) is larger at the nucleus of iron in the neighborhood I than in II.

The ESR spectra of these products indicate the Fe^{3+} existence in a field of high symmetry. In this case five resonance lines usually appear. The only two lines appearing at these polymers are indicative of the following two possible transitions:

(1) When $g\perp$ is characterized by a splitting of the I order on the levels ±2 only the values of about 6 result, as in the case under study. By calculations the transitions between the levels ±5/2 and ±3/2 were found to be forbidden;

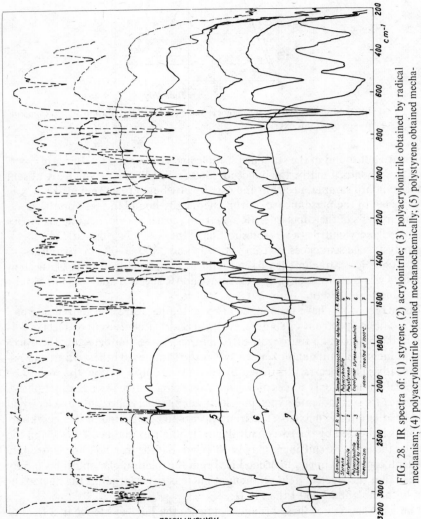

FIG. 28. IR spectra of: (1) styrene; (2) acrylonitrile; (3) polyacrylonitrile obtained by radical mechanism; (4) polyacrylonitrile obtained mechanochemically; (5) polystyrene obtained mechanochemically; (6) copolymer styrene–acrilonitrile obtained mechanochemically; (7) residue obtained from copolymer acrylonitrile–styrene treated at 1000°C.

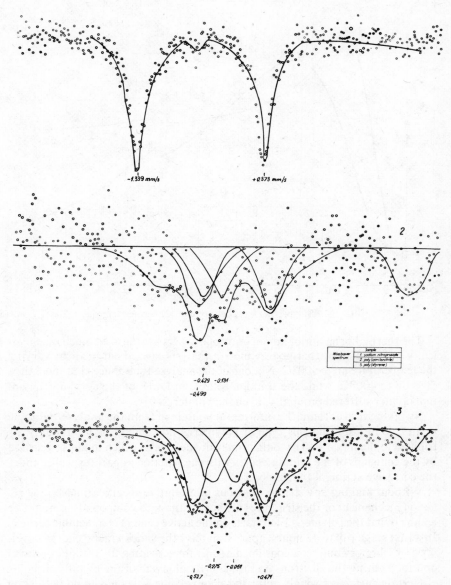

FIG. 29. Mössbauer spectra recorded for: (1) sodium nitroprussiate (standard); (2) polyacrylonitrile; (3) polystyrene.

(2) When the factor g∥—the splitting in null field (D) is large and the resulting frequency is not enough to induce transitions between states with different energies in crystalline fields (Fig. 30).

The values obtained for electrical conductivity and activation energy allow these compounds to be included among semiconducting materials (σ varies between 10^{-2} and 10^{-8} ohm cm^{-1} as a function of temperature and Ea = 1, 8 eV).

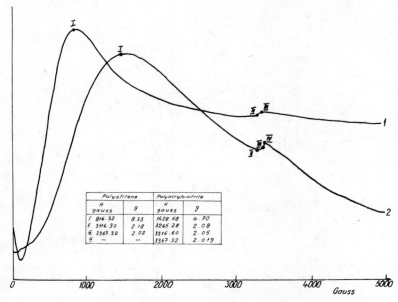

Polystirene		Polyacrylonitrile	
H gauss	g	H gauss	g
I 816.32	8.33	1428.58	4.70
II 3116.30	2.18	3265.28	2.08
III 3367.32	2.02	3316.60	2.05
IV —	—	3367.32	2.019

FIG. 30. ESR spectra of: (1) polystyrene; (2) polyacrylonitrile.

The thermal behavior of homo- and copolymers synthesized mechanochemically was studied by thermogravimetric analysis carried out in air by varying the temperature up to 900°C. Not one of the analyzed substances lost more than 20% up to 400°C, while the fraction soluble in DMF of the acrylonitrile homopolymer suffered practially no change under 240°C.

By taking into account the information obtained from the analysis data, the main role in initiating this type of polymerization should be indubitably attributed to the metal wall or to that finely divided metal which might be released by the collisions of the milling bodies with the reaction apparatus walls, under the corrosive action of the monomer.

Without making any strictly rigorous or definitive specifications on the reaction mechanism or the structures of thus synthesized compounds, it might be assumed that the polymerization starts on the active centers of a metallic surface. In a first stage (up to 48 hours, approximately) the shock kinetic energy developed by vibratory milling is consumed mainly for activating the metallic particles and for exciting the electrons on the superficial layers of the metallic atom.

Certain surfaces strongly activated then become able on contact with vinyl monomer to attach it and thus to initiate the polymerization.

The escape of the excited electrons from the metallic surface should also be accepted, the electron streams formed being capable in their turn to initiate the polymerization.

The obtained data evidence the significance of the —C≡N group in the polymerization development.

Finally, the appearance of the obtained curve is indicative of competitive mechano-destructive processes along with the synthesis reaction.

CONCLUSIONS

1. The mechanochemical transformations proceed in stages: mechano-activation, mechano-degradation, and supermolecular structure reorganization representing fracturing phenomena. In each stage, active centers are formed capable of initiating different types of mechanochemical syntheses.

2. The number and intensity of the stages of the mechano-destructive process depend on the polymer structure, nature of the medium, temperature, and the applied mechanical regime.

3. The radical mechanism is the dominant one in mechanochemical transformation. On its basis, different types of syntheses were studied, especially the mechanochemical grafting, block copolymerization, and polycondensation.

4. Under the conditions of mechanical activation the synthesis initiation is also possible by complexing.

5. The action of the mechanical energy released by vibratory milling in metallic equipment allows the initiation of polymerization of vinyl monomers, the vessel walls and the milling bodies having a catalytic role; the metal is chemically bound in the structure unity of the forming macromolecular compound which is crystalline.

REFERENCES

[1] W. F. Watson, *Ind. Eng. Chem.*, *47*, 1281 (1955).
[2] D. J. Angier and W. F. Watson, *J. Polym. Sci.*, *18*, 129 (1955).
[3] D. J. Angier and W. F. Watson, *J. Polym. Sci.*, *25*, 1 (1957).
[4] R. J. Ceresa and W. F. Watson, *J. Appl. Polym. Sci.*, *1*, 101 (1959).
[5] D. J. Angier, R. J. Ceresa, and W. F. Watson, *J. Polym. Sci.*, *34*, 699 (1959).
[6] J. Mullins and W. F. Watson, *J. Appl. Polym. Sci.*, *1*, 245 (1959).
[7] D. J. Angier and W. F. Watson, *Trans. I.R.I.*, *33*, 22 (1957).
[8] D. J. Angier, E. D. Farlie and W. F. Watson, *Trans I.R.I.*, *34*, 8, (1958).
[9] W. F. Watson, *Trans. I.R.I.*, *34*, 237 (1958).
[10] R. J. Ceresa and W. F. Watson, *Trans. I.R.I.*, *35*, 19 (1959).
[11] D. J. Elliot and W. F. Watson, *Trans. I.R.I.*, *35*, 63 (1959).
[12] R. J. Ceresa, *Plastics Inst. Trans.*, *28*, 178 (1960).
[13] R. J. Ceresa, *Polymer*, *1*, 72 (1960).
[14] R. J. Ceresa., *Trans. I.R.I.*, *36*, 211 (1960).
[15] R. J. Ceresa, *Rev. Gen. Caout.*, *37*, 1323 (1960); *ibid.*, *37*, 1331 (1960).
[16] R. J. Ceresa, *Block and Graft Copolymers*, Butterworths, London, 1962, chap. 5.
[17] D. J. Angier, W. F. Watson, and R. J. Ceresa, British Patent, 832,193.
[18] W. F. Watson, *Trans. Faraday Soc.*, *49*, 1369 (1953).
[19] J. Scanlow and W. F. Watson, *Rubber Chem. Tech.*, *33*, 1201 (1960).
[20] R. J. Ceresa, *Polymer*, *1*, 477, 488 (1960).
[21] R. J. Ceresa, *J. Polym. Sci.*, *53*, 9 (1961).
[22] R. J. Ceresa, *Trans. I.R.I.*, *36*, 45 (1960).
[23] R. J. Ceresa, *Rubber Plastics Age*, *42*, 517 (1961).
[24] G. Ayrey, C. G. Moore, and W. F. Watson, *J. Polym. Sci.*, *19*, 1 (1956).
[25] D. J. Angier, W. F. Chambers, and W. F. Watson, *J. Polym. Sci.*, *25*, 129 (1957).
[26] W. F. Watson, *Rubber Chem. Tech.*, *33*, 80 (1960).
[27] W. F. Watson, in *Chemical Reactions of Polymers*, E. M. Fettes, Ed., Interscience, New York, 1964, p. 1085.
[28] W. F. Watson and D. Wilson, *J. Sci. Inst.*, *31*, 98 (1954).

[29] H. Grohn, K. Bisdrof, M. Cösche, and K. Moeckel, *Plaste Kaut., 8,* 593 (1961).

[30] H. Grohn and K. Bisdrof, *Wiss. Z. Techn. Hochs. Chem. Leunna-Merseburg, 4,* 153, (1961/62).

[31] K. Bisdrof, *Rev. Chim. (Bukarest), 13,* 205 (1962).

[32] H. Grohn, K. Bischof, and H. Hensinger, *Plaste Kaut., 9,* 180 (1962).

[33] K. Bischof and R. Korn, *Plaste Kaut., 10,* 28 (1963).

[34] K. Bischof and R. Korn, *Plaste Kaut., 10,* 80 (1963).

[35] H. Grohn and F. Krause, *Plaste Kaut., 11,* 2 (1964).

[36] H. Grohn and K. Bischof, *Chem. Tech., 11,* 384 (1959).

[37] H. Grohn and Cl. Vasiliu Oprea, *Analele Univ. Bucureşti, Seria St. Nat, 14,* 1, 27 (1965).

[38] H. Grohn and Cl. Vasiliu Oprea, *Rev. Roum. Chim., 11(11),* 1297 (1966).

[39] H. Grohn and Cl. Vasiliu Oprea, *Plaste Kaut., 13,* 385 (1966).

[40] N. K. Baramboim, "Mechanochimia Polimerov," *Izd.nauc'no-tehniceskoi literaturî,* R.S.F.S.R Moskva (1961).

[41] N. K. Baramboim, *Legkaia Prom., 10,* 4, 22 (1950).

[42] N. K. Baramboim, *Nauk trudî Mosk. Tek. Inst. Legk. Prom., 4,* 104 (1954).

[43] N. K. Baramboim, *Nauk trudî Mosk. Tek. Inst. Legk. Prom., 6,* 45 (1945).

[44] N. K. Baramboim, *Nauk. trudî Mosk. Tek. Inst. Legk. Prom., 9,* 87 (1957).

[45] N. K. Baramboim, *DAN, SSSR, 114,* 568 (1957).

[46] N. K. Baramboim, *J. Fiz. Himii, 32,* 806 (1958).

[47] N. K. Baramboim, *J. Fiz. Himii, 32,* 1049 (1958).

[48] N. K. Baramboim, *J. Fiz. Himii, 32,* 1248 (1958).

[49] N. K. Baramboim, *Uspehi Himii, 28,* 877 (1959).

[50] N. K. Baramboim and V. A. Sviridova, *Vysokomol. Soedin., 2,* 1193 (1960).

[51] N. K. Baramboim and V. A. Santin, *Vysokomol. Soedin., 2,* 1196 (1960).

[52] N. K. Baramboim and M. A. Zaharova, *Nauk trudî Mosk. Tek. Inst. Legk. Prom., 16,* 78 (1960).

[53] N. K. Baramboim and V. I. Popov, *Nauk trudî. Mosk. Tek. Inst. Legk. Prom., 19,* 54 (1961).

[54] N. K. Baramboim, *Nauk trudî Mosk. Tek. Inst. Legk. Prom. 19,* 54 (1961).

[55] N. K. Baramboim and N. K. Sokolova, *Nauk trudî Mosk. Tek. Inst. Legk. Prom., 25,* 146 (1962).

[56] N. K. Baramboim, *Vysokomol. Soedin., 4,* 109 (1962).

[57] N. K. Baramboim, *Nauk. Trudî Mosk. Tek. Inst. Legk. Prom., 29,* 127 (1964).

[58] N. K. Baramboim and Yu. S. Simakov, *Nauk trudî Mosk. Tek. Inst. Legk. Prom., 30,* 188 (1966).

[59] N. K. Baramboim and Yu. S. Simakov, *Vysokomol. Soedin., 8,* 235 (1966).

[60] N. K. Baramboim and Yu. S. Simakov, *Vysokomol. Soedin., 8,* 235 (1966).

[61] S. A. Komissarov and N. K. Baramboim, *Polym. Sci., URSS (Eng), 11,* 1189 (1969).

[62] A. A. Berlin, *DAN SSSR, 110,* 401 (1956).

[63] A. A. Berlin, G. S. Petrov, and V. F. Prosvirkina, *Him. Nauka Prom., 2,* 522 (1957).

[64] P. Yu. Butiaghin, A. A. Berlin, and L. A. Bljumfeld, *Vysokomol. Soedin., 1,* 865 (1959).

[65] A. A. Berlin, G. S. Petrov, and V. F. Prosvirkina, *J. Fiz. Himii, 32,* 2656 (1958).

[66] A. A. Berlin and I. M. Gilman, *Kaut. rezina, 19,* 12, 1 (1960).

[67] A. A. Berlin and V. F. Prosvirkina, *Plast. Masy., 5,* 4 (1964).

[68] P. Yu. Butiaghin, *Vysokomol. Soedin, 7,* 1410 (1965).

[69] G. L. Slonimsky, *Him. Nauka Prom., 1,* 73 (1959).

[70] G. L. Slonimski and E. V. Restsova, *J. Fiz. Himii, 33,* 480 (1959).

[71] V. A. Karghin and N. A. Plate, *Vysokomol. Soedin., 1,* 330 (1959); *J. Polym. Sci., 52,* 155 (1961).

[72] P. Goodman, *J. Polym. Sci., 25,* 325 (1957).

[73] P. Goodman and A. B. Bestul, *J. Polym. Sci., 28,* 235 (1955).

[74] A. B. Bestul, *J. Appl. Phys., 25,* 1069 (1954).

[75] A. B. Bestul, *J. Chem. Fiz., 25,* 1069 (1954).

[76] A. B. Bestul, *J. Chem. Phys., 24,* 1196 (1956).

[77] A. B. Bestul, *J. Phys. Chem., 61,* 418 (1957).
[78] H. S. White and H. V. Belher, *J. Res. NBS, 60,* 215 (1958).
[79] A. B. Bestul, *J. Chem. Phys., 32,* 350 (1960).
[80] R. Arisawa and R. S. Porter, *J. Appl. Polym. Sci., 14,* 879 (1970).
[81] R. S. Porter, M. J. R. Cantow, and J. F. Johnson, *J. Polym. Sci., C 16,* 1 (1967).
[82] R. S. Porter, M. J. R. Cantow, and J. F. Johnson, *Polymer, 8,* 87 (1967).
[83] R. S. Porter, R. F. Klaver, and J. F. Johnson, *Rev. Sci. Inst., 36,* 1846 (1965).
[84] E. M. Barall, R. S. Porter, and J. F. Johnson, *J. Chromato., 11,* 177 (1963).
[85] R. S. Porter and J. F. Johnson, *Chem. Rev., 66,* 1 (1966).
[86] R. S. Porter, W. J. Macknight, and J. F. Johnson, *Rubber Chem. Techn., 41,* 1 (1968).
[87] R. S. Porter and J. F. Johnson, *J. Phys. Chem., 63,* 202 (1959).
[88] R. S. Porter and J. F. Johnson, *J. Appl. Phys., 35,* 3149 (1964).
[89] A. Casale, R. S. Porter, and J. F. Johnson, *Rubber Chem. Tech., 44,* 2, 534 (1971).
[90] Y. Go, K. Kondo, E. Gambe, and T. Okamura, *Kobun. Kagaku, 25,* 477 (1968).
[91] Y. Minoura, T. Kasuya, S. Kawamura, and A. Nakano, *J. Polym. Sci., A1, 5,* 43 (1967).
[92] J. Minoura, T. Kasuya, S. Kawamura, and A. Nakamo, *J. Polym. Sci., A2, 5,* 125 (1967).
[93] K. Goto and H. Fujii, *Chem. High. Polymers (Japan), 13,* 305 (1956).
[94] K. Goto, *Chem. High. Polymers (Japan), 13,* 305 (1956).
[95] K. Goto and H. Fujii, *Kobun. Kagaku, 14,* 644 (1957).
[96] K. Goto, *Kobun. Kagaku, 14,* 327 (1957).
[97] K. Goto and H. Fujiwara, *Kobun. Kagaku, 21,* 716 (1964).
[98] Cr. Simionescu and Cl. Vasiliu Oprea, *Mat. plast., VII, 8,* 385 (1970); *Plaste Kaut., 18,* 7, 484 (1971).
[99] Cr. Simionescu and Cl. Vasiliu Oprea, *Cell. Chem. Technol., 3,* 361 (1969).
[100] Cr. Simionescu and Cl. Vasiliu Oprea, *Cell. Chem. Technol., 2,* 155 (1968).
[101] Cr. Simionescu and Cl. Vasiliu Oprea, *Mecanochimia compusilor macromoleculari,* Ed. Acad., 1967, şi Ed. Mir, 1971.
[102] Cl. Vasiliu Oprea, Cl. Neguleanu, and Cr. Simionescu, *Eur. Polym. J., 6,* 181 (1970).
[103] Cl. Vasiliu Oprea, Cl. Neguleanu, and Cr. Simionescu, *Plaste Kaut., 17,* 9, 639 (1970).
[104] Cr. Simionescu, Cl. Vasiliu Oprea, Cl. Neguleanu, and M. Popa, *Plaste Kaut.,* in press.
[105] Cl. Vasiliu Oprea and Cr. Simionescu, *Plaste Kaut., 19,* 12, 897 (1972).
[106] Cl. Vasiliu Oprea and Cr. Simionescu, *Plaste Kaut., 20,* 3, 179 (1973).
[107] H. Grohn and Cl. Vasiliu Oprea, *Rev. Roum. Chim., 9,* 11, 757 (1964).
[108] Cl. Vasiliu Oprea, Doctoral thesis, Technische Hochschule für Chemie Leunna-Merseburg, R.D.G., 1965.
[109] Cr. Simionescu, Cl. Vasiliu Oprea, and Cl. Neguleanu, in press.
[110] Cl. Vasiliu Oprea, Cl. Neguleanu and Cr. Simionescu, *Die Makromol. Chem., 126,* 217 (1969).
[111] Cr. Simionescu, Cl. Vasiliu Oprea, and Cl. Neguleanu, *Bul. Inst. Polit. Iaşi, Fasc. 3/4,* 45 (1970).
[112] Cl. Neguleanu, Cl. Vasiliu Oprea, and Cr. Simionescu, *Die Makromol. Chem., 175,* 2, 371 (1974).
[113] Cr. Simionescu and Cl. Vasiliu Oprea, in "Polymerization Kinetics and Technology", N. Platzer, Ed., Adv. in Chemistry Series, *128,* 68–101 (1973).
[114] Cl. Vasiliu Oprea, Cl. Neguleanu, and Cr. Simionescu, *Die Makromol. Chem., 176,* 133 (1975).
[115] Cr. Simionescu and Cl. Vasiliu Oprea, *Comunicare la Conf. IUPAC:* "Transformările chimice ale polimerilor", Vol. III, Iunie, 22–24, 1971, Bratislava, RSC.
[116] Cl. Vasiliu Oprea, Comunicare la seminarul din 29 XI/1974, Deutschen Kunststoffinstitut Darmstadt, R.F.G.
[117] Cl. Vasiliu Oprea, Cr. Simionescu, and Cl. Neguleanu, Comunicare la cel de-al 39-lea Congres Internaţional de Chimie Industrială, Vol. IV, Bucureşti, Sept., 1970.
[118] Cr. Simionescu, Cl. Vasiliu Oprea, and Cl. Neguleanu, Patent român, Nr. 69,467 (1972).
[119] Cr. Simionescu, Cl. Vasiliu Oprea, and Cl. Neguleanu, *Die Makromol. Chem., 163,* 75 (1973).

[120] Cl. Vasiliu Oprea, Cl. Neguleanu, and Cr. Simionescu, *Die Makromol. Chem., 176*, 1335 (1975).

[121] Cl. Neguleanu, Lucrare de doctorat. Institutul Politehnic Iaşi, 1974. "Cercetări în domeniul policondensării şi complexării mecano-chimice. Poliesteri şi poliamide modificate."

[122] Cr. Simionescu, Cl. Vasiliu Oprea, and Cl. Neguleanu, *Angew. Makromol. Chem., 44*, 17 (1975).

[123] Cr. Simionescu, Cl. Vasiliu Oprea, and Cl. Neguleanu, *Rev. Roum. Chim., 19*, 2, 259 (1974).

[124] Cl. Vasiliu Oprea, I. Avram, and R. Avram, *Angew. Makromol. Chem.*, in press.

[125] Cl. Vasiliu Oprea and Cl. Neguleanu, in press.

[126] Cl. Vasiliu Oprea and M. Popa, in press.

[127] V. R. Regel, A. M. Teksovski, A. I. Slutsker, and V. P. Tamuj, *Mech. Polim., 4*, 597 (1972).

[128] S. N. Jurkov, A. Ia. Savostin, and E. E. Tomasevski, *DAN SSSR, 159*, 303 (1964).

[129] S. N. Jurkov and E. E. Tomasevski, *Proc. Conf. Phys. Yield Fracture*, Oxford, 200 (1966).

[130] K. L. Devries, D. K. Roylance, and N. L. Williams, *J. Polym. Sci., Part A-1, 1*, 237 (1970).

[131] T. Sawakima, S. Himada, and H. Kashiwabara, *Polymer J., 5(2)*, 136 (1973).

[132] B. Crist and A. Peterlin, *Die Makromol. Chem., 171*, 211 (1973).

[133] S. N. Jurkov, V. A. Marihin, and A. I. Stutsker, *Fiz. tverdogo tela, 1*, 1159 (1959).

[134] S. N. Jurkov, V. A. Marihin, and A. I. Stutsker, *Fiz. tverdogo tela, 11*, 296 (1969).

[135] S. N. Jurkov, V. S. Kuksenko, and A. I. Stutsker, *Proc. 2nd Intern. Conf. Fracture*, Brighton, 531 (1969).

[136] A. I. Stutsker and V. S. Kuksenko, *Mehan. Polim. 1*, 84 (1975).

[137] S. N. Jurkov, V. I. Vettegren, V. I. Korsukov, and V. E. Novak, *Fiz. tverdogo tela, 11*, 290 (1968).

[138] S. N. Jurkov and V. E. Korsikov, *Fiz. tverdogo tela, 5*, 2071 (1973).

[139] V. E. Korsikov, V. I. Vettegren, and V. I. Novak, *J. Polym. Sci., C*, 847 (1973).

[140] S. N. Jurkov and V. E. Korsikov, *J. Polym. Sci., 12*, 385 (1974).

[141] U. Gafirov, *Mech. Polim., 4*, 649 (1971).

SOME RELATIONSHIPS BETWEEN PHYSICAL PROPERTIES AND MORPHOLOGY OF HETEROGENEOUS POLYMERIC SYSTEMS

P. H. LINDENMEYER

165 Lee Street, Seattle, Washington 98109

SYNOPSIS

The effect of morphology on the physical properties of heterogeneous polymeric systems may be quantitatively expressed by a single parameter called the contiguity parameter. The formalism of small systems thermodynamics holds considerable promise of providing the means for a quantitative theoretical understanding of the very complicated morphological structures which exist in polymer solids.

INTRODUCTION

The physical properties of polymeric materials are functions of not only the chemical structure of the polymer molecule which is essentially fixed during polymerization but they are also functions of the size, shape, packing geometry, and connectivity of any heterogeneous or anisotropic regions. When viewed on a small enough scale, even a homopolymer is anisotropic and hence for some purposes can be considered as a heterogeneous system where the various regions differ only by their molecular orientation. When one considers block or graft copolymers the chemical differences make it almost certain that the morphology of the various heterogeneous regions (whether or not a true phase separation occurs) will play an important role in determining the physical properties. It is the purpose of this paper to emphasize the importance of morphology in determining physical properties and to present (a) a practical means of estimating the effect of morphology in two phase systems, and (b) a theoretical tool for relating the thermodynamics and kinetics which combine to determine the morphological structure.

When one considers the physical properties of a heterogeneous system, one is always faced with the problem of how to distribute the stress among the different regions. Exact calculations can only be made by assuming that either (a) all regions are subjected to the same strain (Voigt average), or (b) all regions have the same stress (Reuss average). Both of these assumptions have in common the fact that they cannot be strictly correct since the assumption of constant strain requires a discontinuity in the stress at the boundary between regions, whereas the assumption of homogeneous stress requires a similar discontinuity

Journal of Polymer Science: Polymer Symposium 64, 181–187 (1978)
© 1978 John Wiley & Sons, Inc.   0360-8905/78/0064-0181$01.00

in the strain. However, these assumptions represent the two extreme cases and it is certain that any real material lies between these two extreme bounds. Unfortunately, in many cases the upper (Voigt average) and the lower (Reuss average) bounds differ by one to two orders of magnitude so that some means of extrapolation between them is required.

The Tsai–Halpin Equation

Our model for such an extrapolation is to borrow the Tsai–Halpin equation [1] from composite theory and to generalize it to include the orientation of anisotropic regions. This equation was originally proposed as a way of predicting the stress in a composite of discontinuous fibers. It contains a single parameter, ξ, sometimes called the contiguity parameter, which permits one to extrapolate from the Reuss to the Voigt average as the parameter varies from zero to infinity. In the case of a fibrous composite the contiguity parameter can be identified with the length-to-diameter ratio of the fiber, a continuous fiber requires $\xi = \infty$ which corresponds to the assumption of homogeneous strain whereas a spherical particle has $\xi = 1$ which does not differ greatly from $\xi = 0$ which represents the constant stress case. Halpin and Kardos [2] have applied this equation to the estimation of the elastic moduli of semicrystalline polymers where the ξ parameter is taken as a measure of morphology (i.e., the length-to-width ratio of a polymer crystal in the direction of the chain). In all these cases it has been tacitly assumed that either the materials are isotropic or that only the modulus in the direction of the fiber or polymer chain need be considered. In a recent paper [3] we have shown how these same general principles can be applied to an exact calculation of the elastic moduli of semicrystalline polyethylene using all nine of the theoretical crystal moduli and allowing for the orientation of the anisotropic crystal. In that paper we introduced the following arbitrary relationships as a convenient means of estimating the value of a mechanical property, P, of the phase, α, with an average orientation characterized by the orientation function, f.

$$P_\alpha(f) = \frac{P_\alpha^R(f)\, P_\alpha^V(f)[1 + \xi_P]}{P_\alpha^V(f) + \xi_P P_\alpha^R(f)} \tag{1}$$

The term $P_\alpha^R(f)$ represents the value of the property, P, averaged over the various regions using the assumption of homogeneous stress (Reuss average) and $P_\alpha^V(f)$ represents the corresponding average using the assumption of homogeneous strain (Voigt average). The quantity ξ_P is the contiguity parameter associated with the particular stress (or strain) distribution: the subscript, P, indicates that the particular choice of ξ_P may depend upon the property, P, under consideration.

As in the Tsai–Halpin relationship, a value of $\xi_P = 0$ yields the Reuss average (constant stress) and a value $\xi_P = \infty$ yields the Voigt average (constant strain). Intermediate values of ξ_P generate averages between these extremes. In the case that the Voigt and Reuss averages are equivalent (e.g., for $f = 1$) $P_\alpha(f)$ is independant of ξ. With this modification the Tsai–Halpin relationship takes on the following form:

$$P(f) = P_\alpha(f)[(1 + \xi_P\chi V_\beta)/(1 - \chi V_\beta)] \qquad (2)$$

where V_β is the volume fraction of the phase and

$$\chi = [P_\beta(f) - P_\alpha(f)]/[P_\beta(f) + \xi_P P_\alpha(f)] \qquad (3)$$

with $P_\beta(f) \geqslant P_\alpha(f)$ and defined as in eq. (1).

These relationships provide an arbitrary, but nontheless useful, means by which a wide range of micromechanical models can be employed to analyze the behavior of heterogeneous systems. For example, the values of ξ associated with specific micromechanical models [4] can be estimated as illustrated in Figure 1. One may ask what advantage this particular model for the distribution of stress has over other models which have been proposed. Since at the present time the origin of the relationships is arbitrary rather then fundamental, their principal justification must be their usefulness as well as their generality. They involve only a single parameter and are applicable to anisotropic as well as isotropic materials.

The Contiguity Parameter

In all of these cases the contiguity parameter, ξ, is interpreted as a characteristic of the interval stress (or strain) distribution. In the case of a fibrous composite, ξ is approximated by the length-to-diameter ratio of the fiber [1]. In a semicrystalline polymer it has been taken [2] as the ratio between the crystal thickness in the chain direction and the width perpendicular to this direction. Note that lamellar crystal would have $\xi \ll 1$. In both of the above examples it is clear that ratios would be only approximations to the true value of the conti-

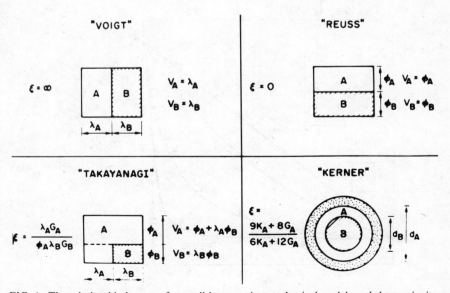

FIG. 1. The relationship between four well-known micromechanical models and the contiguity parameter.

guity since adhesion of the fiber to the matrix and connectivity of the crystal to the amorphous regions would also play a role. In fact the major advantage of the contiguity parameter is that it lumps together in a single parameter all factors which influence the distribution of stress in a heterogeneous system—the size, shape, and packing geometry of the various regions, as well as their connectivity and adhesion. In other words the contiguity parameter is a quantitative measure of all features which have been called morphology and which play any role in the resulting physical properties. Thus the contiguity parameter represents the only additional piece of information which is required along with the composition and overall orientation of any anisotropic regions in order to determine the physical properties of a blend or copolymer from the known properties of its components.

If, as will usually be the case, the contiguity parameter is not known, the measurement of the physical properties of a known composition and a measurement of the orientation will permit one to calculate the contiguity parameter. Figure 2 shows a typical example of two isotropic materials. The straight line connecting the values for the two pure components represents the homogeneous strain or Voigt average—sometimes called the rule of mixtures—while the lower line represents the Reuss or homogeneous stress average. Between them we see various dotted lines representing the values of ξ between 0 and ∞. Clearly if a particular heterogeneous material has a high contiguity there is little that one can do to improve its properties by varying the morphology. On the other hand if the value of ξ is low, then one can estimate the amount of improvement which is potentially possible by changing morphology. Note that in the isotropic case

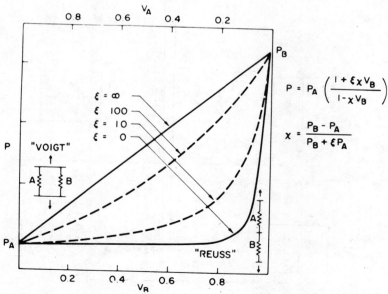

FIG. 2. The use of the Tsai–Halpin relationship for interpolating between two extreme models homogeneous strain (Voigt) and homogeneous stress (Reuss).

in Figure 2 the line corresponding to $\xi = 1$ would be very near the Reuss average or $\xi = 0$ line. This is not the case when one considers the averages obtained for an anisotropic material as a function of the orientation parameters, f, as shown in Figure 3. The example shown in this figure is for the Youngs modulus of crystalline polyethylene and illustrates the results of averaging over a heterogeneous material composed of a single anisotropic phase and a particular orientation distribution. One must use care in averaging over the tensor quantities involved in applying these equations to anisotropic materials; for details concerning Figure 3 please refer to the original work [3]. For our purpose here we merely wish to illustrate that a contiguity of unity, $\xi = 1$, for the anisotropic case—in contrast to the isotropic—is almost an equal division between the upper and the lower averages. Furthermore, Figure 3 illustrates clearly the difference between orientation and contiguity. Increasing the orientation factor, f, increases the Youngs modulus as is well known; however, even when $f = 0$ a substantial increase is still possible if the contiguity could be changed from $0 \rightarrow \infty$. The potential of increasing the physical properties of a heterogeneous material composed of a single anisotropic phase by changing its morphology (i.e., contiguity) has not been generally recognized. Figure 3 clearly illustrates the increase in modulus which could be achieved in an unoriented polyethylene ($f = 0$) if the contiguity increased from $0 \rightarrow \infty$, in other words if, for example, polyethylene crystals were needlelike in the chain direction instead of lamellalike.

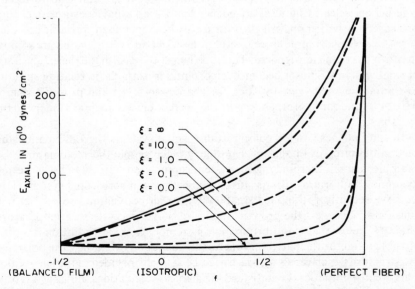

FIG. 3. The use of the contiguity parameter for interpolating between the Voigt and the Reuss averages for the Youngs modulus of crystalline polyethylene as a function of the orientation factor f (see [3]).

Small Systems Thermodynamics

Turning now to the more theoretical aspects of how to predict the conditions which will produce different morphologies, we wish to call attention to a relatively new theoretical approach known as the thermodynamics of small systems. According to T. L. Hill [4] who introduced the formalism of small systems thermodynamics, a thermodynamic system is considered "small" when its extensive thermodynamic functions vary with size in a nonlinear fashion by an amount which is significant for the purpose for which they are to be used. For example, the energy of any thermodynamic system will vary in a nonlinear manner near its surface; however, this surface effect is normally so small in comparison to the bulk of the system as to be completely negligible. But as the system becomes very small, surface effects become increasingly important and at some point the average energy of the system will deviate significantly from being a linear function of size. At this point it becomes necessary to introduce the size of the system as an explicit parameter in the thermodynamics. In effect this introduces an extra degree of freedom into all of the thermodynamic relationships including the phase rule. Thus if one wishes to deal theoretically with the small regions that make up the morphology of heterogeneous polymer systems, it is necessary to use the modified thermodynamics of small systems according to Hill [4]. In addition to the added degree of freedom caused by the explicit introduction of size into the thermodynamic equation, small systems thermodynamics differs from macroscopic thermodynamics by the fact that fluctuations about average values may become nonnegligible. For example, in an ordinary (macroscopic) thermodynamic system it makes no difference whether one selects volume as an independent variable and measures an average pressure, or whether one selects pressure as the independent variable and measures an average volume since fluctuations about an average value are so small they can be completely ignored. This is no longer the case in the thermodynamics of small systems. Thus it becomes important, for example, in relating structure to thermodynamics via statistical mechanics to select the proper ensemble partition function corresponding to the correct environmental (independant) variables.

In spite of these added complications, small systems thermodynamics has certain definite advantages over ordinary macroscopic thermodynamics. For example, a small thermodynamic system is much more likely to be in thermodynamic equilibrium with its surroundings than a macroscopic system. Thus we have recently [5] applied small systems thermodynamics to predict with considerable success the growth kinetics of polymer crystals at temperatures 15–20°C below their equilibrium melting temperatures. In addition it is often possible to factor the partition function in such a way that only the portion of the structure which varies with size need be actively included in the summation. This greatly simplifies the statistical mechanical calculations. Finally, it is often possible to find a single structural parameter which varies from one thermodynamic state to another. By calculating the free energy change as a function of this parameter one can obtain not only the free energy difference between the two states but also the height of the free energy barrier which separates them and thus one can determine the kinetics of transformation from one state to another.

Use of this theoretical tool is still very much in its infancy, yet its success in providing a thermodynamic explanation for chain folding [6] and more recently [5] for fractionation on crystallization and chain folding on an extended chain secondary nucleus provide considerable promise that this approach may prove to be the means of obtaining a quantitative theoretical understanding of the very complicated morphological structures which exist in polymer solids.

It is a pleasure to acknowledge support from Alexander von Humboldt-Stiftung and encouragement and hospitality from Prof. Dr. R. Hosemann and the Fritz-Haber-Institut der Max-Planck-Gesellschaft.

REFERENCES

[1] S. E. Ashton, S. C. Halpin, and P. H. Petet, *Primer on Composite Materials Analysis,* Technomic, Stamford, (1969).

[2] S. C. Halpin and S. L. Kardos, *J. Appl. Phys., 43,* 2235 (1972).

[3] R. L. McCullough, C. T. Wu, J. C. Seferis and P. H. Lindenmeyer, *Poly. Eng. Sci., 16,* 371 (1976).

[4] R. L. McCollough, *Concepts of Fiber-Resin Composites,* Marcel Dekker, New York, 1971.

[5] P. H. Lindenmeyer, H. Beumer, and R. Hosemann, *Polym. Eng. Sci.* (to be published).

[6] P. H. Lindenmeyer, *Polym. Eng. Sci., 14,* 456 (1974).

SOME ASPECTS REGARDING THE STRUCTURE AND THE MECHANISM OF THE PVC SYNTHESIS

A. CARACULACU, E. BURUIANÃ, and GABRIELA ROBILÃ

Institute of Macromolecular Chemistry, "Petru Poni", 6600 Jassy, Romania

SYNOPSIS

The behavior of models of PVC-unsaturated end groups subjected to radical reactions indicates the important role some of these groupings play in the after-polymerization processes. As a result the respective groups lead to new types of abnormal structures in PVC. The PVC head-to-head macroradical is thought to play an essential role in the branching formation and the occurring of the interruption processes leading to labile unsaturated end groups. Based on these results correlated with direct measurements of unsaturated structures in PVC, a reaction mechanism is suggested to explain these phenomena.

Together with other authors the research group from our institute, which is dealing with PVC problems, has been studying abnormal structures of PVC for 15 years.

Since 1953, Cotman has proved that PVC is not a perfectly linear polymer but contains a certain number of branches [1]. Based on this fact it was supposed that, in the branching sites, there are tertiary carbon atoms bearing chlorine atoms (Cl_T) [2]. Taking into account the known reactivity of the tertiary chlorides, it was supposed that these structures can play a decisive role in initiation of PVC chain dehydrochlorination process and therefore on the aging of the polymer.

Our initial papers were directed to the elucidation of the nature of the PVC branching and the role of this structure [3, 4].

After the spectral and chemical characterization of various types of branching in comparison with the normal PVC structure, we succeeded in bringing the first proof that branching in PVC has not a tertiary chloride structure and therefore cannot be the principal cause of dehydrochlorination initiation [3]. In 1970 we succeeded in finding the definitive confirmation [4]. Similar results have been obtained independently of us by Braun and Weiss [5]. Further, these data were more exactly confirmed by Rigo [6] and Abbas [7, 8] who have demonstrated that the branches in PVC have the structure:

$$-CH_2-\underset{\underset{CH_2Cl}{|}}{CH}-CH_2-\underset{\underset{Cl}{|}}{CH}-$$

Journal of Polymer Science: Polymer Symposium 64, 189–208 (1978)
© 1978 John Wiley & Sons, Inc.   0360-8905/78/0064-0189$01.00

In the second stage we have proposed to determine other labile structures with labile chlorine atoms (Cl_L), which are able to start the chain dehydrochlorination reaction. With that end in view we directed our attention to the allylic chlorine atom (Cl_A).

For a more systematic treatment of the problem, first we have proposed a scheme of all the hypothetical unsaturated end groups which can be imagined in PVC [9]. We considered that the models containing five carbon atoms are long enough to give a good representation of this structural defect (Table I).

Also, other mechanisms could be supposed for the formation of all these structures, taking into account the transfer to the monomer by atom extraction followed by various kinds of polymerization with vinyl chloride.

Based on the possibility of the transposition of some of these structures, experimentally verified by us, we extended the study on the transposed models [10] derived from the structures bearing allylic chlorine atoms (Table II).

The models in Table II represent in fact internal unsaturation models. In

TABLE I
Models of Unsaturated End Groups in PVC

	Type	Formation Mode by Disproportionation
I.	$CH_3-CH_2-\underset{\underset{Cl}{\mid}}{CH}-CH=\underset{\underset{Cl}{\mid}}{CH}$	Head-to-tail sequence with H elimination
II.	$CH_3-CH_2-\underset{\underset{Cl}{\mid}}{CH}-\underset{\underset{Cl}{\mid}}{C}=CH_2$	Head-to-head sequence with H elimination
III.	$CH_3-CH_2-\underset{\underset{Cl}{\mid}}{CH}-CH=CH_2$	Head-to-head sequence with Cl elimination
IV.	$CH_3-\underset{\underset{Cl}{\mid}}{CH}-CH_2-CH=\underset{\underset{Cl}{\mid}}{CH}$	Tail-to-tail sequence with H elimination
V.	$CH_3-\underset{\underset{Cl}{\mid}}{CH}-CH_2-\underset{\underset{Cl}{\mid}}{C}=CH_2$	Tail-to-head sequence with H elimination
VI.	$CH_3-\underset{\underset{Cl}{\mid}}{CH}-CH_2-CH=CH_2$	Tail-to-head sequence with Cl elimination

TABLE II
Models of Transposed Unsaturated End Groups in PVC

	Type	Formation Mode
VII.	$CH_3-CH_2-CH=CH-CHCl_2$	Transposition of the model I
VIII.	$CH_3-CH_2-CH=\underset{\underset{Cl}{\mid}}{C}-\underset{\underset{Cl}{\mid}}{CH_2}$	Transposition of the model II
IX.	$CH_3-CH_2-CH=CH-\underset{\underset{Cl}{\mid}}{CH_2}$	Transposition of the model III

addition we also added the internal unsaturation model with allylic chlorine, 4-chlorohexene-2, resulting by incidental dehydrochlorination in PVC or by copolymerization with acetylenic impurities [11]:

$$X. \quad CH_3-CH=CH-\underset{\underset{Cl}{|}}{CH}-CH_2-CH_3$$

To synthesize all these models, most of them unknown in the literature, we were obliged to find original synthesis methods. The synthesis of 2,4-dichloropentene-1 and 1,4-dichloropentene-1 is representative [9]:

$$CH_3-\underset{\diagdown\;O\;\diagup}{CH-CH_2} + CH\equiv CH \xrightarrow[NH_3]{NaNH_2} CH_3-\underset{\underset{OH}{|}}{CH}-CH_2-C\equiv CH$$

$$CH_3-\underset{\underset{OH}{|}}{CH}-CH_2-C\equiv CH \xrightarrow[HgCl_2]{HCl} CH_3-\underset{\underset{OH}{|}}{CH}-CH_2-\underset{\underset{Cl}{|}}{C}=CH_2$$

$$CH_3-\underset{\underset{OH}{|}}{CH}-CH_2-\underset{\underset{Cl}{|}}{C}=CH_2 \xrightarrow{SOCl_2} CH_3-\underset{\underset{Cl}{|}}{CH}-CH_2-\underset{\underset{Cl}{|}}{C}=CH_2$$

and

$$ClCH=CHCl \xrightarrow[NH_3]{NaNH_2} NaC\equiv CCl$$

$$CH_3-\underset{\diagdown\;O\;\diagup}{CH-CH_2} + NaC\equiv CCl \xrightarrow[NH_3]{-33\;°C} CH_3-\underset{\underset{OH}{|}}{CH}-CH_2-C\equiv CCl$$

$$CH_3-\underset{\underset{OH}{|}}{CH}-CH_2-C\equiv CCl \xrightarrow[ether]{LiAlH_4} CH_3-\underset{\underset{OH}{|}}{CH}-CH_2-CH=CHCl \quad trans$$

$$CH_3-\underset{\underset{OH}{|}}{CH}-CH_2-CH=CHCl \xrightarrow[Py]{SOCl_2} CH_3-\underset{\underset{Cl}{|}}{CH}-CH_2-CH=CHCl$$

All the above mentioned models were similarly synthesized [9]. The structure of these substances was demonstrated by physical and chemical methods.

We examined the NMR spectrum of ten models to see if the obtained maxima are characteristic enough to be put in evidence in the NMR spectrum of the PVC [10–12]. The results of these measurements were encouraging. The position of the different signals in comparison with the PVC spectrum is shown schematically in Figure 1.

As one can see, there are generally marked differences with respect to the position and the aspect of signals in the unsaturated region.

Before starting to search for these structures in PVC, we considered it necessary to study the behavior of these unsaturated end groups under polymer-

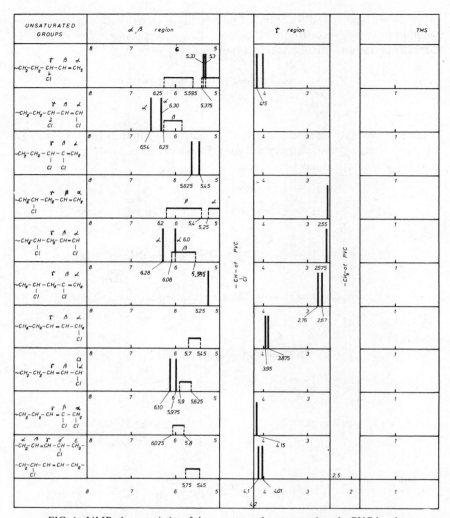

FIG. 1. NMR characteristics of the unsaturated groups against the PVC bands.

ization conditions. This was done in the hope of verifying the possibility of ulterior disappearance of these groups after reaction with different radicals leading to the appearance of other unforeseen structures.

We found that none of these models was able to polymerize in the same conditions as vinyl chloride [13].

2,4-Dichloropentene-1, 3-chloropentene-1 and 2,3-dichloropentene-1, double bond asymmetrically substituted models, in mixture with vinyl chloride up to a certain percentage, are able to copolymerize, except the last one.

By copolymerization of these substances inhibition occurred and the decrease of molecular weight of the obtained products was observed, analogous with those noticed in the vinyl chloride copolymerization with isopropenyl chloride (Table

TABLE III
Copolymerization of Vinyl Chloride (VC) with Isopropenyl Chloride (IPC)

Proof	IPC in Polymerization Mixture (%)	Weight of Polymerization Mixture (g)	Weight of Obtained Polymer (g)	Overall Yield (%)	Molecular Weight (Osmometry)
1	0	53.2	29.66	55.7	225,000
2	5	52.7	4.23	8.0	120,000
3	10	52.8	1.76	3.3	53,000
4	25	52.6	0.56	1.1	24,500
5	50	51.5	0.00	—	—

III). Some years ago Smets [15] and Caraculacu [14] proposed the hypothesis that the incapability of isopropenyl chloride to polymerize as well as the inhibition effect observed, could be explained by degradative transfer by allylic extraction followed by the appearance of a relatively stable radical which is able to give only recombination.

$$R\cdot + CH_2{=}C{-}CH_3 \longrightarrow \overset{\displaystyle\cdot}{\overline{CH_2{-}C{-}CH_2}} + RH$$

$$\underset{Cl}{\quad} \qquad \underset{Cl}{\quad}$$

$$CH_2{=}C{-}CH_2{-}CH_2{-}C{=}CH_2 \qquad CH_2{=}C{-}CH_2{-}R$$

$$\underset{Cl}{\quad}\qquad\underset{Cl}{\quad} \qquad\qquad \underset{Cl}{\quad}$$

There is a marked similarity between the behavior of isopropenyl chloride and the above pentane derivatives.

The behavior of these models was studied under conditions of strong radical attack, generated by the decomposition of a large quantity of the initiator, in the hope of isolating some low molecular products which would aid in explaining the nature of the processes.

The nature of the walls of the reactor vessel (stainless steel or glass) has a strong influence on the reaction [16]. In the first case all ten models reacted, leading to the appearance of a great number of compounds (approximately 50 gas chromatographically observed peaks). In the second case the reactions were more moderate, only three models, 2,4-dichloropentene-1, 3-chloropentene-1, and 2,3-dichloropentene-1, reacted, and the number of compounds decreased markedly.

We used NMR spectroscopy, IR, mass spectrometry, column chromatography, elemental analysis, and molecular weight determination to elucidate the structure of the reaction products, which were very difficult to separate. Some of the measurements were carried out on the raw products and the others on enriched fractions from reaction mixtures.

The reactions were made in the presence of two initiators: benzoyl peroxide (PB) and azobisisobutyronitrile (AIBN). The study of the nature of the formed compounds was easier in the case of the benzoyl peroxide due to the presence of the aromatic ring and the carbonyl group, which were able to give valuable information on the structure of the formed end groups, therefore on the initiation and termination processes.

Taking into account the observed similarity between the behavior of both initiators, we present only the results obtained in the case of the benzoyl peroxide, and also the reaction carried out in glass reactors. The principal results obtained in this study are presented in Table IV.

Based on the behavior similarity we classified the models into groups and subgroups with respect to the position and nature of the substituents at the double bond.

The reaction of 2,4-dichloropentene-1 with benzoyl peroxide was effected at different ratios between monomer and initiator, following the influence of the nature of the termination process. The NMR spectrum of the reaction products (in the case of a molar ratio model:initiator 1:1) indicates the absence of signals in the unsaturated region, corresponding to the saturated products, which contain two initiator fragments for every molecule with two monomer units. Such a product could appear as a result of the starting of a polymerization process, which is immediately stopped either by recombination with a similar radical, or by recombination with an initiator fragment.

Due to the strong hindrance of the macroradical, recombination seems to be of little probability. A more detailed analysis of the NMR spectrum leads to the second possibility, namely recombination with initiator fragments.

Together with increasing of the monomer-to-initiator ratio (Table IV), similar products with higher molecular weight were obtained. However, even under such conditions, similar to a normal polymerization, the polymerization degree does not exceed 9. The polymerization degree was determined based on the integral ratio of NMR aromatic and methyl groups signals. This mode of molecular weight determination was verified in some cases by comparison with values obtained by vapor pressure method, giving a good agreement (Table IV).

The structure of the formed compounds can be represented by the general formula:

$$
C_6H_5COO \Bigg[CH_2-\underset{\underset{\underset{CH_3}{|}}{\overset{\overset{CH_2}{|}}{\underset{CHCl}{|}}}}{\overset{Cl}{\underset{|}{C}}} \Bigg]_n CH_2-\underset{\underset{\underset{CH_3}{|}}{\overset{\overset{CH_2}{|}}{\underset{CHCl}{|}}}}{\overset{Cl}{\underset{|}{C}}}-O-COC_6H_5
$$

$$(C_6H_5-) \qquad\qquad\qquad\qquad (C_6H_5-)$$

$$n = 1\text{-}8$$

If in the case of monomer-to-initiator ratio 1:1 there are no observed signals in the unsaturated region of the NMR spectrum, together with decrease of this ratio, the characteristic maxima of these structures increase progressively. Thus

TABLE IV
Reaction of Model Compounds with Benzoyl Peroxide

Model Type	Model Compound	Model Initiator (mol. ratio)	Polymerization Degree NMR Calc.	Polymerization Degree Exp.[b]	No. of products	Nature of End Groups[a] A Phenyl	A Benzoate	B Unsatu-rated	B Saturated
$R_1R_2C{=}CH_2$	$(R{-}CH_2)(Cl)C{=}CH_2$: $CH_3{-}CH(Cl){-}CH_2{-}C(Cl){=}CH_2$	1/1	2	—	—	55	45	0	0
		1/0.5	3	—	—	18	62	10	10
		1/0.1	5–6	5–6 $M = 647$	6	18	48	17	17
		1/0.05	6	—	—	10	50	20	20
		1/0.01	9	—	8	0	40	30	30
	$(R{-}CHCl)((Cl)H)C{=}CH_2$: $CH_3{-}CH_2{-}CH(Cl){-}CH{=}CH_2$	1/0.5	2	—	—	10	40	25	25
		1/0.1	2	2 $M = 283$	3	10	40	25	25
		1/0.01	2	—	—	0	50	25	25
	$CH_3{-}CH_2{-}CH(Cl){-}C(Cl){=}CH_2$	1/0.01	2	—	—	0	50	25	25
$R_2C{=}CR_2$	1,3-dichloropentene-1; 1,4-dichloropentene-1; 1,1-dichloropentene-2; 1,2-dichloropentene-2; 1-chloropentene-2	Unreacted model compounds and degradation products of the initiator							

[a] A, by initiation and interrupting with initiator fragments; B, by disproportionation reaction.
[b] Determined on the chromatographically isolated principal reaction product

termination by disproportionation occurs, leading to the formation of structures as follows:

$$
C_6H_5COO\!\!-\!\!\left[CH_2\!-\!\underset{\substack{|\\CH_2\\|\\CHCl\\|\\CH_3}}{\overset{\substack{Cl\\|}}{C}}\!\!-\!\!CH_2\!-\!\underset{\substack{\|\\CH\\|\\CHCl\\|\\CH_3}}{\overset{\substack{Cl\\|}}{C}}\right]_n \;+\; C_6H_5COO\!\!-\!\!\left[CH_2\!-\!\underset{\substack{|\\CH_2\\|\\CHCl\\|\\CH_3}}{\overset{\substack{Cl\\|}}{C}}\!\!-\!\!CH_2\!-\!\underset{\substack{|\\CH_2\\|\\CHCl\\|\\CH_3}}{\overset{\substack{Cl\\|}}{CH}}\right]_n
$$

$(C_6H_5\!\!-\!\!)$ $(C_6H_5\!\!-\!\!)$

$n = 2\text{-}8$ $n = 2\text{-}8$

As can be seen in Table IV, the number of products identified by high pressure liquid chromatography is eight. Their nature differs probably by the type of initiator fragments at the ends of the molecules as well as by the type of the chain breaking.

One of the important conclusions which can be drawn from the results presented above refers to the transfer to monomer, which, in our opinion, is not capable of taking place.

In the case of the molar ratio 1:1, the intensities of the NMR aromatic bands indicate the existence of two aromatic ends per molecule. If a transfer to monomer exists, then one aromatic end per molecule at the most should be present.

The missing of olefinic NMR signals in this region led us to affirm, also, that a degradative transfer could not be taken into account (at least in this range of monomer:initiator ratio).

A process which should take place by abstraction of an atom from the allylic position should be:

$$
R\!\cdot + \; CH_2\!\!=\!\!\underset{\substack{|\\CH_2\\|\\CHCl\\|\\CH_3}}{\overset{\substack{Cl\\|}}{C}} \;\longrightarrow\; RH \;+\; CH_2\!\!=\!\!\underset{\substack{|\\CH\cdot\\|\\CHCl\\|\\CH_3}}{\overset{\substack{Cl\\|}}{C}}
$$

From the literature data, this radical should dimerize or react with other radicals from the system. In both cases there should appear maxima in the olefinic region. In the case of a higher monomer:initiator ratio, signals appeared in the olefinic region, but the characteristic signal $\delta = 5.25$ ppm for the end structure $-CCl\!\!=\!\!CH_2$ [12], corresponding to the above mentioned reaction products, was not observed.

Evidently the radical is stabilized before reaction, leading to the appearance of a structure obtained similarly by disproportionation:

$$
\begin{array}{ccc}
& \overset{\displaystyle Cl}{\underset{|}{}} & & \overset{\displaystyle Cl}{\underset{|}{}} \\
CH_2\!\!=\!\!C & & \cdot CH_2\!\!-\!\!C \\
| & \longleftrightarrow & \| \\
CH\cdot & & CH \\
| & & | \\
CHCl & & CHCl \\
| & & | \\
CH_3 & & CH_3
\end{array}
$$

but as we will further see, the presence of such a radical with a chlorine atom in the α position eliminates the chlorine atom, leading to the formation of a double bond. Such reaction products were not detected in the reaction mixture.

We can consider that in this case degradative transfer, which is supposed to exist in similar situations, is missing.

In conclusion, the very low polymerization capacity of 2,4-dichloropentene-1 and its inhibition effect is due to the low reactivity of the formed macroradical which is stabilized by the presence of the chlorine, together with two bulky chains, each of them bearing a chlorine atom in the α position. Recombination with initiator radical or disproportionation diminished the possibility of macromolecular growth.

The reaction of the 3-chloropentene-1 and 2,3-dichloropentene-1 models with benzoyl peroxide was similarly studied (Table IV) but the behavior of these compounds were essentially different. The variation of the monomer:initiator ratio in wide limits from 1:0.5 to 1:0.01 do not lead to changes in the molecular weight of the obtained products. In all cases only dimer formation was observed. The presence of the unsaturated region as well as the aromatic region, quantitatively correlated with the methyl intensity signal led us to conclude that only dimers were formed, stabilized by disproportionation:

$$
\begin{array}{llll}
& \overset{H(Cl)}{\underset{|}{}} & \overset{H(Cl)}{\underset{|}{}} & \\
C_6H_5COO\!-\!CH_2\!-\!C\!-\!CH_2\!-\!C & & + & C_6H_5COO\!-\!CH_2\!-\!C\!-\!CH_2\!-\!CCl \\
\end{array}
$$

$$
\begin{array}{cccc}
& H(Cl) & H(Cl) & \qquad\qquad H(Cl) \qquad H(Cl) \\
& | & | & \qquad\qquad | \qquad\qquad | \\
C_6H_5COO\!-\!CH_2\!-\!C\!-\!CH_2\!-\!C & \;+\; & C_6H_5COO\!-\!CH_2\!-\!C\!-\!CH_2\!-\!CCl \\
(C_6H_5\!-\!) & | & \| & (C_6H_5\!-\!) \qquad | \qquad\qquad | \\
& CHCl & CH & \qquad\qquad CHCl \qquad CHCl \\
& | & | & \qquad\qquad | \qquad\qquad | \\
& CH_2 & CH_2 & \qquad\qquad CH_2 \qquad CH_2 \\
& | & | & \qquad\qquad | \qquad\qquad | \\
& CH_3 & CH_3 & \qquad\qquad CH_3 \qquad CH_3
\end{array}
$$

The presence of a large NMR signal in the olefinic region led us to suppose that disproportionation takes place preferentially by chlorine elimination, which is abstracted by other similar radicals in the system. This supposition is supported also by the fact that in the olefinic NMR region the observed signals correspond to two protons.

Taking into account the above observed behavior, one can conclude that the unsaturated end groups of 3-chloropentene-1 and 2,3-dichloropentene-1 type, which contain allylic chlorine atoms are practically unable to polymerize and exhibit a strong inhibitory effect as a result of a degradative transfer:

$$
\begin{array}{ccc}
 & \text{H(Cl)} & \text{H(Cl)} \\
 & | & | \\
\text{C}_6\text{H}_5\text{COO}-\text{CH}_2-\text{C} & + \text{CH}_2\!\!=\!\!\text{C} & \longrightarrow \\
 & | & | \\
 & \text{CHCl} & \text{CHCl} \\
 & | & | \\
 & \text{CH}_2 & \text{CH}_2 \\
 & | & | \\
 & \text{CH}_3 & \text{CH}_3
\end{array}
$$

$$
\begin{array}{cccc}
\text{H(Cl)} & \text{H(Cl)} & & \text{H(Cl)} \\
| & | & & | \\
\text{C}_6\text{H}_5\text{COO}-\text{CH}_2-\text{C}-\text{CH}_2-\text{C}\bullet & \longrightarrow & -\text{CH}_2-\text{C} \\
| & | & & \| \\
\text{CHCl} & \text{CHCl} & & \text{CH} \\
| & | & & | \\
\text{CH}_2 & \text{CH}_2 & & \text{CH}_2 \\
| & | & & | \\
\text{CH}_3 & \text{CH}_3 & & \text{CH}_3
\end{array}
$$

In the case of AIBN initiator [16] which has another stabilizing effect on the first stage macroradical than PB we succeeded in finding the elimination of α chlorine (with respect to radical) even immediately after the first stage growth.

The unsaturated end group models, substituted on both ends of the double bond and submitted to similar treatment as in the case of previous models, in a glass reactor, remain practically unchanged. One can observe only the appearance of the initiator degradation products which decompose almost totally.

As a consequence, it is to be expected that such end groups do not play a significant role in the postpolymerization processes, and remain practically unchanged.

Search for Unsaturated Groups in PVC

Taking into account the above discussion, we present the following situation on the different types of unsaturated groups in PVC, including the eventual postpolymerization processes (Table V).

Determination of the Unsaturation in PVC by NMR Method

Based on the situation summarized in Table V we started to search for these types of structures in PVC, using Fourier transform integration spectra [17].

We succeeded in finding characteristic signals in the unsaturated region. An example of such a spectrum is shown in Figure 2. The type and concentration of the unsaturation in PVC, using a series of four industrial and laboratory PVC samples, are summarized in Table VI. The difficulties of this kind of measurement and calculation mode led us to find some other method to verify these results.

Determination of the Total Labile Chlorine Content in PVC by Phenolysis

Taking into account that all the unsaturated structures which we found in PVC with the aid of the NMR spectra contain allylic chlorine atoms, we found a chemical method to detect them.

Similar to the behavior of the tertiary chlorine (Cl_T) in reaction with phenol [18], the phenol is able to react, without any catalyst, with all types of labile chlorines including those of an allylic type [19]. In this case different reaction products are obtained, function on the labile structure which acts in the reac-

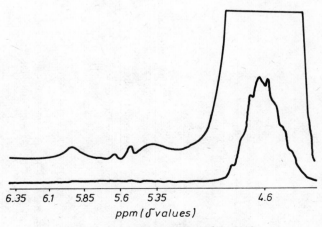

FIG. 2. NMR–PFT spectrum of the PVC.

TABLE V

Unsaturated End Groups and their Possible Transformation Products to Be Found in PVC

Unsaturated Group	Branches	Possible Transformation Products — Abnormal Internal Structures
$-CH_2-CH_2-CH-CH=CH$ with $-Cl$ (on CH) and $-Cl$ (terminal)	—	$-CH_2-C=CH-CH_2-$ (with $-Cl$) and $-CH_2-C-CH-CH_2-$ (with Cl, Cl, CH_2)
$-CH_2-CH_2-CH-C=CH_2$ with $-Cl$ (on CH) and $-Cl$ (on C)	$CH_2-CH-CH_2-$ with $-CHCl-$	
$-CH_2-CH_2-CH-CH=CH_2$ with $-Cl$	—	$-CH_2-CH=CH-CH_2-CH-CH-CH_2-$ (with $-Cl$, $-Cl$, $-Cl$)
$-CH_2-CH-CH_2-CH=CH$ with $-Cl$ (on CH) and $-Cl$ (terminal)	—	—

$$—CH_2—CH—CH_2—C=CH—CH—CH_2—$$
$$\quad\quad\quad |\quad\quad\quad\quad |\quad\quad\quad |$$
$$\quad\quad\quad Cl\quad\quad\quad\; Cl\quad\quad\; Cl$$

$$—CH_2—CH—CH_2—CH=CH—CH— \;+$$
$$\quad\quad\quad |\quad\quad\quad\quad\quad\quad\quad |$$
$$\quad\quad\quad Cl\quad\quad\quad\quad\quad\quad\;\; Cl$$

$$—CH_2—CH—CH_2—CH_2—CH_2—CH—$$
$$\quad\quad\quad |\quad\quad\quad\quad\quad\quad\quad\quad\; |$$
$$\quad\quad\quad Cl\quad\quad\quad\quad\quad\quad\quad\; Cl$$

$$—CH_2—CCl—CH_2—$$
$$\quad\quad\quad\;\; |$$
$$\quad\quad\quad\;\; CH_2—$$

$$—CH_2—CH—CH_2—$$
$$\quad\quad\quad |$$
$$\quad\quad\quad CH_2—$$

$$—CH_2—CH—CH_2—C=CH_2$$
$$\quad\quad\quad |\quad\quad\quad\quad |$$
$$\quad\quad\quad Cl\quad\quad\quad\; Cl$$

$$—CH_2—CH—CH_2—CH=CH_2$$
$$\quad\quad\quad |$$
$$\quad\quad\quad Cl$$

$$—CH_2—CH_2—CH=CH—CH—CHCl_2$$
$$\quad\quad\quad\quad\quad\quad\quad\quad\quad\quad |$$
$$\quad\quad\quad\quad\quad\quad\quad\quad\quad\quad Cl$$

$$—CH_2—CH_2—CH=C—CH_2—$$
$$\quad\quad\quad\quad\quad\quad\quad\quad\; |\quad\; |$$
$$\quad\quad\quad\quad\quad\quad\quad\quad\; Cl\; Cl$$

$$—CH_2—CH_2—CH=CH—CH—CH_2$$
$$\quad\quad\quad\quad\quad\quad\quad\quad\quad\quad |$$
$$\quad\quad\quad\quad\quad\quad\quad\quad\quad\quad Cl$$

$$—CH_2—CH=CH—CH—CH_2—$$
$$\quad\quad\quad\quad\quad\quad\quad\quad |$$
$$\quad\quad\quad\quad\quad\quad\quad\quad Cl$$

TABLE VI

Structure and Concentration of Unsaturated Groups in PVC (PFT–NMR Technique Determination)

Structure	Chemical Shift (ppm, central)	Cl_A concentration, atoms/1000 monomer units			
		Pliovic BL-90	Roumanian PVC	Hostalit 3057	PVC-TTS

$-CH_2-CH-CH=CH_2$
         $\|$
        Cl
	$\delta = 5.35$	1.2	0.7	0.7	2.2

$-CH_2-CH=C=CH_2$
            $\|$ $\|$
           Cl   Cl
	$\delta = 5.55$	—	—	1.0	2.85

and (or)
 $-CH_2-CH-CH=CH-CH_2-$
         $\|$
        Cl					

$-CH_2-CH=C-CH_2-$
            $\|$ $\|$
          Cl   Cl
	$\delta = 5.95$	1.25	1.76	2.7	0.6

and (or)
 $-CH_2-CH=C-CH-$
            $\|$ $\|$
          Cl Cl					

Weak signals	$\delta = 5.68$ $\delta = 5.78$	0.2–0.5 0.15–0.45	— —	0.2–0.7 0.15–0.45	— —
Total		2.45	2.46	4.4	5.05

tion, as shown:

Cl_T

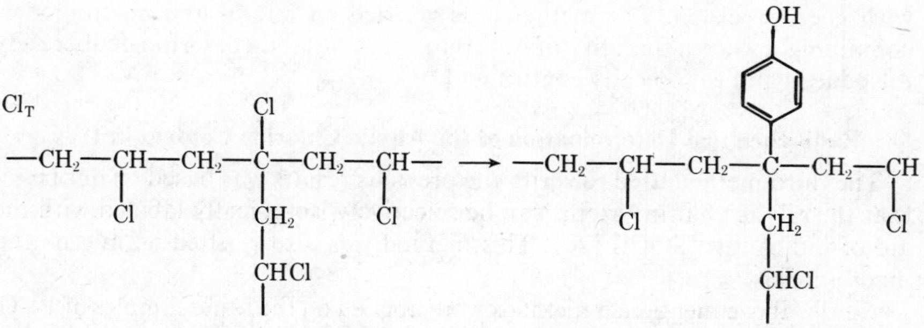

Cl_{A_1}

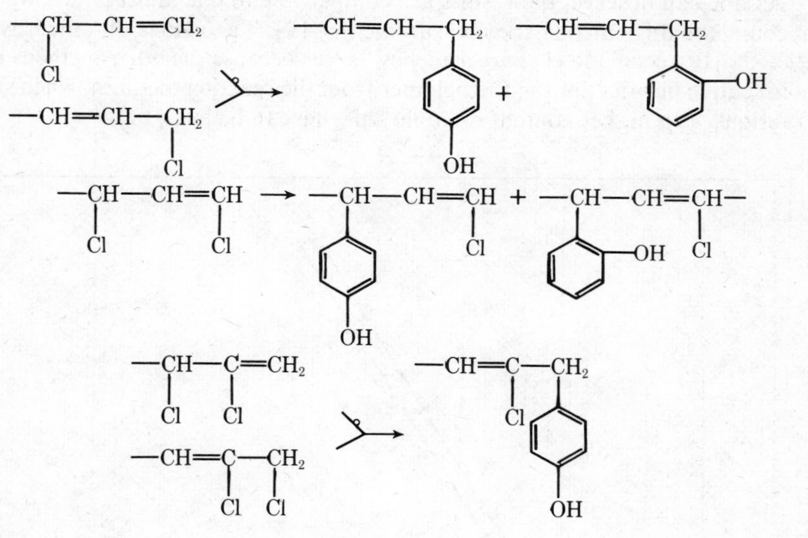

Cl_{A_2}

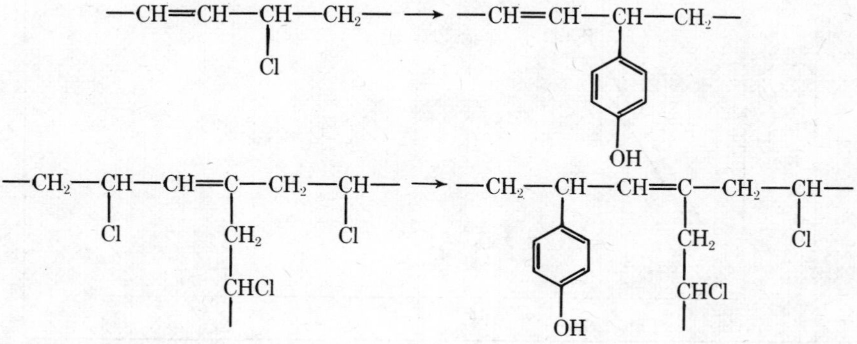

As one can see, all the compounds obtained contain an aromatic ring attached to the aliphatic chain in the place of the labile chlorine atoms.

Based on the approximate equality of the UV extinction coefficients of these compounds (Fig. 3) it was possible to establish a quantitative analytical method with good precision. The method was verified on micro- and macromodels containing a known quantity of different types of labile chlorine deliberately introduced and isotopically confirmed [19].

Radiochemical Determination of the Allylic Chlorine Content in PVC

The third method used to verify the previous results was based on our belief that the allylic chlorine atoms can be selectively isotopically labeled with the aid of radioactive $SOCl_2^*$ [20]. This method was also verified on micro- and macromodels.

All the above mentioned methods were applied on the same samples of PVC. The comparative results are given in Table VII.

As one can observe, the results are comparable to one another, taking into account the difficulty of the measurements [19]. The values obtained by the PFT–NMR technique [17] are somehow larger because the other methods need consecutive purifications of the polymer from the reaction medium, when small fractions with higher content of labile chlorine can be lost [19].

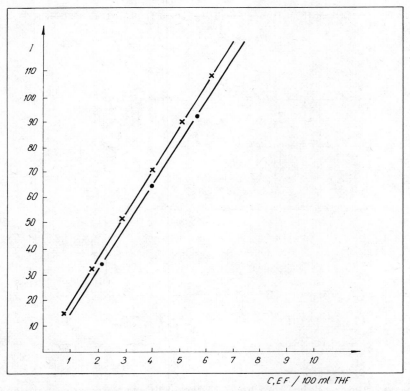

FIG. 3. Calibration curves for phenolysis method: ●, phenol; ×, 3-(p-hydroxyphenyl)-pentene-1.

TABLE VII
Comparison of the Values Obtained for Unsaturated Groups by Different Methods

Method	Cl$_L$ (%)			
	Sample 1	Sample 2	Sample 3	Sample 4
Phenolysis	0.14	0.095	0.12	0.17
NMR	0.245	0.246	—	—
Radiochemically	—	—	0.16	0.16

If we have a look at the nature of the unsaturation in PVC found by the NMR method, the results seem to be initially unexpected. In this connection we should make some comments.

Some Problems of the Vinyl Chloride Polymerization

There are still a lot of uncertainties regarding the polymerization of vinyl chloride in spite of the great number of papers published on the subject.

As is well known, the macroradical formed during the polymerization reaction is continually growing by reaction with new monomer molecules. Statistically the collision may take place at any part in the molecule (leaving aside the shielding effect of the chlorine atom). However studies on the PVC structure have shown that most of this macromolecule is a result of a head-to-tail collision [1]:

The existence of a small number of head-to-head structures has been suggested earlier [21].

A logical explanation of this situation would be the fact that the process is essentially determined by the energy state of the reaction products.

Structure (1) represents a radical stabilized on one side by the bond C—C, on the other side by the $-I_s$ effect of chlorine atom thus inducing a much lower energy state as compared with structure (2), which somewhat resembles the polyethylene macroradical, a very active radical as it is well known.

Thus the efficiency of these two types of collisions is not the same, the frequency of structure (2), of high energy, being much lower. However it must be taken into consideration.

Recent results enable us to suppose that these two types of radicals further polymerize quite differently. The radical (1) is stable enough to exist as such before a new collision with another monomer molecule or radical takes place, thus the normal growing process is continued. The radical (2) of high energy level close to the reactivity of polyethylene radical tries to get stabilized through all types of transformations which occur during ethylene polymerization, i.e., transposition, stabilization through α-atoms eliminations (Cl⁻ or H⁺) and, reaction with the monomer. The scheme of all these transformations is as follows:

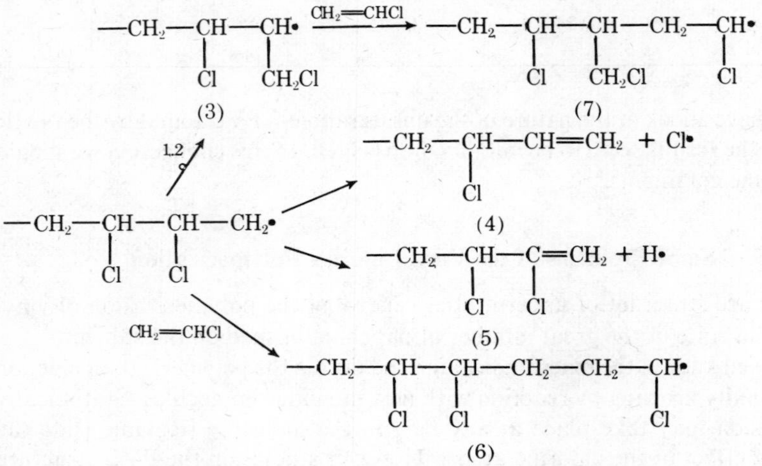

The appearance of radical (3) and of the PVC bearing —CH_2Cl branches (structure 7) was established by Abbas [7, 8] by NMR measurements, and by Rigo [6] by the analysis of degradation products of reduced PVC irradiated with γ-rays.

A process leading to structure (4) is strongly supported by our observation on the behavior of unsaturated end group model compounds having allylic chlorine atoms. These models give radicals which easily eliminate the α-chlorine atom [16]. We have made evident by PFT–NMR measurements [17] the structures of type (4) and (5) corresponding to some chain breaking processes and representing unsaturated end groups (Table VI).

Finally the macroradical (2) can normally react with vinyl chloride leading to a polymer having head-to-head units bearing vicinal chlorine atoms. Although there is some evidence for the existence of such structures [21], we think this problem is yet unsolved.

In our previous paper [17] we presented, besides the peaks corresponding to the unsaturated groups (4) and (5), a third type of such groups (8), whose appearance could be explained starting from the head-to-head macroradical (2):

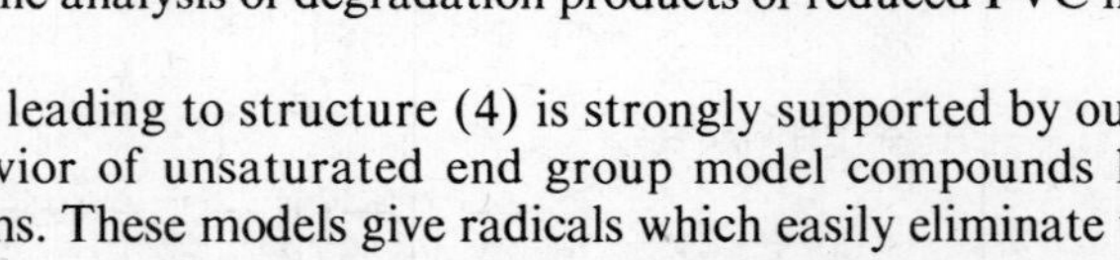

We have earlier experimentally noted the possibility of isomerization of structure (5) to (8) [10].

Considering our somewhat unexpected results on the nature of the PVC unsaturation in the light of the above statement, the situation becomes explainable. The difference we find in the ozonolysis [22] and NMR measurements [17] of the double bonds is due to the different principles the two methods are based upon, as well as to the processing of the samples. As a result one cannot exactly compare the results obtained by these two methods. However, we consider that the difference between the values obtained by these methods enables us to presume that in the case of PFT–NMR signal at $\delta = 5.55$ ppm there is an accumulated contribution of both types of double bonds, internal and terminal.

The internal double bond could appear through another process as the copolymerization with acetylenic impurities [11] (as we have demonstrated), by an after-polymerization process [16] of the unsaturated end groups, or by dehydrochlorination. That a hydrogen atom is extracted from the —CH$_2$— group in PVC through a radical mechanism was demonstrated by other authors [23]. As a result of this process, a radical with structure (9) is formed. As we have earlier experimentally proved, such a radical easily eliminates the chlorine atom resulting in the appearance of an internal double bond and not of branching [16]:

$$-CH_2-CH-CH-CH- \xrightarrow{-Cl\cdot} -CH_2-CH-CH=CH-$$

(9)

From the above, one can conclude that the head-to-head structure, of high energy level, appearing under certain conditions, plays an essential role both in the forming of branches and the appearance of chain breaking processes. This completely agrees with the well known argument that the lower the polymerization temperature, the lower the number of the branches and the higher the molecular weight of the obtained PVC [24].

Considering the much disputed transfer to monomer, the absence of unsaturated end groups of type IV, V and VI (Table I) led us to think that an existing transfer to monomer would be realizable only by extracting an atom from a high energy structure as (2), the monomer being the acceptor.

This is to be expected since, while elimination from the structure (2) leads to more stable forms (4) and (5), elimination from the monomer molecule, as some authors suppose [25], leads to structures of a much higher energy level as compared with initial state (vinyl chloride), (10), (11), and (12), and much more improbable:

$$CH_2=C\cdot \qquad CH=CH\cdot \qquad CH_2=CH\cdot$$

(10)   (11)   (12)

REFERENCES

[1] J. D. Cotman, *Ann. New York Acad. Sci.*, *57*, 417 (1953).

[2] E. Parker, *Kunststoffe*, *47*, 443 (1957).

[3] A. A. Caraculacu, *J. Polym. Sci.*, *A-1*, *4*, 1839 (1966).

[4] A. A. Caraculacu, E. C. Bezdadea, and G. Istrate, *J. Polym. Sci.*, *A-1*, *8*, 1239 (1970).

[5] D. Braun and F. Weiss, *Angew. Makromol. Chem.*, *13*, 55 (1970).

[6] A. Rigo, G. Palma, and G. Talamini, *Makromol. Chem.*, *153*, 219 (1972).

[7] A. Abbas, F. Bovey, and F. Schilling, *Makromol. Chem.*, *Suppl. 1*, 227 (1975).

[8] F. Bovey, K. Abbas, F. Schilling, and W. Starness, *Macromolecules*, *8*, 437 (1975).

[9] E. C. Bezdadea and A. A. Caraculacu, *Rev. Roum. Chim.*, *14*, 1171 (1969).

[10] E. C. Bezdadea, E. C. Buruiană, G. Robilă, and A. A. Caraculacu, *Eur. Polym. J.*, *9*, 445 (1973).

[11] E. C. Buriuană, E. C. Bezdadea, G. Robilă, and V. T. Bărbînţă, *Eur. Polym. J.*, in press.

[12] E. C. Bezdadea, E. C. Buruiană, and A. A. Caraculacu, *Eur. Polym. J.*, *7*, 1649 (1971).

[13] A. A. Caraculacu, E. C. Bezdadea, E. C. Buruiană, and F. V. Dobre, *Bull. Polytech. Inst.*, *Jassy*, in press.

[14] A. A. Caraculacu, *J. Polym. Sci.*, *A-1*, *4*, 1829 (1966).

[15] M. Aelterman and G. Smets, *Bull. Soc. Chem. Belges*, *60*, 459 (1951).

[16] E. C. Buruiană, G. Robilă, and A. A. Caraculacu, unpublished data.

[17] A. A. Caraculacu and E. C. Bezdadea, *J. Polym. Sci.*, in press.

[18] A. A. Caraculacu, Thesis.

[19] G. Robilă, E. Buruiană, and A. Caraculacu, *Eur. Polym. J.*, in press.

[20] E. C. Buruiamă, V. T. Bărbînţă, and A. A. Caraculacu, *Eur. Polym. J.*, in press.

[21] S. Enomoto, *J. Polym. Sci.*, *A-1*, *7*, 1255 (1969).

[22] A. Michel, E. Castaneda, and A. Guyot, Second International Symposium on PVC, July 5–9th, 1976, Lyon-Villeurbanne, France.

[23] S. Sobajima, W. Tagaki, and H. Watade, *J. Polym. Sci.*, *A-2*, *6*, 223 (1968).

[24] G. Boccato, A. Rigo, G. Talamini, and F. Zilio-Grandi, *Makromol. Chem.*, *108*, 216 (1967).

[25] I. W. Breitenbach, O. F. Olaj, H. Reelf, and A. Scindler, *Makromol. Chem.*, *122*, 51 (1969).

ISOMERIC COPOLYMERS OF ACETYLENES

C. SIMIONESCU, SV. DUMITRESCU, and V. PERCEC

"P. Poni" Institute of Macromolecular Chemistry, Jassy, 6600, Romania

SYNOPSIS

The triple bond opening in the *cis* position during Ziegler polymerization of acetylenes does not result however in pure *cis* polymers. The heat of polymerization and the exothermal effect of catalyst destroying has been found to be responsible for the variable *cis* content due to *cis–trans* thermal isomerization. The *cis–trans* thermal isomerization is accompanied by cyclization and scission of the polymer chains. These side reactions lead to copolymer structures instead of homopolymer ones. These processes probably determine the low thermal stability of the polymers (even at room temperature), and at the same time allow the interpretation of the order–disorder phenomenon found by Ehrlich in the case of high crystalline polyphenylacetylene at about 120°C.

On the polymerization of acetylenic derivatives, a great number of papers have been published on the structure and properties of the corresponding polymers. Although the acetylenic monomers have been polymerized by all classical techniques (radical, ionic, and coordinative), their polymerization behavior differs from that of corresponding vinylic monomers and the polymerization mechanisms and structure of polyacetylenes present yet a number of unsolved aspects.

Although the first trans-polyacetylene was synthesized by Natta et al. in 1958 [1] by polymerization of acetylene with Ziegler catalysts (i.e., $AlEt_3/TiCl_4$), and Watson et al. [2] showed in 1961 that by polymerization of acetylene with $Al(i\text{-}Bu)_3/TiCl_4$ catalytic system a *cis*-polyacetylene may be obtained, only in 1968 was the synthesis of a *cis*-rich polyacetylene reported by using a $Ti(OBu)_4/AlEt_3$ system at low temperature [3] (Table I). One year later, Kleist and Byrd [4] reported the synthesis of *cis*-rich polyacetylene (having a 60–70% *cis* content), by polymerization of acetylene with $AlEt_3/Fe(dmg)_2\cdot2Py$.

Shirakawa et al. [5, 6, 7] prepared polyacetylene by using the $Ti(OBu)_4/AlEt_3$ system over a wide temperature range (ca. $-100°–180°C$). They reported that the spectral data of polyacetylenes are best interpreted on the basis of an all *cis* structure for the polymers prepared at temperatures lower than $-78°C$, and an all *trans* structure for the polymers prepared at temperatures higher than 150°C. The physical properties of *cis*-polymers are different from those of *trans*-polymers [7] (Table I). Regarding the polymerization of arylacetylenes, the results were published with emphasis on phenylacetylene (PA).

Journal of Polymer Science: Polymer Symposium 64, 209–227 (1978)
© 1978 John Wiley & Sons, Inc.   0360-8905/78/0064-0209$01.00

TABLE I

Synthesis of *Cis* and *Trans* Isomers of Polyacetylene (PA) and Polyphenylacetylene (PPA)

No.	Polymer	Structure	Catalyst	Polymerization temperature(°C)	Literature	Observation
1.	PA	trans	$AlEt_3/TiCl_4$	room temperature	1	-
2.	PA	cis	$Al(iBu)_3/TiCl_4$	20 - 25	2	low cis content
3.	PA	cis	$AlEt_3/Ti(OBu)_4$	20	3	rich cis content
4.	PA	cis	$AlEt_3/Fe(dmg)_2 \cdot 2Py$	25	4	60-70% cis content
5.	PA	all cis	$AlEt_3/Ti(OBu)_4$	-78	5,6,7	λ_{max}=594 nm (clear red)
6.	PA	all trans	$AlEt_3/Ti(OBu)_4$	150	5,6,7	λ_{max}=700 nm (deep blue)
7.	PPA	trans	$AlEt_3/TiCl_4$	70	8	-
8.	PPA	trans	$AlEt_3/M(acac)_n$	80	9	-
9.	PPA	cis	$AlH(iBu)_2/Fe(acac)_3$	room temp.	10	-
10.	PPA	cis	$AlEt_3/TiCl_3$	50	11	-
11.	PPA	cis	$AlEt_3/Fe(dmg)_2 \cdot 2Py$	25	12	-
12.	PPA	cis or trans	WCl_6, $MoCl_5$	30	13	polar solvents → trans structure;nonpolar solvents → cis structure.

Berlin et al. [13] reported a *trans*-polyphenylacetylene (PPA) obtained by polymerization of PA with $AlEt_3/TiCl_4$. Simionescu et al. [9] reported a *trans*-PPA obtained by polymerization of PA with other Ziegler system, i.e., $AlEt_3/M(acac)_n$. Later on, Kern [10], Berlin et al. [11], and Simionescu et al. [12], synthesized also *cis*-PPA with Ziegler catalysts (Table I).

Finally, Masuda et al. [13] showed in 1975 that the polymerization of PA with WCl_6 and $MoCl_5$ leads to a polymer whose *cis* content depends on the polar or the nonpolar character of the solvent, i.e., the polymers obtained in polar solvents showed higher *trans* contents than the polymers produced in nonpolar solvents (Table I).

Ehrlich et al. [14–17] pointed out an order–disorder transition at about 120°C in insoluble and crystalline PPA obtained with Ziegler catalysts. The unpaired spin concentration (N_s) of PPA goes through a maximum at this temperature. This order–disorder transition is accompanied by a decrease of the crystallinity degree and by an increase of the polymer solubility. The polymer could not be recrystallized by cooling. The nature of this order–disorder transition is not yet elucidated.

Also, Ehrlich et al. [18] reported that when a PPA solution (obtained by heating the crystalline PPA in *o*-dichlorobenzene at 145°C) was heated at more than 160°C, the polymer undergoes a thermal degradation, accompanied by a decrease of the molecular weight.

Masuda et al. [13] showed that when the PPA obtained by PA polymerization with WCl_6 and $MoCl_5$ was left standing at room temperature, in light, for 2–6 months, the molecular weight decreased to about half of the original sample (Table II).

In the present paper we will try to answer a few unsolved problems, which appear from the literature data presented above.

(1) Why both *cis* and *trans* polymeric isomers could be obtained when acetylenes are polymerized with the same catalysts (i.e., Ziegler catalysts) and in the same reaction conditions?

TABLE II
Stability of Polyphenylacetylene[a]

	Standing period, month			
	0	2	4	6
$\overline{Mn}$, Room temperature, in light	7100	4500	4100	4100
D_{870}/D_{910}, Room temperature, in light	0.602	-	0.549	0.565

[a] According to reference [13].

(2) Why the polyphenylacetylene synthesized by Masuda (having a low *cis* content) is not stable even at room temperature?

(3) What is the meaning of the order–disorder transition in crystalline polyphenylacetylene at about 120°C, as pointed out by Ehrlich?

In some previous works we studied the coordinative, cationic, and thermal polymerizations of PA [12, 19], α-ethynylnaphthalene (α-EN) [20, 21], β-ethynylnaphthalene (β-EN) [22, 23], N-ethynylcarbazole (NEK) [24, 25], 9-ethynylanthracene (9-EA) [26], and 3-ethynylphenanthrene (3-EPh) [27], and the structures of the obtained polymers.

The four following structures are possible from the theoretical point of view for a monosubstituted polyacetylene (Fig. 1): *trans*-transoidal, *trans*-cisoidal, *cis*-cisoidal, and *cis*-transoidal.

Three fractions were separated from the polymers obtained with Ziegler catalysts: one insoluble in benzene (fraction I), one soluble in benzene but in-

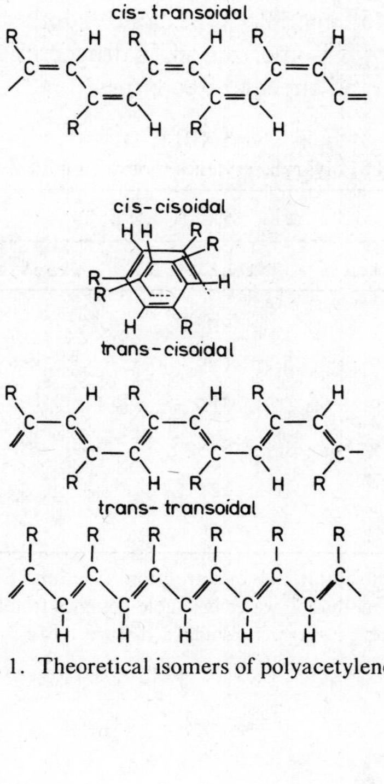

FIG. 1. Theoretical isomers of polyacetylenes.

soluble in methanol (fraction II), and in a few cases, a polymer fraction soluble in methanol, but insoluble in a 3/1 methanol–water mixture (fraction III). Polymer structures according to these fractions are presented in Table III. In all cases the polymerizations were carried out at 20°C, the monomer was added in the catalytic system at −78°C, and subsequent to polymerization reaction, the catalytic systems were destroyed with 10% methanolic HCl at −78°C.

The cationic polymerizations lead in all cases, except 9-EA, to *trans*-cisoidal structures. By thermal or radical polymerization, trans-cisoidal polymer structures or in some cases *cis–trans* copolymers with low *cis* content were obtained. Both linear polymers and cyclic trimers were obtained by thermal polymerization of arylacetylenes.

As we can see in Table III, by polymerization of β-EN and 3-EPh, fractions II with *trans* structures were obtained. Fractions III with *trans* structures were obtained by polymerization of α- and β-ethynylnaphthalene with Ziegler catalysts.

The reproducibility of synthesis regarding the *cis* content of the polymer is low, even if the synthesis was carried out at the same temperature [19]. Also, if the monomers were introduced in the reaction mixture at room temperature and/or the catalyst was destroyed also at room temperature, usually instead of polymers with high *cis* content, *trans* polymers or *cis–trans* copolymers with low *cis* content were obtained. The same behavior was observed in the case of acetylene polymerization with Ziegler catalysts [7]. In order to elucidate the *cis–trans* isomerization phenomenon which probably can take place during the polymerization, the *cis–trans* thermal isomerization of polymers was studied.

The thermal analysis of acetylene polymers presents the following characteristics [28]. Thermograms of *cis-poly*acetylene revealed the existence of two exothermic peaks at 145° and 325°C and one endothermic peak at 420°C which were assigned to *cis–trans* isomerization, hydrogen migration accompanied with crosslinking reaction, and thermal decomposition, respectively [28]. The en-

TABLE III
Structures of Polyarylacetylenes Obtained with Ziegler Catalysts

MONOMER	POLYMER STRUCTURE[a]			litera-ture
	fraction I	fraction II	fraction III	
PA	c–c	c–t	–	12,19
−EN	c–c	c–c	t–c	20,21
−EN	c–c	t–c	t–c	22,23
NEK	c–c/c–t	c–c/c–t	–	24,25
9-EA	c–c	c–c	–	26
3-EPh	c–c	t–c	–	27

[a] Fraction I = benzene-insoluble polymer fraction; fraction II = methanol-insoluble polymer fraction; fraction III = 3/1 methanol–water insoluble polymer fraction; c-c = *cis*-cisoidal structure; c-t = *cis*-transoidal structure; t-c = *trans*-cisoidal structure; c-c/c-t = *cis*-cisoidal and *cis*-transoidal stereoblocks structure.

thalpy change associated with the exothermic phenomenon at around 145°C was found to be 1.85 kcal/mole, calibrated to 100% *cis*-polyacetylene. Taking into account the fact that both *cis*- and *trans*-polyacetylenes are crystalline polymers, but with different XRD spectra, the exothermic phenomenon at around 145°C should include *cis–trans* isomerization (exothermic phenomenon), in-plane disorder of molecular chains (endothermic phenomenon), and rearrangement (exothermic phenomenon).

The thermograms of various polyarylacetylenes are presented in Figure 2 [29]. All *cis* polymers present an exothermic peak in a temperature range depending on the polymer nature [22, 25, 29, 31]. No peaks have been observed in the cases

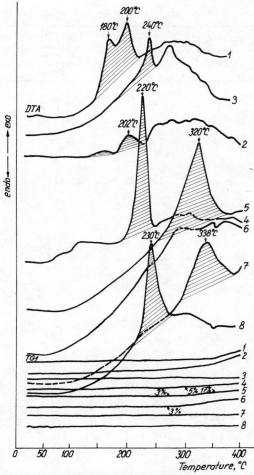

FIG. 2. DTA and TG curves of: PPA with *cis*-transoidal structure (1), PPA with *cis*-cisoidal structure (2), Pα-EN with *cis*-cisoidal structure (3), Pβ-EN with *cis*-cisoidal structure (4), PEK with *cis*-cisoidal/*cis*-transoidal stereoblock structure (5), PEK with *trans*-cisoidal structure (6), P9-EA with *cis*-cisoidal structure (7), and P3-EPh with *cis*-cisoidal structure (8).

of the *trans* polymers [30]. The exothermic effects are not accompanied by weight losses, except in the cases of *cis*-PEK and P-9EA. The weight losses represent 1–10% of the sample weight. It has to be mentioned that in the cases of *trans*-PEK and P-9EA this first step of thermal degradation is not accompanied by caloric effects. IR, UV, NMR, and XRD spectroscopy data showed that these exothermic peaks are connected to a *cis–trans* isomerization, which changes the *cis* insoluble or soluble, crystalline or amorphous polymer (Table IV) into a *trans*-cisoidal amorphous and soluble polymer. In the case of PEK only, the exothermic peak appears in a temperature range of 200°–330°C. The temperature value is a function of the *cis*-transoidal vs. *cis*-cisoidal stereoblock ratio.

TABLE IV
Cis–Trans Isomerization Heats of Different Polyacetylene Isomers

Polymer structure	Crystallinity degree (%)	Isomerization range(°C)	Temp. of exothermal phenomenon (°C)	Isomerization heat (kcal/mole)
–CH = CH–	–	47–227	47–227	$1.6{-}1.7^{a)}$
–CH = CH–	–	145	145	$1.85^{b)}$
–C = CH–⬡	5	130–210	180–200	$6.35^{c)}$
–C = CH–⬡	60	135–225	135–225	$1.45^{d)}$
–C = CH–⬡⬡	80	220–250	240	2.30
–C = CH–⬡⬡	60	190–240	220	5.08
–C = CH–⬡N⬡	60	275–360	320	19.80
–C = CH–⬡⬡⬡	amorphous	266–370	323	12.40
–C = CH–⬡⬡⬡	amorphous	190–265	230	12.56

[a] According to Ref. [4], 60–70% *cis* content, corrected for 100% *cis* content.
[b] 88% *cis* content, corrected for 100% *cis* content [28].
[c] *Cis*-cisoidal structure (94.45% *cis* content, corrected for 100% *cis* content).
[d] *Cis*-cisoidal structure (90.10% *cis* content, corrected for 100% *cis* content).

The enthalpy changes associated with the exothermic phenomena are presented in Table IV. Almost in all cases, the exothermic peaks include the exothermic phenomenon of *cis–trans* isomerization and the endothermic phenomenon of in-plane disorder of molecular chains.

The *cis–trans* isomerization of *cis* P9-EA and P3-EPh with amorphous structures is not accompanied by the disordering of molecular chains of a crystalline structure into an amorphous one, but by a conformational change from a *cis*-cisoidal conformation into a *trans*-cisoidal conformation.

At the present time it is impossible to determine independently the heat of isomerization from the apparent enthalpy change. This is the reason why the enthalpy changes obtained for *cis*-transoidal PPA (5% crystallinity degree), P9-EA and P3-EPh, both having an amorphous *cis*-cisoidal structure, are close to *cis–trans* heat of isomerization.

The difference between enthalpy changes of the transformation of *cis*-cisoidal and *cis*-transoidal PPA structures into *trans*-cisoidal structures, and that between *cis*-cisoidal structures of Pα-EN and Pβ-EN can be explained by differences in the crystallinity degree of the corresponding polymers (Table IV).

The enthalpy changes at *cis–trans* isomerization temperature are increasing with the size of the acetylene substituent. The highest caloric effect was found in the case of PEK and could be attributed to the rigidity of polymeric chains.

In order to elucidate the *cis–trans* isomerization phenomenon of polyarylacetylenes, a few structural details have to be taken into consideration. The IR spectra of *cis*-cisoidal and *cis*-transoidal structure of PPA are identical (Fig. 3), and present at 740, 870, and 1380 cm^{-1} the peaks characteristic of *cis* structure, while the IR spectra of *trans*-cisoidal structures are void of these bands and present three characteristic bands at 910, 970, and 1265 cm^{-1}. The ratios of the bands at 870 and 910 cm^{-1}, which are also characteristic of *cis* and *trans* structures, respectively, were used for the estimation of PPA *cis* content [13].

The NMR spectrum of the *cis*-transoidal structure of PPA presents the best resolution at 70°C (Fig. 4) having three protonic resonances centered at $\delta = 5.82$ ppm (one *cis* polyenic proton), $\delta = 6.7$ ppm (one aromatic proton), and $\delta = 6.85$ ppm (four aromatic protons).

The *cis* content in *cis*-transoidal PPA was determined from NMR spectra (at 70°C), using the area of the peak from $\delta = 5.82$ ppm ($A_{5.82}$) and the total area of the spectrum (A_t) according to eq. (1):

$$\% \ cis = \frac{A_{5.82} \times 10{,}000}{A_t \times 16.66} \tag{1}$$

The contents of *cis* sequences lower than 70% could not be determined from the NMR spectrum, due to the superposition of the $\delta = 5.82$ ppm signal with the resonance of aromatic protons (Fig. 5). The percentages of cis sequences determined from NMR spectra of PPA, were compared with the value of D_{760}/D_{740} ratio from their IR spectra (Fig. 6). The intensity of the absorbance at 760 cm^{-1} should be constant (C-H out of plane deformation vibration from

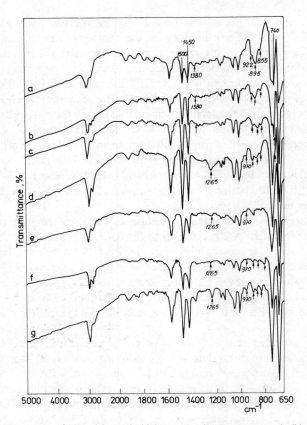

FIG. 3. Typical IR spectra of PPA: (a) *cis*-cisoidal structure (91.7% *cis* content); (b) *cis*-transoidal structure (94.45% *cis* content); (c) *cis*-transoidal structure (46.30% *cis* content); (d) *trans*-cisoidal structure obtained by polymerization with Ziegler catalyst; (e) *trans*-cisoidal structure obtained by polymerization with $(PPh_3)_2 \cdot PdCl_2$ at 140°C; (f) *trans*-cisoidal structure obtained by polymerization with $TiCl_4$ at 20°C; (g) *trans*-cisoidal structure obtained by thermal polymerization at 140°C.

monosubstituted benzene ring). Plotting the *cis* content determined by NMR vs. the D_{760}/D_{740} ratio from IR spectra, the *cis* content for both *cis*-transoidal soluble and *cis*-cisoidal insoluble structures can be determined from the obtained straight line (Fig. 6).

The lowest limit of *cis* content which can be determined from this plot is dependent on the presence of the 740 cm^{-1} band in the IR spectrum. It seems that *cis* contents lower than 40% could not be determined in this way, because these polymers exhibit no band at 740 cm^{-1}, and consequently these sequences can not be determined by IR measurements. In this way it is possible to explain the insoluble *trans* structure which may be obtained by the polymerization of PA. This polymer is insoluble perhaps due to a low content of *cis*-cisoidal sequences, undetectable by this method.

The comparative study of NMR spectra of *cis*-transoidal PPA with *trans*-cisoidal PPA obtained by thermal polymerization (from which the cyclic trimers

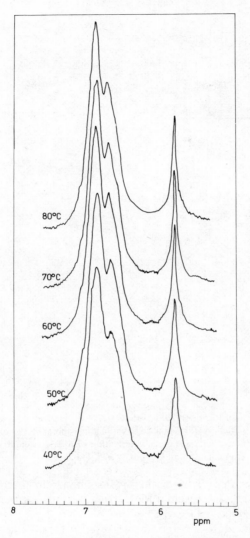

FIG. 4. NMR spectrum of *trans*-cisoidal structure of PPA (94.45% *cis* content, solvent CCl$_4$, reference TMS).

were extracted) pointed out the following differences. Between $\delta = 0$ and 5.5 ppm, the NMR spectrum of *cis*-transoidal PPA does not present protonic resonances (Fig. 7). The resonances of polyenic and aromatic protons appear at $\delta = 5.5$–7.6 ppm. The NMR spectrum of *trans*-cisoidal PPA obtained by thermal polymerization presents at $\delta = 2$–4.5 ppm the methinic protons resonance of 1,3 and 1,4 cyclohexadiene structures, between $\delta = 4.5$ and 6 ppm the resonance of double bonds protons from cyclohexadiene structures, at $\delta = 6$–7.6 ppm the resonance of aromatic monosubstituted protons and of *trans*-polyenic protons, and at $\delta = 7.6$–8.4 ppm the proton resonances from polyphenylene structures.

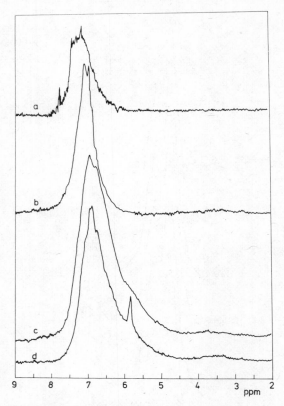

FIG. 5. NMR spectra of PPA: (a) *trans*-cisoidal structure obtained by thermal polymerization at 140°C; (b) *trans*-cisoidal structure obtained by cationic and phosphine complexes polymerizations; (c) *trans*-cisoidal structure obtained by polymerization with Ziegler catalyst; (d) *cis*-transoidal structure (46.30% *cis* content); solvent CCl₄, reference TMS.

The NMR spectra of *trans*-cisoidal PPA obtained by thermal isomerization of *cis*-cisoidal and *cis*-transoidal PPA structures present the same resonance as the thermally obtained PPA.

In order to elucidate the appearance of such copolymer structure during thermal or Ziegler polymerizations, we studied the structural changes which take place during *cis–trans* thermal isomerization of *cis*-PPA. The IR spectra of *cis*-PPA were run as a function of temperature (Fig. 8). Above 120°C, the bands from 740, 895, and 1380 cm⁻¹ diminished and a new band appeared at 1265 cm⁻¹. The absorption of the band from 920 cm⁻¹ increases in intensity. Increasing the temperature to 170°C, the first three bands, characteristic of structures, vanished, and the IR spectrum of the parent polymer changed to a spectrum similar to that of *trans*-polymers obtained by thermal or Ziegler polymerizations.

The NMR spectrum of the *cis*-transoidal structure was also recorded at different temperatures (under argon in hexachlorobutadiene) (Fig. 9). The resolution of the spectrum increased up to 70°C. At higher temperatures than 70°C

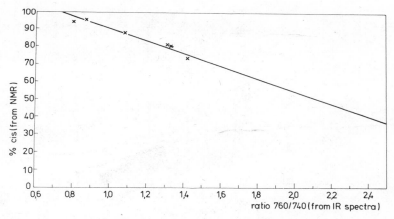

FIG. 6. The dependence of *cis* content of PPA determined by NMR, versus 760 cm^{-1}/740 cm^{-1} ratio.

the area of the peaks at δ = 5.82 ppm and δ = 6.70 ppm start to decrease. At the same time the resonance of methinic protons from cyclohexadiene structures appears between δ = 2–4.5 ppm. Subsequently, the superposition of signal at δ = 5.82 ppm (*cis*-polyenic protons) with the protons from the double bonds of cyclohexadiene structures (situated at δ = 4.6–6.2 ppm) starts to occur. At higher temperatures than 120°C, a new signal increases at δ = 7.6 ppm resembling the signal from the NMR spectrum of thermal *trans*-cisoidal PPA. The appearance of this signal proves that the thermal isomerization of the polymer is accompanied to a very small extent by destructive processes which give some cyclic derivatives (i.e., 1,3,5-triphenylbenzene). The rescanned NMR spectrum of the purified isomerized polymer is void of signals from δ = 7.6 ppm. So we can conclude that thermal *cis–trans* isomerization is accompanied by scission of the polymeric chain which isolates low amounts of cyclic compounds (i.e., 1,3,5-triphenylbenzene), and by appearance of cyclohexadiene and polyphenylene structures in the polymeric chain.

The kinetic study of *cis–trans* thermal isomerization of polyacetylene showed that in the solid state at room temperature the rate of isomerization is very low. Kleist and Byrd [4] found that the *cis*-rich polyacetylene isomerized to the *trans* polymer at room temperature with a rate of *cis* to *trans* conversion of 5–6% per day. Ito et al. [28] showed that a polyacetylene synthesized and purified carefully at −78°C contained·98% *cis* isomer, whereas a polymer obtained at room temperature had about 60% *cis* content. In order to obtain a polymer containing the same composition as the polymer synthesized at room temperature, the *cis*-rich polymer needs a thermal treatment of 300 minutes at 75°C. The apparent activation energy for the *cis–trans* isomerization is 17.0 kcal/mole for the polymer containing 88% *cis* configuration and it increased with increasing *trans* content up to 38.8 kcal/mole for 80% *trans* content.

Since the ease of *cis–trans* isomerization must be influenced by the electron density at the double bond, the activation energy for the *cis–trans* isomerization

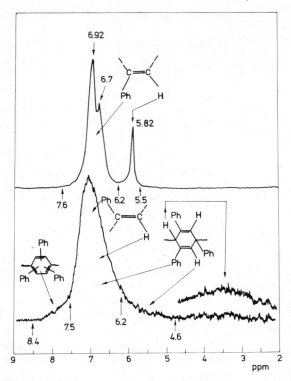

FIG. 7. NMR spectra of: *cis*-transoidal structure of PPA (94.45% *cis* content) (up), and *trans*-cisoidal structure obtained by thermal isomerization or thermal polymerization (down); the cyclic trimers from the *trans* polymer were removed (solvent CCl$_4$, reference TMS).

of a conjugated polyene system may be lower than those of unconjugated compounds. In the longer conjugated polyenes the decreased double bond order should result in lower activation energy. In contrast to the low value observed for *cis*-rich polyacetylene, however, the activation energy increased with increasing *trans* content, although the *trans*-polyacetylene has a higher conjugation than the *cis* one.

In the case of *cis*-transoidal PPA having 85% *cis* content, the thermal isomerization at room temperature in solution takes place (Fig. 10) with a 5–7% rate per day, and at a *cis* content lower than 60%, the rate is 2–4% per day [32]. The *cis*-PPA isomerizes at room temperature both in solution and in the solid state in the presence of UV or even visible radiation [32]. In order to reduce the reaction of chain scission, the kinetics of *cis–trans* isomerization of *cis*-transoidal PPA were studied at temperatures lower than 110°C (Fig. 11, Table V). Table V presents the dependence of *cis–trans* activation energy in the case of *cis*-transoidal PPA vs. *cis* content of the polymer. In the case of PPA with *cis* contents lower than 80%, the determinations (in solution) become more difficult due to the superposition of the peak at $\delta = 5.82$ ppm (*cis* polyenic proton) on the signal between $\delta = 4.5$ and 6.0 ppm (protons from the double bonds of cy-

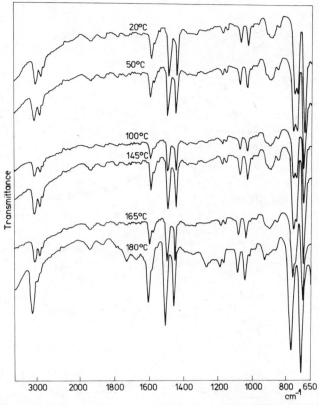

FIG. 8. IR spectrum of *cis*-transoidal or *cis*-cisoidal structures of PPA as a function of temperature (KBr pellet).

clohexadiene structures). In this case we can observe also an increase of activation energy of *cis–trans* thermal isomerization with the decrease of polymer *cis* content. In the case of PPA there are more parameters which influence the value of *cis–trans* activation energy. Because of the steric repulsion between hydrogen and phenyl groups located in every other double bond along the chains, it may give rise to a twist around the *cis* double bond to some extent. So the structure of polyarylacetylenes is built up from conjugated blocks of which sizes depend on the configuration and conformation of the polymer [21]. The *trans*-PPA presents the longest conjugated blocks but at the same time they are polydisperse in length, by contrast with *cis*-PPA which presents homogeneous conjugated blocks but of shorter length. At the same time the chain of *trans* polymer contains cyclohexadienes and polyphenylene sequences. Because of this steric problem, the *cis* polymers seem to be thermodynamically more unstable than the *trans* polymers. These are of course only a few factors which complicate the *cis–trans* isomerization phenomenon and retard the reverse reaction of *trans–cis* isomerization.

In these conditions we may consider that *cis* opening of the triple bond occurs

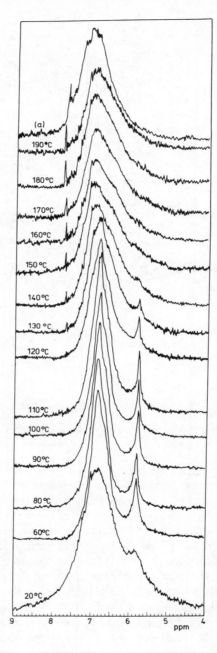

FIG. 9. NMR spectrum of *cis*-transoidal structure of PPA (94.45% *cis* content) as a function of temperature (solvent hexachlorbutadiene, reference HMDS); (a) typical NMR spectrum of *trans*-cisoidal PPA obtained by thermal isomerization (120°C, time 6 hours) of *cis*-transoidal or *cis*-cisoidal structures of PPA (solvent hexachlorbutadiene, reference HMDS).

TABLE V

The Activation Energies for *Cis–Trans* Thermal Isomerization of *Cis-Transoidal* Polyphenylacetylenes with Different *Cis* Content

Cis content (%)	Activation energy	
	Solution (kcal/mole)	Solide state (kcal/mole)
96.0	13.10	28.99
89.5	16.36	–
86.0	20.78	32.77
81.0	15.57	23.31

at any polymerization temperature with Ziegler catalysts, to yield *cis* conjugated double bonds. *Trans* double bonds in the polymer must be formed by *cis–trans* isomerization of the *cis* conjugated double bonds. Since the rate of *cis–trans* thermal isomerization of the polymers is too low at room temperature as we have shown above, we can conclude that thermal isomerization takes place, first during the propagation reaction under the action of heat of polymerization, and second due to thermal effects which accompanied the working up of the polymer (i.e., isomerization of the polymer in solution or in the solid state).

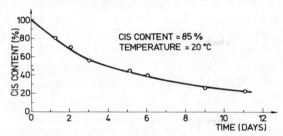

FIG. 10. Kinetics plot for the thermal *cis–trans* isomerization of *cis*-transoidal structure of PPA in solution (solvent hexachlorbutadiene).

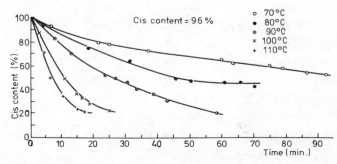

FIG. 11. Kinetics plots for the thermal *cis–trans* isomerization of *cis*-transoidal PPA in solution (solvent hexachlorbutadiene).

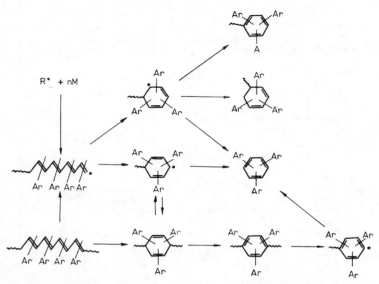

FIG. 12. The reactions which accompany the *cis–trans* thermal isomerization of PPA.

Consequently, we consider that by coordinative polymerization of arylac-etylene monomers, we may obtain homopolymers with *cis* structure, or co-polymers with *cis, trans,* cyclohexadiene and polyphenylene sequences.

In Figure 12, the reactions which take place during radical polymerization of arylacetylene monomers [33] and which accompanied also the *cis–trans* thermal isomerization are presented.

The cyclic trimers may be obtained when we lead the polymerization into a *cis*-cisoidal configuration, or during the *cis–trans* thermal isomerization of the

TABLE VI

$\overline{M}n$ of *Cis* and *Trans* Polyphenylacetylenes

Catalyst	Structure	$\overline{M}n$	Literature
AlEt$_3$/Fe(acac)$_3$	cis	4900	10
AlEt$_3$/Fe(acac)$_3$	trans	4200	10
AlEt$_3$/Co(acac)$_3$	cis–97.5%	4150	19
AlEt$_3$/Fe(dmg)$_2$·2Py	cis–79.2%	3800	19
AlEt$_3$/Co(acac)$_3$	cis–78.5%	3750	19
AlEt$_3$/Co(acac)$_3$	trans	3500	19
WCl$_6$(solvent,C$_6$H$_6$)	$D_{870}/D_{910}=0.746$	11800	13
WCl$_6$(solvent,CH$_2$Cl$_2$)	$D_{870}/D_{910}=0.568$	6100	13
MoCl$_5$(solvent,C$_6$H$_6$)	$D_{870}/D_{910}=1.27$	5900	13
MoCl$_5$(solvent,CH$_2$Cl$_2$)	$D_{870}/D_{910}=0.909$	5600	13

TABLE VII

ESR Characteristics of Polyarylacetylene Isomers

No.	Polymer	Polymer Structure	$\overline{M}n$	Softening Temperature (°C)	Unpaired spins concentration vs. gram
1.	PPA	cis-cisoidal (90% cis content)	5000[a]	230-257	6.00×10^{17}
2.	PPA	cis-cisoidal (89% cis content)	5500[a]	238-243	2.00×10^{17}
3.	PPA	cis-transoidal (94% cis content)	4150	205-213	2.5×10^{19} (6.00×10^{18})[b]
4.	PPA	cis-transoidal (80% cis content)	3750	185-195	9.30×10^{19} (5.00×10^{18})[b]
5.	PPA	trans-cisoidal	850	120-124	3.02×10^{19}
6.	Pα-EN	cis-cisoidal	-	244-250	3.00×10^{19}
7.	Pα-EN	trans-cisoidal	764	158-161	6.00×10^{19}
8.	Pβ-EN	cis-cisoidal	-	246-250	7.40×10^{18}
9.	Pβ-EN	trans-cisoidal	1435	194-196	8.00×10^{19}
10.	PEK	c-t/c-c[c]	-	338-342	7.25×10^{19}
11.	PEK	c-t/c-c[c]	2700	265-280	9.09×10^{19} (5.05×10^{19})[b]
12.	PEK	trans-cisoidal	1935	250-255	3.70×10^{19}

[a] Molecular weight determined after thermal isomerization of the *cis*-cisoidal insoluble polymer.

[b] Determined as 10% solution in diphenyl ether.

[c] *Cis*-transoidal/*cis*-cisoidal stereoblock structure.

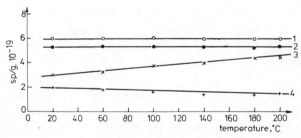

FIG. 13. Temperature dependence of unpaired spins concentration for *trans*-cisoidal structures of: (1) Pα-EN; (2) Pβ-EN; (3) PPA; (4) PEK; heating time, 5 minutes.

polymer. In these conditions we can not neglect the possibility of *cis* opening of the triple bonds also in the case of thermal polymerization.

Table VI presents the molecular weight of *cis* and *trans* PPA, both prepared using the same catalyst. The *cis* polymers present higher molecular weight than the *trans* polymers. The *cis–trans* isomerization reaction which takes place during polymerization being accompanied by scission of polymer chain, explains the decrease of $\overline{M}n$ of *trans* polymers, comparatively with that of *cis* polymers. Based on these reactions we may explain the degradation of PPA observed by

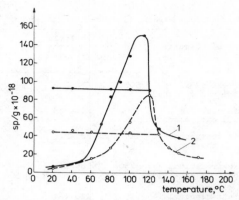

FIG. 14. Temperature dependence of unpaired spins concentration for *cis*-transoidal structure of PPA (10% solution in diphenyl ether): (1) 94% *cis* content; (2) 80% *cis* content; heating time, 5 minutes; the straight lines represent the recordings on cooling.

Masuda (when the PPA was stored in light at room temperature) or by Ehrlich (at higher temperature than 160°C).

In order to elucidate the order–disorder phenomenon observed by Ehrlich in the case of crystalline PPA, the ESR spectra of all polymers were determined as a function of temperature [34]. The unpaired spins concentration of polymers are presented in Table VII. In the cases of PPA and Pβ-EN only, we found a difference of two and one magnitude order, respectively, between the N_s of crystalline and amorphous structures.

All polymers with *trans*-cisoidal structure obey the Curie law (Fig. 13) in a temperature range between 20° and 200°C, i.e., up to above their softening temperatures. Both in solution and in the solid state, the unpaired spins concentration of the *cis* polymers goes through a maximum at the *cis–trans* isomerization temperature. For *cis* polymers the Curie law is obeyed only for temperatures lower than that at which the polymer configuration is changed.

Evidently, the *cis–trans* thermal isomerization takes place via an excited state (the double bond is excited to a higher energy level) in which the π electrons are no longer involved in bond formation (the so-called antibonding state), so that free rotation is then possible about the remaining single σ-bond connecting the carbon atoms of the original double bond. When the antibonding state drops to the ground state with release of its excitation energy, the double bond is formed again, but with the prevalence of the *trans* conformation.

Consequently, we believe that at the isomerization temperature, the number of unpaired π orbitals is maximal and at the same time it depends on the amount of *cis* units in the polymer, as shown in Fig. 14.

Consequently, the order–disorder phenomenon observed by Ehrlich in the case of crystalline PPA and evidenced for all *cis* polymers [34], represents in fact a structural *cis–trans* irreversible isomerization.

REFERENCES

[1] G. Natta, G. Mazzanti, and P. Corradini, *Rend. Accad. Nazl. Lincei Rend. Classe. Sci. Fis. Mat. Nat., 8, 25*, 3–12 (1958).

[2] W. H. Watson Jr., W. C. McMordie Jr., and L. G. Lands, *J. Polym. Sci., 55*, 137–144 (1961).

[3] M. Hatano, K. Shimamura, S. Ikeda, and S. Kambara, *Preprts. Progr. Polym. Phys. Japan, 11*, 123 (1968).

[4] F. D. Kleist and N. R. Byrd, *J. Polym. Sci., A-1, 7*, 3419–3425 (1969).

[5] H. Shirakawa and S. Ikeda, *Polym. J., 2*, 231–244 (1971).

[6] H. Shirakawa, T. Ito, and S. Ikeda, *Polym. J., 4*, 460–462 (1973).

[7] T. Ito, H. Shirakawa and S. Ikeda, *J. Polym. Sci., Polym. Chem. Ed., 12*, 11–20 (1974).

[8] A. A. Berlin, M. I. Cherkashin, I. P. Cernisheva, E. I. Assev, E. I. Barkan, and P. P. Kisilitsa, *Vysokomol. Soedin (A), IX*, 1840–1846 (1967).

[9] C. Simionescu and Sv. Dumitrescu, *Eur. Polym. J., 6*, 635–654 (1970).

[10] R. J. Kern, *J. Polym. Sci., A-1, 7*, 621–631 (1969).

[11] A. A. Berlin and M. I. Cherkashin, *Vysokomol. Soedin (A), XIII*, 2298–2308 (1971).

[12] C. Simionescu, Sv. Dumitrescu, I. Negulescu, V. Percec, M. Grigoraş, I. Diaconu, M. Leancă, and L. Goraş, *Vysokomol. Soedin, (A) XVI*, 790–800 (1974).

[13] T. Masuda, N. Sasaki and T. Higashimura, *Macromolecules, 8*, 717–721 (1975).

[14] P. Ehrlich, R. J. Kern, E. D. Pierron and T. Provder, *J. Polym. Sci., Part B, 5*, 911–915 (1967).

[15] P. Ehrlich, *J. Macromol. Sci.-Phys., B2(2)*, 153–162 (1968).

[16] G. M. Holob, P. Ehrlich, and R. D. Allendoerfer, *Macromolecules, 5*, 569–573 (1972).

[17] S. H. C. Chang, E. C. Mertzlufft, P. Ehrlich and R. D. Allendoerfer, *Macromolecules, 8*, 642–644 (1975).

[18] B. Biyani, A. J. Campagna, D. Daruwalla, C. M. Shirastava, and P. Ehrlich, *J. Macromol. Sci.-Chem., A9(3)*, 327–339 (1975).

[19] C. Simionescu, V. Percec and Sv. Dumitrescu, *J. Polym. Sci., Polym. Chem. Ed.,* in press.

[20] C. Simionescu, Sv. Dumitrescu, V. Percec, I. Negulescu, and I. Diaconu, *J. Polym. Sci., Polym. Symp., 42*, 201 (1973).

[21] C. Simionescu, Sv. Dumitrescu, and V. Percec, *Polym. J., 8*, 313–317 (1976).

[22] C. Simionescu, Sv. Dumitrescu, and V. Percec, *IVth Symposium Polymer,* Varna, October 4–6 (1973), Preprint Vol. I, p. 466.

[23] C. Simionescu, Sv. Dumitrescu, and V. Percec, *Polym. J., 8*, 139–149 (1976).

[24] C. Simionescu, Sv. Dumitrescu, and V. Percec, *IUPAC, International Symposium on Macromolecules,* Madrid, 1974, Preprint I, 5–13.

[25] Sv. Dumitrescu, V. Percec, and C. Simionescu, *J. Polym. Sci., Polym. Chem. Ed.,* in press.

[26] C. Simionescu, Sv. Dumitrescu, and M. Grigoraş, *IVth Symposium Polymer,* Varna, October 4–6 (1973), Preprint Vol. I, p. 479.

[27] C. Simionescu, Sv. Dumitrescu, M. Grigoraş, and I. Diaconu, *IUPAC, International Symposium on Macromolecules,* Madrid, 1974, Preprint I, 5–14.

[28] T. Ito, H. Shirakawa, and S. Ikeda, *J. Polym. Sci., Polym. Chem. Ed., 13*, 1943–1950 (1975).

[29] C. Simionescu, Sv. Dumitrescu, V. Percec, and M. Grigoras, *Rev. Roum. Chim.,* in press.

[30] C. Simionescu, Sv. Dumitrescu, S. Vasiliu, and V. Percec, *J. Thermal. Anal., 6*, 389–399 (1974).

[31] C. Simionescu, V. Percec, and Sv. Dumitrescu, *IUPAC, International Symposium on Macromolecules,* Jerusalem, 1975, Abstract I.1, p. 71.

[32] I. M. Radionov, Ph.D. Thesis, Moskwa, 1975.

[33] C. Simionescu, Sv. Dumitrescu, and V. Percec, unpublished data.

[34] Sv. Dumitrescu, V. Percec, and C. Simionescu, *Rev. Roum. Chim.,* in press.

POLYMERIZATION OF VINYLPYRIDINIUM SALTS. X. COPOLYMERIZATION STUDIES OF CATIONIC–ANIONIC MONOMER PAIRS

J. C. SALAMONE, A. C. WATTERSON, T. D. HSU, C. C. TSAI, M. U. MAHMUD, A. W. WISNIEWSKI, and S. C. ISRAEL

Polymer Science Program, Department of Chemistry, University of Lowell, Lowell, Massachusetts 01854

SYNOPSIS

The mechanism of the spontaneous polymerization of vinylpyridinium salts in aqueous and acidic solution has been actively investigated in recent years. Recent reports have suggested that these reactions are free radical in nature. In order to investigate the involvement of free radical initiation in the spontaneous polymerization reactions, an investigation of ion pair copolymers between the cationic 4-vinylpyridinium moiety and the anionic vinylsulfonate, 2-acrylamido-2-methylpropanesulfonate, and *p*-styrenesulfonate has been undertaken. The results of spontaneous polymerization suggest that for 4-vinylpyridinium vinylsulfonate, only the cationic moiety is polymerized, whereas for 4-vinylpyridinium 2-acrylamido-2-methylpropanesulfonate, two homopolymers or a block copolymer apparently result, and for 4-vinylpyridinium *p*-styrenesulfonate, an alternating copolymer appears to be obtained.

INTRODUCTION

In recent years there has been considerable interest in the spontaneous polymerization of vinylpyridinium salts to their corresponding vinyl polymers. A certain degree of confusion has ensued with regard to their mechanisms of polymerization, since three monomers have been studied (i.e., 2-, 4-, and 2-methyl-5-vinylpyridine) under conditions of acylation, alkylation, protonation, and changing monomer concentration in aqueous solution [1–3]. There is general agreement that in alkylation reactions both 2- and 4-vinylpyridinium salts can homopolymerize by a zwitterionic mechanism which involves the attack of a neutral vinylpyridine upon the β-position of the double bond of a vinylpyridinium ion to generate a resonance stabilized zwitterion. Studies of the spontaneous alkylation polymerizations of 2- and 4-vinylpyridine in the presence of other comonomers such as styrene, acrylonitrile, and vinyl ether, have shown that none of the comonomers is incorporated in the zwitterionic polymerization system.

In regard to the mechanisms of polymerization of all three vinylpyridinium monomers in acid or in aqueous solution, it has been suggested by Kabanov et al. that a zwitterionic growth reaction occurs through counterion attack on the

Journal of Polymer Science: Polymer Symposium 64, 229–243 (1978)
0360-8905/78/0064-0229$01.00

vinylpyridinium double bond [1]. In regard to Kabanov's assertion of an anionic polymerization, which is characteristic of the polymerization by alkylating agents, it has been shown that the demonstrated reagents of anionic growth of (vinyl)pyridine (for initiation) and a resonance stabilized pyridone methide (for propagation) have no effect whatsoever on the spontaneous polymerization of salts from 2-methyl-5-vinylpyridine, although such compounds rapidly polymerize salts of 2-vinylpyridine and 4-vinylpyridine [3]. Interestingly, no resonance stabilized structures can be drawn for 2-methyl-5-vinylpyridinium salts upon anionic attack in the β-position of the double bond. The major point of significance of this work is that while 2-methyl-5-vinylpyridinium salts cannot undergo anionic initiation and propagation, GPC results and rate of polymerization measurements confirm that in strong acid all pyridine monomers behave similarly with regard to rate, molecular weight, and molecular weight distribution [4]. In addition, when spontaneous polymerization reactions of 2-vinylpyridine are conducted directly in fuming sulfuric acid, either a high molecular weight poly-2-vinylpyridinium salt or a low molecular weight addition compound of 2-hydroxy-2-(2'-pyridyl)-ethylsulfonic acid (I) is obtained [5]. The structure of the addition compound was verified by mass spectroscopy and ^{1}H NMR spectroscopy. The polymer that is isolated from fuming sulfuric acid can also be of narrow molecular weight distribution; for example, at a molar ratio of fuming sulfuric acid to 2-vinylpyridine of 1, GPC results revealed $\overline{M}_n = 1.57 \times 10^6$, $\overline{M}_w = 1.80 \times 10^6$ and $\overline{M}_w/\overline{M}_n = 1.15$ [5]. An anionic polymerization reaction, although capable of giving a narrow molecular weight distribution, is difficult to conceive in such a strongly acidic medium.

Further to the isolation of I given above, the concept of a pseudocationic mechanism was also considered. Using both the sulfonic acid I and the related sulfuric acid derivative of 2-(2'-pyridyl)ethyl bisulfate (II) [6] in the attempted initiation of 2-vinylpyridinium bisulfate, no significant initiation behavior was observed. These results would indicate that the pseudocationic mechanism is not valid in these spontaneous polymerization reactions.

Ringsdorf et al., on the other hand, have recently suggested that the spontaneous polymerization of vinylpyridinium salts in aqueous solution above a certain critical concentration is free radical in nature. This mechanism appears to have more validity as strong inhibition to the polyvinyl structure has been reported with typical radical inhibitors [3]. In support of the possibility of a radical mechanism, we have noted that the ^{13}C NMR spectra of several poly-2-vinylpyridinium salts obtained through spontaneous polymerization in acid solution were related to those obtained through the utilization of free radical initiators [8].

HO—CH—CH$_2$—SO$_3$H   CH$_2$—CH$_2$—OSO$_3$H

I   II

In this work are reported studies of the spontaneous polymerization behavior of a new class of ionic monomeric salt. These salts are cationic–anionic monomer pairs of the form III in which both cationic and anionic moieties of the salt are capable of vinyl polymerization[8]. These salts should be considered ion pair comonomers which contain no nonpolymerizable counterions. Through an investigation of the polymerization behavior of cationic–anionic monomer pairs, a more detailed understanding of how the spontaneous polymerization process is related to a free radical mechanism could hopefully be obtained depending on whether one or both components of the comonomer structure are incorporated into the polymer structure.

EXPERIMENTAL

Monomer Preparations

Cationic–Anionic Monomer Pairs

The preparations of 4-vinylpyridinium vinylsulfonate (IVa), 4-vinylpyridinium 2-acrylamido-2-methylpropanesulfonate (IVb), and 4-vinylpyridinium *p*-styrenesulfonate have recently been reported [8].

4-Vinylpyridinium Methanesulfonate (X)

To 4.8 g (50 mmol) of methanesulfonic acid in 60 ml of THF at $-10°C$ was added with stirring a solution of 4.5 ml (45 mmol) of freshly distilled 4-vinyl-pyridine in 60 ml of THF. The crystals were filtered, washed with THF and with ether and then dried *in vacuo,* giving a quantitative yield of 4-vinylpyridinium methanesulfonate, mp 109.5–110.5°C. NMR (D_2O) showed aromatic protons at $\delta 8.0$ (*d*, 2*H*, *J* = 7 Hz), $\delta 8.7$ (*d*, 2*H*, *J* = 7 Hz), the vinyl protons at $\delta 6.0$ (*d*, 1*H*), $\delta 6.5$ (*d*, 1*H*), $\delta 7.0$ (*d* of *d*, 1*H*), J_{gem} = 0 Hz, J_{cis} = 11 Hz, J_{trans} = 17 Hz, and the methyl protons at $\delta 2.8$ (*s*, 3*H*).

ANAL.: *Calcd.* for $C_8H_{11}NO_3S$: C, 47.75%; H, 5.51%; N, 6.96%. *Found:* C, 47.64%; H, 5.68%; N, 6.88%.

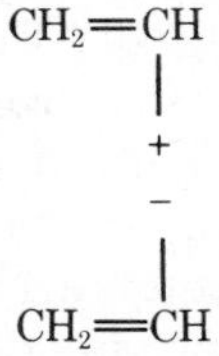

$$CH_2{=\!=}CH$$
$$|$$
$$+$$
$$-$$
$$|$$
$$CH_2{=\!=}CH$$
III

Sodium 2-Acrylamido-2-Methylpropanesulfonate (XI)

2-Acrylamido-2-methylpropanesulfonic acid (Lubrizol Corporation) (10 g) was dissolved in 200 ml methanol and passed through a cationic ion exchange column containing Amberlite IRA-120 in the sodium form. The eluate was evaporated under reduced pressure to obtain sodium 2-acrylamido-2-methyl-propanesulfonate. Recrystallization from 2-propanol/methanol (1:1 v/v) solution gave crystals of the monomer, mp 123°C with polymerization. NMR (D$_2$O) of the sodium salt showed δ6.2 (m, 3H, CH$_2$=CH—CO—NH—), δ5.7 (d of d, 1H, CH$_2$=CH—CO—), δ3.4 (s, 2H, methylene protons), and δ1.5 (s, 6H, methyl protons).

ANAL.: Calcd. for C$_7$H$_{12}$NO$_4$SNa: C, 36.68%; H, 5.28%; N, 6.11%. Found: C, 36.46%; H, 5.50%; N, 6.18%.

Polymer Preparations

Spontaneous Polymerizations of 4-Vinylpyridinium Vinylsulfonate, Nitrate, and p-Styrenesulfonate

The required amount of water was added to the required amount of 4-vinyl-pyridinium salt. The reaction was allowed to proceed at room temperature for a specified length of time. The polymer from 4-vinylpyridinium p-styrenesul-fonate precipitated and the solid content was separated by filtration, and then washed with water. The solubility property of this compound suggested that it was a copolymer. The polymers from 4-vinylpyridinium vinylsulfonate and 4-vinylpyridinium nitrate remained in solution and were precipitated by pouring into a large amount of acetone. The NMR spectra showed that they were ionenes [10] and the intrinsic viscosities of the products obtained (from 0.9M monomer solutions) were 0.137 dl/g and 0.034 dl/g in 0.1M KBr solution, for poly-1,4-pyridiniumdiylethylene vinylsulfonate and nitrate, respectively.

Spontaneous Polymerization of 4-Vinylpyridinium 2-Acrylamido-2-Methylpropanesulfonate

The spontaneous polymerization reactions of 4-vinylpyridinium 2-acrylamido-2-methylpropanesulfonate were conducted at 25°C while the monomer concentrations, solvents, and length of reaction time were varied (Table I). To each polymerization ampule, 10 ml of a freshly prepared solution of the monomeric salt was added. The solution was degassed, the ampule sealed and then placed in a thermostated bath at 25°C. After the appropriate time, the viscous solution was diluted with 100 ml of deionized water and divided into two equal parts. One part was precipitated into 500 ml of acetone, filtered, and washed with methanol to remove unreacted monomer. The existence of ionene polymer was examined by ^{1}H NMR. The other part was neutralized with excess sodium carbonate resulting in a polymer precipitate. This base insoluble polymer was obtained by centrifugation and purified by exhaustive dialysis (as it was water

TABLE I

Spontaneous Polymerization of 4-Vinylpyridinium 2-Acrylamido-2-Methylpropanesulfonate at 25.0°C

Solvent	Conc., M	Time, h	Structures Obtained	Total % Conv.	% Base Insoluble Polymer	% Base Soluble Polymer
Water	0.5	10	Ionene + Vinyl polymers	19.0	23	0
Water	1.0	4	Ionene + Vinyl polymers	24.1	62	27
Methanol	1.0	24	Ionene + Vinyl polymers	21.4	85	0
Dimethyl-formamide	0.5	39	Ionene + Vinyl polymers	22.8	60	30

soluble) against deionized water, followed by lyophilization. The remaining sodium carbonate solution was also exhaustively dialyzed against deionized water and lyophilized to obtain the base-soluble fraction. Infrared spectra of both base-soluble and base-insoluble portions were similar, with the latter showing a greater preponderance of polyvinylpyridine residues. The percentage products obtained and the elemental analyses of the base-soluble and insoluble portions are given in Tables I and II, respectively. For comparison purposes, the elemental analyses of poly-4-vinylpyridine and sodium poly-2-acrylamido-2-methylpro-panesulfonate are given below.

ANAL. (for poly-4-vinylpyridine): *Calcd.* for $(C_7H_7N)_x$: C, 79.97%; H, 6.71%; N, 13.32%. (For sodium poly-2-acrylamido-2-methylpropanesulfonate): *Calcd.* for $(C_7H_{12}NO_4SNa)_x$: C, 36.68%; H, 5.28%; N, 6.11%.

TABLE II

Elemental Analyses of Base-Soluble and Insoluble Fractions from the Spontaneous Polymerization of 4-Vinylpyridinium 2-Acrylamido-2-Methylpropanesulfonate

Solvent	Solubility in Base	% Conv. Component	Elemental Analyses %C	%H	%N
Water [a]	Soluble	15	57.25	6.77	9.65
Water [a]	Insoluble	7	59.08	6.46	9.75
Dimethyl Formamide [b]	Soluble	14	58.36	6.54	9.52
Dimethyl Formamide [b]	Insoluble	7	60.22	6.66	9.94
Methanol [c]	Soluble	18	57.80	6.17	9.26

[a] Monomer concentration 1.0M.

[b] Monomer concentration 0.5M.

[c] Monomer concentration 1.0M.

Poly-1,4-Pyridiniumdiylethylene Vinylsulfonate (VIII)

To 0.97 g (9 mmole) of vinylsulfonic acid in 20 ml of nitromethane at room temperature was added dorpwise with stirring a solution of 1.05 g (10 mmole) of 4-vinylpyridine in 50 ml of nitromethane. The precipitate was filtered after 24 hr, washed with ether, reprecipitated from acetone, and then dried *in vacuo,* giving a quantitative yield of ionene vinylsulfonate complex. NMR (D$_2$O): pyridinium protons at $\delta 8.04–8.86$ (*m, 4H*), ethylene peaks at $\delta 3.74$ (*t, 2H*) and $\delta 4.82$ (*t, 2H*), vinyl protons of the vinylsulfonate moiety at $\delta 5.75$ (*d, 1H*), $\delta 5.98$ (*d, 1H*), and $\delta 6.66$ (*2d, 1H*), $J_{gem} = 0$ Hz, $J_{cis} = 10.0$ Hz, and $J_{trans} = 16.8$ Hz; $[\eta] = 0.029$ dl/g in $0.1M$ KBr solution.

ANAL.: *Calcd.* for C$_9$H$_{11}$NO$_3$S: C, 50.70%; H, 5.16%; N, 6.57%. *Found:* C, 50.81%; H, 5.17%; N, 6.45%.

Free Radical Polymerization of 4-Vinylpyridinium 2-Acrylamido-2-Methylpropanesulfonate

In an ampule, 2.5 g (8 mmol) of 4-vinylpyridinium 2-acrylamido-2-methyl-propanesulfonate and 7.2 mg (0.026 mmol) of 4,4′-azobis-4-cyanovaleric acid were dissolved using an appropriate amount of distilled water to make 8 ml of solution. The solution was degassed, sealed, and placed in a thermostated bath at 70°C. After 4 hr, the seal was broken and the pastelike material was dissolved in 100 ml of a 1% sodium carbonate solution. This solution was exhaustively dialyzed against deionized water, followed by lyophilization to obtain the co-polymer: NMR (D$_2$O + D$_2$SO$_4$) $\delta 8.5$ (broad, pyridine ring), $\delta 7.0$ (broad, pyridine ring), $\delta 3.6$ (*s,* —CH$_2$—SO$_3$H), $\delta 1.5$ (broad, backbone of polymer chain), $\delta 1.4$ (*s,* methyl groups).

ANAL.: *Calcd.* for —(C$_7$H$_8$N)$_1$—(C$_7$H$_{12}$NO$_4$S)$_1$—: C, 53.83%; H, 6.45%; N, 8.97%; S, 10.30%. *Calcd.* for —(C$_7$H$_7$N)$_1$—(C$_7$H$_{12}$NO$_4$SNa)$_1$—: C, 50.29%; H, 5.37%; N, 8.38%, S, 9.59%. *Found* (water-insoluble portion): C, 53.41%; H, 6.75%; N, 8.79%; S, 10.28%. (Base-soluble portion): C, 50.77%; H, 6.28%; N, 8.19%; S, 9.23%.

Free Radical Copolymerization of Sodium 2-Acrylamido-2-Methylpropanesulfonate with N-Vinylpyrrolidone

All copolymerization reactions were run in a standard manner differing only in length of the reaction time. Quantities of monomers, times, conversions, and mole fraction of monomers in the copolymers are recorded in Table III.

To a weighed test tube which had been constricted to allow ease of sealing were weighed the appropriate amount of sodium 2-acrylamido-2-methylpro-panesulfonate, 14 mg of 4,4′-azobis-4-cyanovaleric acid, 9.00 g of distilled water, and the appropriate amount of N-vinylpyrrolidone. The solution was degassed, sealed, and placed in a thermostated bath at 50°C. After the appropriate time, polymerization was stopped by breaking the seal and diluting the solution with 50 ml of cold distilled water. The solution was then exhaustively dialyzed against

TABLE III
Copolymerization of Sodium 2-Acrylamido-2-Methylpropanesulfonate (M_1) with N-Vinylpyrrolidone (M_2) in Water

M_1 (mmol)	M_2 (mmol)	time (min)	% Conv.	%S	F_1*
0.944	9.559	60	2.97	6.87	0.319
1.931	7.847	23	4.11	8.05	0.397
2.954	6.670	41	7.83	8.76	0.448
5.327	6.292	25	9.32	9.52	0.508
6.203	3.870	17	6.80	10.38	0.582
6.914	3.042	20	7.12	11.00	0.641
7.831	2.178	13	5.35	10.72	0.614
8.960	1.026	11	4.52	11.72	0.715

[a] F_1 = mole fraction of M_1 in copolymer.

deionized water followed by lyophilization to obtain polymer. From the mole fraction of sodium 2-acrylamido-2-methylpropanesulfonate in the copolymer (Table III), reactivity ratios were determined by the Fineman–Ross method [9] to be 0.174 for sodium 2-acrylamido-2-methylpropanesulfonate and 0.078 for N-vinylpyrrolidone, respectively. These reactivity ratios yield Alfrey–Price Q–e values [11] of sodium 2-acrylamido-2-methylpropane sulfonate of $Q = 0.17$ with $e = +0.93$, and $Q = 19.0$ and $e = 3.21$. The former values are considered more acceptable.

RESULTS AND DISCUSSION

The preparation of three ion pair comonomer salts has been described through the protonation of 4-vinylpyridine with vinylsulfonic acid, 2-acrylamido-2-methylpropanesulfonic acid, and p-styrenesulfonic acid. The resulting salts of 4-vinylpyridinium vinylsulfonate (IVa), 2-acrylamido-2-methylpropanesulfonate (IVb), and p-styrenesulfonate (IVc) were all obtained in crystalline form, although with varying degrees of stability. It is apparent that such divinylic salts are capable of a variety of polymerization reactions. Using 4-vinylpyridinium vinylsulfonate (IVa) as an example, the polymer structures shown in Scheme I could be anticipated depending on whether one or both vinyl groups are polymerizable under the conditions employed (Scheme I). In this Scheme, reactions 1 and 2 lead to a polymerization of both vinylic moieties in which either a copolymer (V, alternating or random, reaction 1) or a polyelectrolyte complex (VI, two individual homopolymerization, reaction 2) result. Reactions 3 and 4 consider the polymerization of only the vinylpyridinium moiety, with the vinylsulfonate moiety remaining unchanged. In reaction 3, spontaneous polymerization at a monomer concentration less than $1M$ in water could lead to ionene formation

$$CH_2=CH$$

IV a. X^- -= $CH_2=CH-SO_3^-$

b. x^- = $CH_2=CH-\overset{\overset{O}{\|}}{C}-NH-\overset{\overset{CH_3}{|}}{\underset{\underset{CH_3}{|}}{C}}-CH_2-SO_3^-$

c. x^- = $CH_2=CH-\langle\bigcirc\rangle-SO_3^-$

of a poly-1,4-pyridiniumdiylethylene vinylsulfonate salt (VII) by a step-growth, hydrogen-transfer polymerization [10, 12]. On the other hand, for reaction 4 at a monomer concentration of greater than $1M,$ a chain growth polymerization of the vinylpyridinium moiety can occur giving the polyvinylpyridinium backbone of poly-4-vinylpyridinium vinylsulfonate (VIII). The separate polymerization of the vinylic anion with no polymerization of the vinylpyridinium moiety has not been considered because of the facile ease of polymerization of the latter salts. It can be seen that through such a variety of polymerization reactions, a greater elucidation of the spontaneous polymerization process of vinylpyridinium salts in aqueous solution in the presence of free radically polymerizable counterions could be obtained.

4-Vinylpyridinium Vinylsulfonate

The monomer of 4-vinylpyridinium vinylsulfonate was the most stable toward spontaneous polymerization of the three monomeric salts prepared. However, it was observed that during the determination of the melting point of this salt, several interésting observations were noted. By the capillary melting point technique, the melting point could not be easily ascertained as the interval between when the monomer melted and when it polymerized was extremely short. However, the melting point could be determined by differential scanning calorimetry (DSC) at a rapid heating rate of 80°C/min to be 99–100°C. An endothermic peak appeared immediately before an exothermic peak which indicated that the monomer polymerized upon melting. If the monomer was heated at a slower heating rate in the calorimeter, only an exothermic peak was produced. It is expected that a heating rate of 80°C/min cannot be considered to give an accurate melting point. However, it is doubtful that many of the melting points which have been reported for labile vinylpyridinium salts are accurate. For example, it was originally reported that the melting point of 4-vinylpyridinium nitrate was 155°C at an unspecified heating rate [11]. However, a rein-

SCHEME I

Possible Polymerization Reactions of 4-Vinylpyridinium Vinylsulfonate in Water

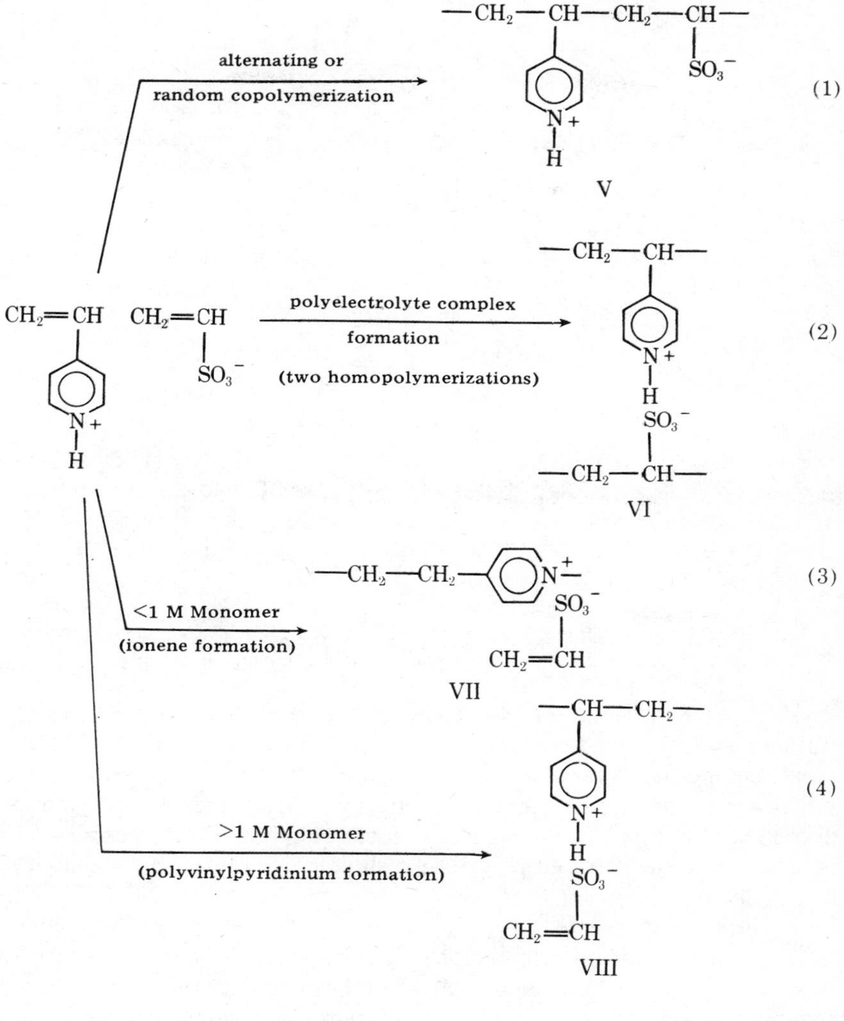

vestigation of this compound by hot-stage polarized microscopy at a heating rate of 10°/min gave a melting point of 112–113°C whereas by DSC at a heating rate of 80°C/min, a melting point of 137°C was obtained, followed immediately by an exothermic peak indicating that polymerization had occurred.

When the melting point of 4-vinylpyridinium vinylsulfonate was inspected by hot-stage polarized microscopy, the rapid transition from monomeric salt to polymeric salt could be observed as the birefringent crystals of the monomeric salt yielded other birefringent crystals of the same dimensions as the original monomeric crystal structure. An inspection of the ^{1}H NMR structure of the resulting polymer showed it to be that of poly-1,4-pyridiniumdiylethylene

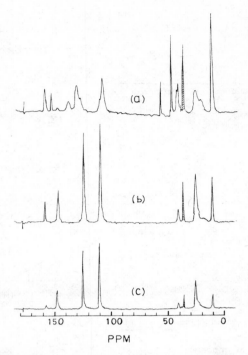

PPM

FIG. 1. ^{13}C NMR of (a) free radical copolymer, (b) polyelectrolyte complex, and (c) spontaneous polymerization product. Values are given in ppm downfield from high field peak of DSS.

vinylsulfonate (VII), i.e., an essentially solid state conversion of the 4-vinylpyridinium moiety into a linear ionene structure with an unpolymerized vinylsulfonate counterion. A powder X-ray spectrum of VII prepared thermally in the solid state did not appear more crystalline than that of the same polymer produced from solution. The monomer 4-vinylpyridinium nitrate was also found to yield the ionene polymer, poly-1,4-pyridiniumdiylethylene nitrate, above its melting point.

It is interesting to note that upon thermally induced spontaneous polymerization, 4-vinylpyridinium vinylsulfonate yielded solely the ionene structure VII. A similar phenomenon was observed when this monomer was attempted to be spontaneously polymerized in aqueous solution at a monomer concentration of $0.9M$ in which an ionene with an intrinsic viscosity of 0.14 dl/g in $0.10M$ KBr resulted. Above $1M$, however, the polyvinylpyridinium structure VIII resulted with an intrinsic viscosity of 1.40 dl/g in $0.10M$ KBr solution. Whereas it is not particularly surprising the vinylsulfonate was not incorporated into the latter structure because of its presumed overall lower reactivity relative to that of a vinylpyridinium salt (in free radical copolymerization), a similar result was obtained when the water-soluble azo initiator 4,4'-azobis-4-cyanovaleric acid (ACVA) was added to the $0.9M$ aqueous solution of 4-vinylpyridinium vinylsulfonate. Since vinylsulfonate salts can be difficult to polymerize and may necessitate high monomer concentrations to effect polymerization [13], the free radical initiation of this monomer by a variety of other radical and redox initiators at monomer concentrations $>2.3M$ was also studied. However, in these cases only the product poly-4-vinylpyridinium vinylsulfonate (VIII) resulted as there was no observable reaction of the vinylsulfonate ion. Under similar conditions sodium vinylsulfonate was readily polymerized. Consequently, through the use of this monomer no distinction could be made between the products (of a chain growth nature) obtained by spontaneous polymerization and by free radically initiated polymerization.

4-Vinylpyridinium 2-Acrylamido-2-Methylpropanesulfonate

The monomer 4-vinylpyridinium 2-acrylamido-2-methylpropanesulfonate was also found to be relatively stable (in the absence of moisture) and its melting could be determined by the capillary melting technique at slow heating to give a sharp melting point of 84.5–85.0°C with polymerization [8].

A preliminary ^{1}H NMR study of $1M$ monomer in D_2O indicated that the vinyl protons of the vinylpyridinium moiety disappeared faster than those of the acrylamido moiety to a polyvinyl structure. This initial observation was significant in that it indicated both cationic and anionic components were in fact polymerizable, although apparently at different rates. Further spontaneous polymerization studies of 4-vinylpyridinium 2-acrylamido-2-methylpropanesulfonate were then conducted in distilled water, a protic organic solvent (methanol), and an aprotic solvent (dimethyl formamide). For each reaction a mixture of polymer products was obtained. These included a water-soluble ionene polymer of poly-1,4-pyridiniumdiylethylene 2-acrylamido-2-methylpropanesulfonate (IX), a base-insoluble fraction, and a base-soluble fraction (Tables I and II). The determination of the ionene structure was confirmed by ^{1}H spectroscopy from extraction of the combination of products by D_2O. In the presence of base, the ionene structure is decomposed, allowing for more facile determination of the other components. An analysis of the infrared spectra of both base-soluble and base-insoluble components showed the existence of both polyvinylpyridine and

IX

polyacrylamido-2-methylpropanesulfonate units, and that the base-insoluble
fraction had a greater portion of polyvinylpyridine as would be expected.

A result such as obtained above did suggest a vinyl copolymerization reaction,
which had not previously been observed in the spontaneous polymerizations of
vinylpyridinium salts although this effect had previously been investigated [14].
Since the anionic sulfonate gegenion has previously been shown not to contribute
significantly to the spontaneous polymerization process [15], and since the
spontaneous polymerization of 4-vinylpyridinium methanesulfonate (X) could
not be effected even up to monomer concentrations up to $4M$, it was decided to
determine if the 2-acrylamido-2-methylpropanesulfonate moiety could be involved in a charge transfer interaction with the electron-deficient vinylpyridinium
moiety [16, 17]. This possibility did appear remote, but various substituted
acrylamides have been reported with negative e values in the Alfrey–Price Q–e
scheme [18, 19]. However, upon copolymerization of the sodium salt of 2-ac-
rylamido-2-methylpropanesulfonic acid (XI) with N-vinylpyrrolidone in water
at 50°C with ACVA, its Q–e values were found to be $Q = 0.17$ and $e = +0.93$.
By the high positive e value of sodium 2-acrylamido-2-methylpropanesulfonate,
a charge transfer interaction with the vinylpyridinium ion was considered un-
likely. Furthermore, an inspection of the UV spectra of various ratios of 4-
vinylpyridinium hydrochloride and sodium 2-acrylamido-2-methylpropane-
sulfonate showed no evidence for such a charge transfer complex.

In an attempt to clarify the polymerization mechanism of the spontaneous
polymerization of 4-vinylpyridinium 2-acrylamido-2-methylpropanesulfonate,
it was decided to initiate polymerization free radically in aqueous solution at
70°C by ACVA and to study the resulting polymer. A cloudy paste was obtained
of water-insoluble polymer, which was then separated from the unreacted
monomer by washing with additional water. It was found that the resulting
polymer was soluble in sodium chloride solution, in basic solution, or in highly
acidic solution. Solubility characteristics of this type indicated that the free
radically prepared polymer was amphoteric in structure, such as that of XII.
These solubilities could be rationalized by the fact that in water charge neutralization leads to insolubility, in base the neutralization of the pyridinium
residues renders the polymer anionic, and in strong acid a partial protonation
of the sulfonate sites renders the polymer cationic. The solubility of XII in salt
solution is related to a greater hydration of the polyampholyte with reduced
intramolecular and intermolecular ionic interactions [19, 20]. In a related study,
a polyelectrolyte complex of poly-4-vinylpyridinium poly-2-acrylamido-2-
methylpropanesulfonate was found to be insoluble in water but could be dissolved
in aqueous salt solutions.

The solubility results obtained through free radical polymerization did not
appear to conform with those obtained by spontaneous polymerization. Consequently, an investigation of the products obtained by the spontaneous and free
radical polymerizations of 4-vinylpyridinium 2-acrylamido-2-methyl pro-
panesulfonate and by the polyelectrolyte complex formation of poly-4-vinyl-
pyridinium poly-2-acrylamido-2-methylpropanesulfonate was undertaken by
^{13}C NMR in D_2SO_4. The results showed (Fig. 1), very surprisingly, that the

XII

spectrum of the product from the spontaneous polymerization reaction was identical to that of the polyelectrolyte complex, whereas that of the free radically obtained polymer exhibited behavior characteristic of a random copolymer.

These results were indeed unusual in that they indicated two separate spontaneous homopolymerizations for 4-vinylpyridinium 2-acrylamido-2-methylpropanesulfonate, or that a block copolymer formed. The spontaneous polymerization of the anion is not unusual, as the monomer is capable of spontaneous polymerization in certain acid pH ranges [21].

It is significant, however, that while the 2-acrylamido-2-methylpropanesulfonate moiety is readily susceptible to free radical polymerization in contrast to that of the vinylsulfonate moiety, the spontaneous polymerization of 4-vinylpyridinium 2-acrylamido-2-methylpropanesulfonate yields either two homopolymers or a block copolymer whereas a random copolymer structure appears to result in the presence of a free radical initiator.

4-Vinylpyridinium p-Styrenesulfonate

The monomer 4-vinylpyridinium p-styrenesulfonate was the most difficult of the three monomers prepared to recrystallize, isolate, and characterize because of its facile tendency for spontaneous polymerization. This cationic–anionic monomer pair was of considerable interest since the individual components would be aromatic structures, but with strongly differing polarities.

Indeed, during the preparation of this monomer pair, a considerable amount of polymeric product of the polyvinyl structure was always produced. Elemental analyses indicated that the water-insoluble product was equimolar in 4-vinylpyridinium and p-styrenesulfonate moieties, thus indicating the possibility of a charge-transfer polymerization.

In attempting to determine the melting point of the 4-vinylpyridinium p-styrenesulfonate crystals, a similar problem to that of 4-vinylpyridinium vinylsulfonate was encountered, i.e., a melting point could only be obtained by DSC at a high heating rate. By this technique a melting point of 137–138°C was obtained at a heating rate of 80°/min followed immediately by polymerization. In contrast to the polymerizations of 4-vinylpyridinium nitrate and 4-

vinylpyridinium vinylsulfonate upon heating to their respective ionene polymers, no ionene polymer of poly-1,4-pyridiniumdiylethylene p-styrenesulfonate was observed by ^{1}H NMR spectroscopy in the thermal polymerization of 4-vinylpyridinium p-styrenesulfonate. Instead, only the polyvinyl structures could be found of which an elemental analysis indicated an equimolar copolymer.

In addition, an inspection of the melting behavior of crystals of this monomer revealed that birefringence was lost upon thermal polymerization. Of all the products obtained with equimolar compositions of 4-vinylpyridinium and p-styrenesulfonate moieties, the solubility characteristics were found to be those of an ampholytic polymer. That is, the product was water insoluble in the pH range 0–6, but could be dissolved in basic solution (pH > 6.5) and in strongly acidic solution (pH < 0). In addition, the water-insoluble polymer could be dissolved upon addition of sodium chloride. These solubility characteristics clearly correspond to an ampholytic polymer and are quite distinct from a polyelectrolyte complex of poly-4-vinylpyridinium poly-p-styrenesulfonate which is soluble only in a ternary solvent system of water, a water-soluble organic compound, and an inorganic electrolyte [22].

Based on the above results, it seems likely that an alternating copolymer resulted through the spontaneous polymerization of the cationic and anionic moieties of opposite polarities. In order to help substantiate such a charge-transfer copolymerization reaction, the spontaneous polymerization of 4-vinylpyridinium p-styrenesulfonate was conducted in the presence of sodium p-styrenesulfonate of the same molar concentration. An elemental analysis and an investigation of the solubilities of the resulting polymer showed that the copolymer remained 1:1 in regard to the ratio of 4-vinylpyridinium to p-styrenesulfonate units. This result strongly indicates that the spontaneous polymerization of 4-vinylpyridinium p-styrenesulfonate is charge-transfer in nature.

CONCLUSION

The monomer pair of 4-vinylpyridinium p-styrenesulfonate would appear to be the first characterized case of a copolymerization of the vinylpyridinium moiety in aqueous solution. This result could conceivably suggest that while the polymerization may be free radical in nature, it could also incorporate the characteristics of a charge-transfer polymerization in which the anion acts as a donor to the electron deficient vinylpyridinium moiety. When the anion is nonpolymerizable, it may be that a charge-transfer homopolymerization results, such as has been reported with vinylidene cyanide and electron donors [22] and with N-vinylcarbazole and electron acceptors [23]. In order to study these possibilities further, additional studies are currently in progress which involve the electron-deficient vinylpyridinium moiety and various electron-donating vinylic compounds.

The authors gratefully acknowledge the support of the National Science Foundation, Polymers Program, and the donors of the Petroleum Research Fund, administered by the American Chemical Society, for support of this work.

REFERENCES

[1] O. V. Kargina, L. A. Mishustina, V. I. Svergun, G. M. Lukovkin, V. P. Yerdakov, and V. A. Kabanov, *Vysokomol. Soedin., A16,* 1755 (1974).

[2] V. Martin, W. Sutter, and H. Ringsdorf, *Makromol. Chem., 177,* 89 (1976).

[3] J. C. Salamone, D. F. Bardoliwalla, E. J. Ellis, S. C. Israel, and A. W. Wisniewski, *Appl. Polym. Symp., 26,* 309 (1975).

[4] J. C. Salamone, E. J. Ellis, and D. F. Bardoliwalla, *J. Polym. Sci., C, 45,* 51 (1974).

[5] A. W. Wisniewski, M.S. Thesis, Lowell Technological Institute, 1975.

[6] C. C. Tsai, M.S. Thesis, University of Lowell, 1976.

[7] J. C. Salamone, A. C. Watterson, S. C. Israel, E. J. Ellis, and A. Wisniewski, Abstracts, First Chemical Congress of the North American Continent, Mexico City, Nov. 30–Dec. 5, 1975.

[8] J. C. Salamone, A. C. Watterson, T. D. Hsu, C.-C. Tsai, and M. U. Mahmud, *J. Polym. Sci., Polym. Lett. Ed., 15,* 487 (1977).

[9] M. Fineman and S. D. Ross, *J. Polym. Sci., 5,* 259 (1950).

[10] J. C. Salamone, B. Snider, and W. L. Fitch, *Macromolecules, 3,* 707 (1970).

[11] T. Alfrey, Jr. and C. C. Price, *J. Polym. Sci., 2,* 101 (1947).

[12] J. C. Salamone, B. Snider, W. L. Fitch, E. J. Ellis, and P. L. Dholakia, XXIII Int. Conf. Pure Appl. Chem., Boston, 1971; *Macromol. Prepr., II,* 1177 (1971).

[13] D. S. Breslow and A. Kutner, *J. Polym. Sci., 27,* 295 (1958).

[14] V. A. Kabanov, K. V. Aliev, T. I. Patrikeeva, O. V. Kargina, and V. A. Kargin, *J. Polym. Sci., C-16,* 1079 (1967).

[15] J. C. Salamone, B. Snider, and W. L. Fitch, *J. Polym. Sci., Polym. Lett., 9,* 13 (1971).

[16] V. R. Georgieva, V. P. Zubov, V. A. Kabanov, and V. A. Kargin, *Dokl. Acad. Nauk USSR, 190,* 1128 (1970).

[17] J. C. Salamone, P. Taylor, B. Snider, and S. C. Israel, *J. Polym. Sci., Poly. Chem. Ed., 13,* 161 (1975).

[18] L. J. Young, *J. Polym. Sci., 54,* 411 (1961).

[19] L. J. Young, in *Polymer Handbook,* J. Brandrup and E. H. Immergut, Eds., Wiley, New York, 1975.

[20] J. C. Salamone, W. Volksen, A. P. Olson, and S. C. Israel, *Polymer,* in press.

[21] Private communication from Lubrizol Corporation, June 21, 1976, Akron, Ohio.

[22] J. K. Stille, N. Oguni, D. C. Chung, R. F. Tarvin, S. Aoki, and M. Kamachi, *J. Macromol. Sci.-Chem., A9*(5), 745 (1975).

[23] H. Scott, G. A. Miller, and M. M. Labes, *Tetrahedron Lett., 17,* 1073 (1963).

SOME PHYSICOCHEMICAL ASPECTS REGARDING THE SYNTHESIS OF POLYVINYL ALCOHOL

M. DIMONIE, C. CINCU, C. OPRESCU, and GH. HUBCĂ

Polytechnic Institute Bucharest, Faculty of Chemical Engineering, Strada Polizu 1, Sectorul 8, Bucuresti, Romania

SYNOPSIS

The present paper reports some theoretical aspects related to recent technology for the production of polyvinyl alcohol. The procedure involves the polymerization of vinyl acetate in two stages and the alcoholysis of polyvinyl acetate. A model for the calculation of the polymerization degree and grafting degree of vinyl acetate was developed in view of a quantitative evaluation of various possibilities for conducting the polymerization process. The alcoholysis of polyvinyl acetate was studied by conducting the reaction in a Brabender-type mixer. The dependence of the reaction rate on the catalyst or polymer concentration or on the temperature evidences the important role played by the diffusion processes and supports the idea of the cocatalytic effect of OH groups in the alcoholysis reaction of the polyvinyl acetate in the basic catalysis.

INTRODUCTION

Polyvinyl acetate necessary for producing polyvinyl alcohol is usually obtained either by suspension polymerization or by polymerization of vinyl acetate in methanol solution. Polymerization in solution has the advantage of a rather simple technology, and the resulting polymer does not have lateral chains; on the other hand, this procedure does not allow the obtaining of high molecular weights as the transfer reaction with the solvent is quite powerful. Polyvinyl acetate obtained by suspension polymerization leads after the alcoholysis reaction to polyvinyl alcohol with high molecular mass; nevertheless, the final polymer is still branched. Moreover, this technology involves some extra operations, such as drying and dissolution.

The present paper deals with some aspects referring to a new procedure for the preparation of polyvinyl alcohol [1, 2]; this procedure keeps the advantages of the previously quoted technologies, but at the same time eliminates their disadvantages. Our procedures involve the polymerization of vinyl acetate in two distinct stages: first, polymerization in bulk in the absence of solvent; second, polymerization in solution by continuous pouring of the solvent, in a well determined amount, into the system, after a certain period.

Journal of Polymer Science: Polymer Symposium 64, 245–265 (1978)
0360-8905/78/0064-0245$01.00

THEORETICAL

Calculation of the Polymerization Degree

The prediction of the molecular mass, at the same time with the degree of branching as a function of conversion, is of particular interest for assessment of the various possibilities of operating the polymerization according to the present procedure.

Mayo's "classical" equation has a differential character and, accordingly, may only be applied for small conversions (i.e., 5%); the use of this equation for the final degree of polymerization leads, of course, to important errors.

There are several attempts in literature to correlate the degree of polymerization with time or conversion. Bamford [3] started from kinetic considerations and showed that for conversions higher than 5%, namely when the monomer concentration may no longer be considered as constant, Mayo's simple equation should be replaced by the expression:

$$n = \int \frac{d[M]}{dt} \cdot dt \bigg/ \int \left\{ k_m [M] \left[\frac{I}{(k_t + k_t')} \right]^{1/2} + I \left(\frac{k_t}{2} + k_t' \right) \bigg/ (k_t + k_t') \right\} dt$$

in which n is the degree of polymerization; I, rate of initiation; $[M]$, monomer concentration; and k_m, k_t, k_t', rate constants for transfer reaction with the monomer, termination by combination, and termination by disproportionation, respectively.

In order to calculate the degree of polymerization it is necessary to know the dependences $[M] = f(t)$ and $I = f(t)$.

Gee and Melville [4], Harington and Robertson [5], and Saginian and Enicolopian [6] studied the problem of the degree of polymerization as a function of conversion and were concerned with finding equations for the distribution of molecular weights.

Wolf and Burchard [7] have recently developed a statistical theory of polymerization reactions which allows the calculation of the weight average degree of polymerization and the Z-average mean square radius of gyration.

The calculation method proposed by us starts from Mayo's equation and imposes two restrictive conditions, namely the invariance of solvent and initiator concentrations; these conditions are generally satisfied in practice. This method allows a certain simplicity and the use of readily accessible values.

In Mayo's equation

$$\frac{1}{n} = \frac{1}{n_0} + C_s \frac{[S]}{[M]} + C_p \frac{[P]}{[M]} \tag{1}$$

the value $1/n_0$ represents the inverse of the polymerization degree in the absence of transfer reactions with solvent and polymer; it may be written as

$$\frac{1}{n_0} = \frac{(1 + x)k_t^{1/2}}{k_p [M]} \left(\frac{f}{2} \cdot k_d \cdot [I_n] \right)^{1/2} + C_m \tag{2}$$

where x is the fraction of the polymer radicals disappearing by disproportion- ation; $[P]$, $[M]$, $[S]$, the polymer, monomer, and solvent concentrations, re- spectively; k_p, k_t, k_d, the rate constants for propagation, termination and initiator decomposition, respectively; $[I_n]$, the concentration of initiator; C_m, C_s, C_p, the transfer constants with monomer, solvent, and polymer; and f, the fraction of radicals resulting from the decomposition of the initiator which actually initiate the polymerization.

Values x, $(k_t/k_p)^{1/2}$, f, k_d, are constant and known in the literature for the most usual monomers or they may be determined quite easily. The value

$$(1 + x) \frac{k_t}{k_p} \left(\frac{f k_d \cdot [I_n]}{2} \right)^{1/2}$$

may be included in an α constant which may be calculated from the already available data. This value may also be obtained experimentally.

By substituting $1/n_0$ in eq. (1) by its expression and taking into account the obvious relations:

$$[M] = [M_0] \cdot (1 - \eta) \tag{3}$$

$$[P] = [M_0]\eta \tag{4}$$

where $[M_0]$ is the initial monomer concentration and η is the conversion, one obtains:

$$\frac{1}{n_{in}} = \left\{ \frac{\alpha}{[M_0]} + C_s \frac{[S]}{[M_0]} + C_m + (C_p - C_m)\eta \right\} \bigg/ (1 - \eta) \tag{5}$$

The sum

$$\frac{\alpha}{[M_0]} + C_s \frac{[S]}{[M_0]} + C_m$$

assumes the significance of the inverse polymerization degree in the initial conditions and is written as $1/n_{0i}$. Thus eq. (5) becomes

$$\frac{1}{n_i} \alpha = \frac{1/n_{0i} + (C_p - C_m) \cdot \eta}{1 - \eta} \tag{6}$$

The polymerization degree in n_{in} given in Mayo's equation represents the number average of the polymerization degree for instantly formed polymers

$$\bar{n}_{in} = \frac{\sum N_i n_i}{\sum N_i} \tag{7}$$

the weight average being:

$$\bar{n}_{iw} = \frac{\sum N_i n_i^2 \eta}{\sum N_i n_i} \tag{8}$$

N_i represents the number of macromolecules with the polymerization degree n_i.

According to Flory [8, 9], the "most probable distribution" for linear polymers is

$$\bar{n}_{in}/\bar{n}_{iw} = (1 + \zeta)^{-1} = \lambda \tag{9}$$

where

$$\zeta = \frac{k_p[M]}{k_p[M] + k_m[M] + k_s[S] + (Ik_t')^{1/2}} \tag{10}$$

From the expression for ζ, this value is practically independent of the conversion, even at high values (over 95%), provided the polymerization degree is reasonably high ($n \geqslant 150$).

The weight average of the degree of polymerization at higher conversions may be expressed as:

$$\bar{n}_w = \frac{\sum_0^\eta \sum_1^\eta N_i n_i^2}{\sum_0^\eta \sum_1^\eta N_i n_i} \tag{11}$$

A simultaneous consideration of eqs. (8), (9), and (11) gives:

$$\bar{n}_w = \frac{\sum_0^\eta \lambda \bar{n}_{in} \sum_1^\eta N_i n_i}{\sum_0^\eta \sum_1^\eta N_i n_i} \tag{12}$$

$\sum_1^\eta N_i n_i$ represents the polymer amount (ΔP) instantly formed and having the average degree of polymerization, n_i and $\sum_0^\eta \sum_1^\eta N_i n_i$ represents the polymer amount formed by η conversion. Equations (11) and (12) become:

$$\bar{n}_w = \frac{\lambda \sum_0^\eta \Delta P_i \cdot \bar{n}_{in}}{P} \tag{13}$$

$$\bar{n}_n = \frac{\sum_0^\eta \Delta P_i \cdot \bar{n}_{in}}{P} \tag{14}$$

The polymer amounts ΔP_i are not equal to each other, their size depending on the time (conversion) at which they have been produced. The polymer formed in the initial moment of the reaction:

$$\Delta P_0 = K[R\cdot][M_0]\Delta t \tag{15}$$

where $[R\cdot]$ is concentration of the macroradicals. The polymer amount formed at a conversion η is:

$$\Delta P_i = K[R\cdot]([M_0] - [P])\Delta t \tag{16}$$

By dividing the two relations, it becomes:

$$\Delta P_i = \Delta P_0 \left(1 - \frac{[P]}{[M_0]}\right) \tag{17}$$

Taking into account eq. (4) and making $d[P_0] = [M_0]d\eta$ it becomes

$$d[P_i] = [M_0] (1 - \eta)d\eta \tag{18}$$

and

$$\bar{n}_n = \int_0^\eta \frac{(1 - \eta)^2 d\eta}{1/n_{0i} + (C_p - C_m)\eta}\bigg/ \eta \tag{19}$$

Eventually, after integration one obtains

$$\bar{n}_n = \frac{\eta}{2(C_p - C_m)} - \frac{1 + 2 n_{0i}(C_p - C_m)}{n_{0i}(C_p - C_m)^2}$$
$$+ \frac{[1 + n_{0i}(C_p - C_m)]^2}{n_{0i}^2(C_p - C_m)^3 \, \eta} \ln[1 + n_{0i}(C_p - C_m)\eta] \tag{20}$$

$$\bar{n}_w = \lambda \bar{n}_n \tag{21}$$

$\bar{n}_n$ and $\bar{n}_w$ are not real, but fictitious magnitude, nevertheless serving for subsequent calculations. They represent the degrees of polymerization for the unbranched polymer, which would be obtained if the branched polymer remained in the system as linear chains.

The branching degree may be calculated starting from similar considerations as the above presented ones. The equation for the rate information of branching sites is:

$$\frac{d[R^{\cdot *}]}{dt} = k_{tp}[R^{\cdot}][P] \tag{22}$$

Concurrently, for the main polymerization reaction it may be written

$$\frac{d[P]}{dt} = k_p[R^{\cdot}][M] \tag{23}$$

where $[R^{\cdot *}]$ is the concentration of branching centers and k_{tp}, the rate constant of chain-transfer with polymer.

By dividing eqs. (22) and (23) and considering eqs. (3) and (4), we obtain:

$$d[R^{\cdot *}] = C_p[M_0] \frac{\eta}{1 - \eta} d\eta \tag{24}$$

Since the branches appearing as a result of the chain-transfer with polymer have the same polymerization degree as the unbranched polymer simultaneously formed, the concentration of the "instantly" formed graft polymer would be

$$d[P_0] = n_{in}d[R^{\cdot *}] = C_p[M_0]n_{in} \frac{\eta}{1 - \eta} d\eta \tag{25}$$

The average degree of polymerization for the branch may be written as

$$\bar{n}_b = \frac{\text{the amount of branched polymer}}{\text{the number of branching sites}}$$

or

$$\bar{n}_b = \int_0^\eta n_i \frac{\eta}{1 - \eta} d\eta \Big/ \int_0^\eta \frac{\eta}{1 - \eta} d\eta \tag{26}$$

and after integration one obtains

$$\bar{n}_b = \frac{(C_p - C_m)^{-1} - [n_{0i}(C_p - C_m)^2\eta]^{-1} \ln[1 + n_{0i}(C_p - C_m)\eta]}{-[1 + 1/\eta \ln(1 - \eta)]} \tag{27}$$

All the relations obtained verify the initial conditions:

$$\lim_{\eta \to 0} \bar{n}_n \to \bar{n}_{0i}; \lim_{\eta \to 0} \bar{n}_b \to 0$$

The δ ratio between the amount of branched polymer

$$P_b = \sum N_b \cdot n_{ib} \tag{28}$$

and the amount of linear polymer P_1 constituted in main chain

$$P_1 = \sum N_1 \cdot n_{i1} \tag{29}$$

is

$$\delta = \frac{\sum N_b \cdot n_b}{\sum N_1 \cdot n_1} = C_p \left\{ \frac{1}{C_p - C_m} - \frac{\ln[1 + n_{0i}(C_p - C_m)]}{n_{0i}(C_p - C_m)^2 \eta} \right\} \Bigg/ $$
$$1 - C_p \left\{ \frac{1}{C_p - C_m} - \frac{\ln[1 + n_{0i}(C_p - C_m)]}{n_{0i}(C_p - C_m)^2 \cdot \eta} \right\} \tag{30}$$

where N_b represents the number of branched macromolecules with the polymerization degree n_b and N_1 represents the number of linear macromolecules constituted in main chains with the polymerization degree n_1.

Obviously the magnitude $\bar{n}_n$, whose significance was above mentioned, may be expressed as:

$$\bar{n}_n = \frac{\sum N_1 \cdot n_1 + N_b \cdot n_b}{\sum N_1 + \sum N_b} \tag{31}$$

A simultaneous consideration of eqs. (30) and (31) allows the calculation of the branching degree γ (number of branches per one macromolecular chain):

$$\gamma = \frac{\sum N_b}{\sum N_1} \tag{32}$$

$$\gamma = -\bar{n}_n C_p \left[1 + \frac{1}{\eta} \ln(1 - \eta) \right] \Bigg/ 1 + \bar{n}_n C_p \left[1 + \frac{1}{\eta} (1 - \eta) \right] \tag{33}$$

The final degree of polymerization for the branched polymer is:

$$\bar{n}_{nf} = \frac{\sum N_1 n_1 + \sum N_b n_b}{\sum N_1} \tag{34}$$

or

$$\bar{n}_{nf} = \bar{n}_1 + \gamma \bar{n}_b \tag{35}$$

where

$$\bar{n}_1 = \frac{\sum N_1 n_1}{\sum N_1} \tag{36}$$

Its expression is obtained from a simultaneous consideration of eqs. (28), (29), (32), (34), and (36):

$$\bar{n}_{nf} = \frac{1 + \delta}{\delta} \cdot \gamma n_b \tag{37}$$

If $C_p \simeq C_m$, the instant polymerization degree is:

$$\bar{n}_i = \bar{n}_{0i}(1 - \eta) \tag{38}$$

and eqs. (20) and (27) become, respectively,

$$\bar{n}_n = \bar{n}_{0i}(1 - \eta + \tfrac{1}{3}\eta^2) \tag{39}$$

and

$$\bar{n}_b = \frac{\bar{n}_{0i} \cdot \eta}{-2[1 + 1/\eta \ln(1 - \eta)} \tag{40}$$

Stein [10] and very recently Wolf and Bruchard [7] admit that in the case of vinyl acetate polymerization the main mechanism which produces branching is not the ordinary transfer reaction with the polymer, but the addition of the growing radical to the terminal double bonds of the macromolecular chains. Such double bonds may appear as a result of the transfer reaction with the monomer.

According to the determinations performed [10], the reactivity of the terminal double bond towards the polymer radicals is high, the rate constant for this reaction is close to the propagation reaction ($k_n/k_p = 0.8$) (k_n is the rate constant of the macroradical with double bond). In turn, the transfer constant with polymer, found by the same authors, has a particularly low value of $C_p = 1.8 \times 10^{-4}$. In our calculation method this assumption has been also considered.

As it is now accepted for vinyl acetate polymerization, double bonds are formed as a result of the chain transfer reaction with its own monomer (as is shown in equations Ia, Ib) on the other hand, the double bonds are consumed in reaction with growing radicals (see equation II).

$$\text{\textsmall{www}}\overset{|}{\underset{|}{C}} \cdot + CH_2 = CH - COOCH_3 \xrightarrow{k_m} \text{\textsmall{www}}\overset{|}{\underset{|}{C}}H + CH_2 = CH - OCOCH_2^{\cdot}$$

$$\tag{Ia}$$

$$CH_2 = CH - OCOCH_2^{\cdot} + nM \xrightarrow{k_p} CH_2 = CH - OCOCH_2 \text{\textsmall{www}} \cdot$$

$$\tag{Ib}$$

$$\text{\textsmall{www}}CH_2COO - CH = CH_2 + R\cdot \rightarrow \text{\textsmall{www}}CH_2 - COO - \dot{C}H - CH_2R$$

$$\tag{II}$$

Consequently, the rate of formation of terminal double bonds may be written

$$\frac{d[N]}{dt} = k_m[R\cdot][M] - K_n[R\cdot][N] \tag{41}$$

where $[N]$ represents the concentration of terminal double bond. By dividing eq. (41) with that of the polymer formation:

$$\frac{d[P]}{dt} = k_p \, [R\cdot][M] \tag{42}$$

the following equation results:

$$\frac{d[N]}{d} = C_m[M_0] - \frac{k_n}{k_p} \frac{[N]}{1 - \eta} \tag{43}$$

whose solution leads to

$$[N] = \frac{C_m[M_0]}{1 - k_n/k_p} \tag{44}$$

According to similar considerations as those mentioned above, the following relations have been readily deduced:

The concentration of "instantly" formed branching sites by both processes (polymer transfer and addition to the terminal double bond)

$$d[R\cdot^*] = [M_0] \left(C_p + \frac{C_p \, k_n/k_p}{1 - k_n/k_p} \right) \frac{\eta}{1 - \eta} \, d\eta \tag{45}$$

The ratio between the branched polymer amount and that constituted in main chains

$$\delta = \left(C_p + C_m \frac{k_n/k_p}{1 - k_n/k_p} \right) \left\{ \frac{1}{C_p - C_m} - \frac{\ln[1 + n_{0i}(C_p - C_m)\eta]}{n_{0i}(C_p - C_m)^2 \eta} \right\} \Big/$$

$$1 - \left(C_p + C_m \frac{k_n/k_p}{1 - k_n/k_p} \right) \left\{ \frac{1}{C_p - C_m} - \frac{\ln[1 + n_{0i}(C_p - C_m)\eta]}{n_{0i}(C_p - C_m)^2 \eta} \right\} \tag{46}$$

The branching degree (number of branches on one polymer chain)

$$\gamma = -\bar{n}_n \left(C_p + C_m \frac{k_n/k_p}{1 - k_n/k_p} \right) \left[1 + \frac{1}{\eta} \ln(1 - \eta) \right] \Big/$$

$$1 + \bar{n}_n \left(C_p + C_m \frac{k_n/k_p}{1 - k_n/k_p} \right) \left[1 + \frac{1}{\eta} \ln(1 - \eta) \right] \tag{47}$$

It can be easily noticed that the values δ and γ for both calculations are similar; the first variant represents a particular case ($k_n = 0$) of the second one. The equation for the average degree of polymerization for branch $\bar{n}_b$, as well as the magnitude $\bar{n}_n$, are identical in both cases.

RESULTS AND DISCUSSION

Polymerization of Vinyl Acetate

The experimental data and the results obtained using eqs. (30), (33), (37), (46), and (47) are illustrated in Fig. 1. The polymerization degree was determined by the viscometric method [11]. The molecular weight increases in time. This is mainly due to the branching process.

Studies on the kinetics of vinyl acetate polymerization [12, 13] have shown that the mutual reaction between polyvinyl acetate radicals is negligible compared to the reaction of radicals with the monomer of the solvent, except for the case in which the polymerization rate is very high.

The decrease of the termination constant due to the increase in the medium viscosity (diffusion effect) can have only a limited effect on the increase of the molecular weight.

There is relatively good agreement among the values for transfer constants with monomer and solvent (methanol), determined by various authors, but at the same time a certain confusion still exists in the literature with regard to the transfer constant with polymer. A list of such values is given in the monographs by Finch [14] and Ham [15].

For the values [10, 16] $C_p = 1.x \times 10^{-4}$, $C_m = 1.7 \times 10^{-7}$, $k_n/k_p = 0.8$ the deduced eqs. (37), (46), and (47) largely satisfy our results as well as those reported by Stein [10].

Results in good agreement with the experimental data can be obtained by calculation, assuming for the transfer constant with polymer and other values, for instance, that determined by Clarke [17] ($C_p = 7 \times 10^{-4}$; $k_n = 0$).

The above calculation method was used as a valuable tool for the selection of polymerization conditions in order to obtain a polyvinyl acetate with the desired properties.

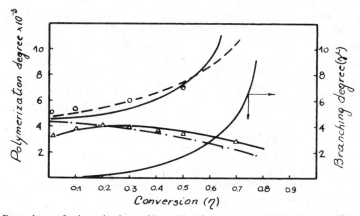

FIG. 1. Dependence of polymerization and branching degree on conversion for polyvinyl acetate:—, calculated; - - -, experimental; O, Stein's experimental data; and for polyvinyl alcohol: - - -, calculated; △, experimental.

As shown in the Introduction, we carried out the vinyl acetate polymerization in the absence of solvent in the first stage; this was necessary to prepare a polymer with a high polymerization degree.

Nevertheless, it has been unanimously accepted that, unlike other vinyl monomers, bulk polymerization of vinyl acetate cannot be achieved on an industrial scale, due to some specific features of this process, discussed below.

The relatively reduced volatility of vinyl acetate, the low thermal stability of polyvinyl acetate melt, as well as the existence of some intense polymer transfer processes, are reasons for which the removal of monomer traces from the reaction medium becomes an unsolvable issue in industrial conditions. In this case the unreacted monomer cannot be removed either by evaporation or by raising the temperature at the end of the process, to achieve the total monomer consumption in reaction.

Another major obstacle is the removal of the polymerization heat. An elementary thermal calculus shows that at reaction rates high enough to satisfy the productivity requirements, only a small part of the reaction heat may be removed through the wall of the reactor (below 10% in the case of industrial reactor). To avoid difficulties related to the thermal transfer, in the period in which polymerization reaction takes place in absence of the solvent, we performed the process under a certain controlled depression, so that the monomer boils at the polymerization temperature.

In this manner the polymerization heat is eliminated as latent evaporation heat; the caloric transfer is achieved without any difficulty, as the value of the transfer constant of condensed organic vapors is higher by 1–2 orders of magnitude compared to the liquid–liquid transfer constant (in the case of high viscous liquids).

A typical feature of the polymerization process conducted under the above described conditions is a tendency towards autoacceleration, which appears around 40% conversion (Fig. 2); this phenomenon is produced regardless of the nature of the initiator or of its concentration.

Autoacceleration is accompanied by an important increase in temperature of the system (Fig. 3).

Two factors contribute directly to the autoacceleration effect, both due to the increasing viscosity of the medium. The first is related to the decrease of the termination rate, due to the well known diffusion effect. It seems that the second factor, namely the marked decrease in the evaporation rate of monomer, has a more important role; the reaction heat cannot be eliminated completely and this leads to the augmentation of the temperature of the system.

Therefore in our work, the solvent is introduced into the system as soon as the autoacceleration effect becomes evident; the polymerization is continued until maximum conversion has been reached.

The necessary amount of solvent is decided on one hand by the need to ensure a sufficient thermal agent to take over the reaction heat and on the other by the requirements imposed by the desired polymerization degree. In the absence of the solvent the polymerization takes place without any control and may develop explosively; the polymer eventually obtained has a tridimensional structure.

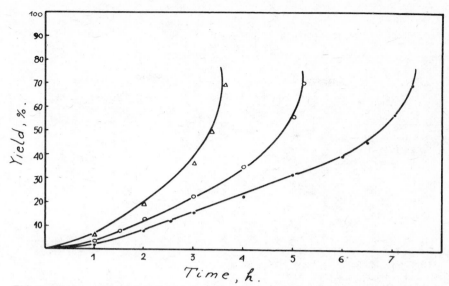

FIG. 2. Yield as a function of time for vinyl acetate polymerization in the absence of solvent. Temperature (T) = 60°C; pressure (P) = 490 mm Hg; concentration of initiator [AIBN] = 0.1% ($\triangle$); 0.03% ($\bigcirc$); 0.015% ($\bullet$).

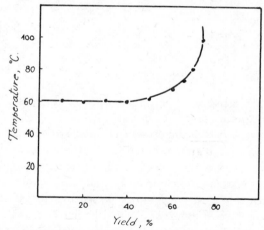

FIG. 3. Temperature against conversion for the vinyl acetate polymerization in the absence of solvent. P = 490 mm Hg; [AIBN] = 0.04%.

The main factors which influence the polymerization degree and the polydispersity of molecular masses of the final obtained polymer, are (1) the time in which the polymerization takes place in the absence of the solvent; (2) the total amount of solvent subsequently introduced; (3) the manner of introducing the solvent (continously or in portions); and (4) the variation of the solvent flow in time.

The variation in these parameters represents a practical possibility for ob-

TABLE I

The Influence of Reaction Conditions on the Degree of Polymerization of Polyvinyl Acetate and Polyvinyl Alcohol[a]

	Polymerization Degree			
Reaction Conditions	Polyvinyl Acetate		Polyvinyl Alcohol	
	Exp.	Calcd.	Exp.	Calcd.[c]
Solution polymn. $\eta = 0.95$[b]	1650	1423	720	650
Bulk polymn. $\eta = 0.1$ Solution polymn. $\eta = 0.95$	1920	1860	1050	820
Bulk polymn. $\eta = 0.2$ Solution polymn. $\eta = 0.95$	2500	2150	—	1120
Bulk polymn. $\eta = 0.4$ Solution polymn. $\eta = 0.95$	3800	3658	2200	1470

VAc/MeOH = 80/20 (g/g); AIBN/VAc = 0.00035/1 (g/g).

[a] VAc, vinyl acetate; MeOH, methanol; AIBN, azobisizobutyronitrile. Solvent was continuously poured.

[b] η, conversion.

[c] Calculated from $\bar{n}_n$ values.

taining different assortments of polyvinyl acetate and, eventually of polyvinyl alcohol. Some results are presented in Table I.

Due to its characteristics, the present technology enables the synthesis of polyvinyl alcohol with a high molecular mass and a low degree of branching; another major advantage of this procedure consists in the possibility of obtaining a large variety of types in terms of hydrolysis degree and molecular weight.

This procedure might present noteworthy perspectives for the polyvinyl alcohol fiber industry; we have in mind, of course, the use of redox systems for the polymerization of vinyl acetate within the above described procedure, but at lower temperatures.

Kawakami [12] and Fridlander [13] have shown that the polyvinyl acetate branching considerably diminished at lower temperatures; this fact may be explained by the larger difference between the values for the activation energy of the propagation ($E_p = 4.4$ kcal/mole) and transfer ($E_t = 12.0$ kcal/mole) reactions, respectively. The above quoted authors have underlined that the decrease of the polymerization temperatures of vinyl acetate is an important method for obtaining polyvinyl alcohol with increased crystallinity.

The Hydrolysis of Polyvinyl Acetate

The polyvinyl acetate solution in methanol obtained by the polymerization process is diluted and subsequently subjected to the alcoholysis reaction in the presence of sodium hydroxide as catalyst.

A well known peculiarity of this alcoholysis reaction is the fact that the viscosity of the medium largely increases as the process goes on, with the formation of an adherent and elastic gel, which shows a certain tendency towards solidification. The formation of the gel is a highly undesired element as the continuous dispersion of the medium involves a huge consumption of energy.

The presence of gel also produced difficulties in the control of the alcoholysis reaction, in view of obtaining polyvinyl alcohol with a desired content of residual acetate groups. The formation of gel hampers the quantitative study of the alcoholysis reaction; this fact is perhaps reflected in the scarce number of papers [18] approaching the problem of the kinetics of alcoholysis at high polymer concentrations.

We performed the alcoholysis by conducting the reaction in a Brabender-type grinder; this has allowed us to measure the necessary effort for homogenization and to obtain reproducible kinetic data.

The shear effort as a function of time is illustrated in Figure 4 and it presents a typical shape which corresponds to two stages.

In the first stage, corresponding to alcoholysis in an homogeneous medium, the mixing effort remains constant for a certain period of time, it reaches a maximum and subsequently returns to the initial value. In the second stage, the average value of the mixing effort also follows a curve with a maximum; the curve is characterized by significant momentary fluctuations, which produce the dented aspect of the curve.

The first maximum, which appears in the stage in which the reaction takes place in an homogeneous medium, corresponds to a conversion of around 60%;

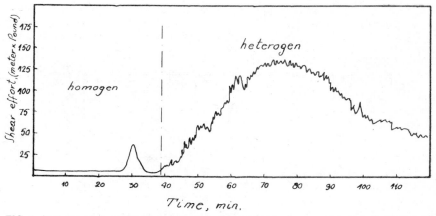

FIG. 4. Variation of shear effort (consistence) in (meter × lb) in the time of alcoholysis of polyvinyl acetate. Concentration of polyvinyl acetate $[PA_cV]$ = 20%; concentration of catalyst $[NaOH]$ = 2×10^2 mole/l; T = 30°C.

accordingly, the moment of its appearance in the alcoholysis reaction depends both on temperature and on catalyst concentration.

The catalyst concentration (Fig. 5), the temperature (Fig. 6) or the addition of solvents (e.g., benzene, methyl acetate) do not practically change the size of this maximum; in contrast, the size of this maximum depends strongly on the polymer concentration (Fig. 7).

For the second maximum, in the heterogeneous stage of the reaction, it was not possible to find a correlation between the moment of its appearance and conversion.

The size of this maximum (mediated value of fluctuations) sensibly decreases with the reduction of the catalyst concentration (Fig. 8) and the increase of the content of methyl acetate (co-solvent) (Fig. 9).

As in the homogeneous stage, the polymer concentration is the most important parameter which determines the value of this maximum (Fig. 10).

The above presented data suggest that the first maximum (homogeneous stage) may be attributed to the configuration of the polymer chain in solution, caused by the substitution of $-COOCH_3$ groups by the $-OH$ groups. It is likely that as the number of $-OH$ groups increases in the polymer chain, intermolecular interactions become important via hydrogen bonds; the result of this process

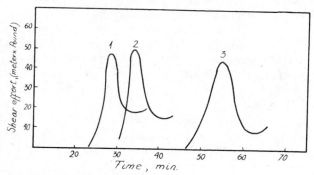

FIG. 5. Dependence of maximum shear effort in homogeneous stage on catalyst concentration and time. (1) $[NaOH] = 3 \times 10^{-2}$ mol/l.; (2) $[NaOH] = 2 \times 10^{-2}$ mol/l.; (3) $[NaOH] = 1.5 \times 10^{-2}$ mol/l.

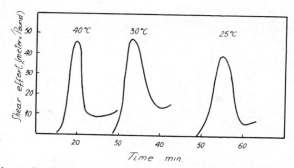

FIG. 6. Dependence of maximum shear effort in homogeneous stage on the temperature and time. $[PA_cV] = 20\%$; $[NaOH] = 2 \times 10^{-2}$ mol/l.

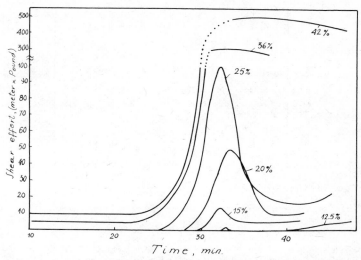

FIG. 7. Dependence of the maximum shear effort of the polyvinyl acetate concentration and time in the homogeneous stage. [PAcV] = 12.5, 15, 20, 25, 36, and 42%.

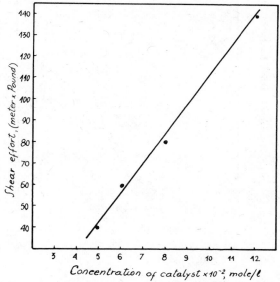

FIG. 8. Dependence of the maximum shear effort on the catalyst concentration, in the heterogeneous stage. [PAcV] = 20%, $T = 30°C$.

is the association of macromolecules and consequently the viscosity increases. It is also possible that the accumulation of –OH groups, beyond a certain point, leads to another change in the macromolecular configuration; the preponderance of intramolecular interactions produces a shrinkage of the chain and accordingly the viscosity again decreases.

The second stage was attributed to the mechanical grinding of polyvinyl alcohol turned insoluble in the reaction medium.

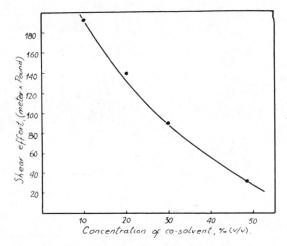

FIG. 9. Dependence of maximum shear effort in heterogeneous stage on the concentration of co-solvent. [PAcV] = 20%; [NaOH] = 2×10^2 mol/l.; $T = 30°C$.

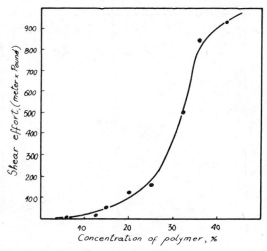

FIG. 10. Dependence of maximum shear effort in heterogeneous stage on the concentration of polyvinyl acetate. [NaOH] = 2×10^{-2} mol/l.; $T = 30°C$.

With regard to the alcoholysis kinetics it may be said that while in the first stage the reaction is rapid, in the second part, which takes place in the heterogeneous phase, the process is slow (Fig. 11).

The initial rate of the alcoholysis process varies linearly with the catalyst concentration (Fig. 12).

Surprisingly, it was noticed that in time the rate for the hydrolysis degree (dx/dt) diminishes as the polymer concentration increases (see Fig. 13).

This fact may be correlated with the autoacceleration effect which is typical for the hydrolysis reaction of the polyvinyl acetate. The above discussed phe-

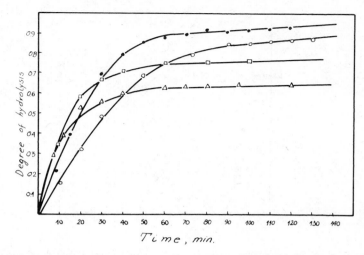

FIG. 11. Degree of hydrolysis as a function of time. $T = 30°$; $[PAcV] = 20\%$; $[NaOH] = 3 \times 10^{-2}$ mol/l. ($\bullet$); $[NaOH] = 2 \times 10^{-2}$ mole/l. ($\bigcirc$); $[NaOH] = 2 \times 10^{-2}$ mole/l.; and concentration of $[H_2O] = (\triangle)$; $[NaOH] = 2 \times 10^{-2}$ mole/l. and $[H_2O] = 2\%$ ($\square$).

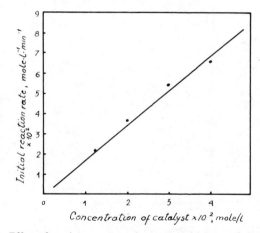

FIG. 12. Effect of catalyst concentration on the initial rate of alcoholysis.

nomenon was put in evidence by Sakurada [19]. According to this author, the reaction rate increases as hydrolysis takes place, due to the absorption of the catalyst on the –OH groups which are in the vicinity of acetate groups. Sakurada [19, 20] determined the equation:

$$\frac{dx}{dt} = k_0(1 + mx)(1 - x)$$

where k_0 is the initial rate constant, x, the degree of hydrolysis, and m, the autoacceleration coefficient of reaction. This equation perfectly verifies our results (Fig. 14).

We have found a value of 1.68 for the autoacceleration coefficient; this value,

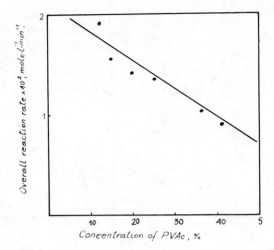

FIG. 13. Effect of concentration of polyvinyl acetate on overall rate of alcoholysis. [NaOH] = 2 × 10⁻² mole/l.; $T = 30°C$.

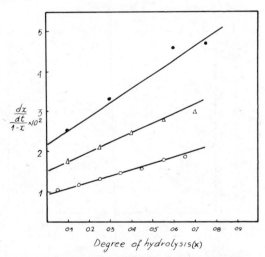

FIG. 14. Dependence of (dx/dt) $(1 - x)$ on the degree of hydrolysis. [PAcV] = 20%; $T = 30°C$; [NaOH] = 1.25 × 10⁻² mole/l. (O); [NaOH] = 2.0 × 10⁻² mole/l. (△); [NaOH] = 3.0 × 10⁻² mole/l. (●).

which is independent of polymer concentration, is much smaller than reported by Sakurada [16].

The origin of this discrepancy may be sought in the difference between the active species in the hydrolysis and alcoholysis reactions.

The decrease in alcoholysis rate at the same time with the polymer concentration may be understood only by admitting the catalyst absorbtion on the polymer. The smaller the concentration of polymer, the higher the content of catalyst absorbed on each macromolecule and accordingly a higher number of reaction sites.

The following simple experiment is illustrative for this reasoning. If a few drops of phenolphtalein are pipetted into the solution at the beginning of the alcoholysis reaction, the whole mass is colored. After gel formation and liquid phase separation, only the solid polyvinyl alcohol particles remain colored. The separated solvent is completely colorless and has a neutral reaction.

The same peculiarity may explain the partial character of block-copolymer which is typical for the partially hydrolyzed polyvinyl alcohol obtained by alcoholysis in basic catalysis.

The sudden drop in the reaction rate as well as the linear increase of conversion in time, in the heterogeneous stage of the process, lead to the concept of diffusive control of the reaction in this stage (see Fig. 11). The fact that the alcoholysis rate in the slow stage of the process depends on the molecular weight (Fig. 15) and hence on the consistence of the medium, is another argument in the same sense.

We have to take into account the diffusion processes and the occlusion of the catalyst, in order to obtain polyvinyl alcohol with a desired content of residual acetate groups.

One of the possibilities is to introduce small amounts of water into the reaction medium. In anhydrous methanol, the alcoholysis reaction occurs according to the following mechanism:

$$CH_3O^- + PV-O-\overset{\overset{\displaystyle O}{\|}}{C}-CH_3 \quad PV-O-\overset{\overset{\displaystyle O^-}{|}}{\underset{\underset{\displaystyle O-CH_3}{|}}{C}}-CH_3 \rightleftharpoons$$

$$PVO^- + CH_3-O-\overset{\overset{\displaystyle O}{\|}}{C}-CH_3 \quad (IIIa)$$

$$PVO^- + CH_3OH \longrightarrow PV-OH + CH_3O^- \quad (IIIb)$$

In the presence of traces of water, OH^- ions may appear, due to the shift towards the right of the protolytic equilibrium:

$$CH_3O^- + H_2O \rightleftharpoons CH_3OH + HO^- \quad (IV)$$

In the presence of hydroxyl ions, the following reactions have also to be taken into account:

$$HO^- + PV-O-\overset{\overset{\displaystyle O}{\|}}{C}-CH_3 \rightleftharpoons PVO-\overset{\overset{\displaystyle O^-}{|}}{\underset{\underset{\displaystyle OH}{|}}{C}}-CH_3 \rightleftharpoons$$

$$PVO^- + CH_3-\overset{\overset{\displaystyle O}{\|}}{C}-OH \quad PVOH + CH_3COO^- \quad (Va)$$

$$HO^- + CH_3COOCH_3 \longrightarrow CH_3COO^- + CH_3OH \quad (Vb)$$

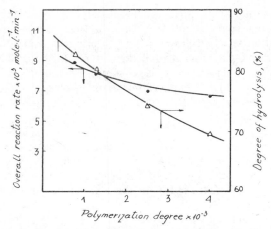

FIG. 15. Influence of molecular weight of polyvinyl acetate on the overall rate of alcoholysis in heterogeneous stage. [PAcV] = 20%; T = 30°C; [NaOH] = 2 × 10⁻² mole/l. (●); [NaOH] = 2 × 10⁻¹ mole/l., and [H₂O] = 2% (△).

The introduction of small amounts of water into the reaction medium leads to a slow consumption of the catalyst as shown in the last scheme; accordingly, the catalyst becomes exhausted before the conversion reaches its maximum.

It must be said, that due to the diffusive control in the presence of the gel as well as to the phenomenon of catalyst occlusion, the mechanism of this reaction is far from simple; the final conversion depends not only on the amount of water introduced, but also on the molecular mass of the polymer (Fig. 15).

In the heterogeneous stage, the rate of the alcoholysis reaction is much more affected by the increase in viscosity by comparison with the rate for the catalyst consumption (see eq. (Vb)). This peculiar feature has as an immediate effect, the fact that the deactivation reactions become predominant as the viscosity of the system increases.

REFERENCES

[1] M. Dimonie, M. Corobea, and Cr. Oprescu, Rom. Patent 58192 (1974).
[2] M. Dimonie, M. Georgescu, and Cr. Oprescu, Rom. Patent 58177 (1971).
[3] C. H. Bamford, W. G. Barb, A. D. Jenkins, and P. F. Onyon, *The Kinetics of Vinyl Polymerization by Radical Mechanisms,* London, 1958, p. 15.
[4] G. Gee and H. W. Melvill, *Trans. Faraday Soc., 40,* 240 (1944).
[5] G. Harington and A. Robertson, *Trans. Faraday Soc., 38,* 490 (1942).
[6] A. A. Shaginyan and N. S. Enicolopian, *Vysokomol. Soedyn., 7,* 1866 (1965).
[7] C. Wolf and W. Burchard, *Makromol. Chem., 177,* 2519 (1976).
[8] P. J. Flory, *J. Am. Chem. Soc., 64,* 2205 (1942).
[9] P. J. Flory, *Ind. Eng. Chem., 38,* 421 (1946).
[10] D. J. Stein, *Makromol. Chem., 76,* 170 (1964).
[11] W. R. Moore and M. Murphy, *J. Polym. Sci., 56,* 519 (1962).
[12] H. Kawakami, N. Mori, K. Kawashima, and M. Sumi, *Kogyo Kagaku Zasshi, 66,* 88 (1963); *C.A.* 4042b (1963).
[13] H. N. Fridlander, H. E. Harris, and J. G. Pritchard, *J. Polym. Sci. A-1, 4,* 649 (1966).

[14] C. A. Finch, *Polyvinyl Alcohol,* Wiley, New York, 1973.

[15] G. E. Ham, *Vinyl Polymerization,* M. Dekker, New York, 1964.

[16] M. Lazar, J. Paulinek, and Z. Manasek, *Coll. Czech. Chem. Commun., 26,* 1380 (1961).

[17] J. T. Clarke, *Kunstoffe-Plastics, 3,* 151 (1956).

[18] V. N. Stepchenko and A. N. Levin, *Soviet. Plastics, 8,* 44 (1961).

[19] I. Sakurada, *Gohsei Seni Kenkyu, 1-1,* 192 (1942); *C.A. 44,* 8161 (1950).

[20] S. Sakaguki, *The Mechanism of Hydrolysis of Polyvinyl Acetate in Polyvinyl Alcohol,* I. Sakurada, Ed., The Society of Polymer Science, Tokyo, 1956, p. 43.

SOLUTE BINDING AND CATALYSIS BY POLY(VINYLBENZO-18-CROWN-6)

J. SMID, S. C. SHAH, A. J. VARMA, and L. WONG

Chemistry Department, State University of New York, College of Environmental Science and Forestry, Syracuse, New York 13210

SYNOPSIS

Poly(vinylbenzo-18-crown-6), abbreviated as P18C6, and a polymer endowed with the cation binding crown ether ligand benzo-18-crown-6 as pendent groups, exhibits in water strong hydrophobic interactions with organic solutes such as picrates and methylorange. The binding can be modified by addition of crown ether complexable cations which superimposes an electrostatic component on the hydrophobic interaction forces. Thermodynamic parameters for the picrate binding have been determined, ΔH being -11.9 kcal/mole and $\Delta S = -16.8$ e.u. This type of binding was utilized to catalyze reactions which are known to be sensitive to the polarity of their micro environment. The P18C6 catalyzed decarboxylation of sodium-6-nitrobenzisoxazole-3-carboxylate in water was studied in detail, and the kinetics were found to resemble that of a micellar catalyzed reaction. The hydrophobic interaction of the carboxylate with P18C6 is weak, but the binding can be greatly enhanced by addition of KCl. The rate constant of decarboxylation of the substrate in the polymer domain was found to be between 2000 to 5000 times larger than that in water, depending on the charge density on the polymer. Finally, the interaction of P18C6 with polyanions such as sodium carboxymethylcellulose is discussed. Formation of insoluble polysalt complexes can be observed in the presence of KCl and CsCl.

Pedersen's [1] original work on the synthesis and cation binding properties of macrocyclic polyethers (crown ethers) induced many researchers to look for new macrocyclic and macropolycyclic ligands capable of forming selective cation inclusion complexes, especially with alkali and alkaline earth cations [2–6]. The lipophilic exterior of the ligands facilitates solubilization of their cation complexes in low polarity media, thus making the ligands useful in ion selective electrodes and as cation carriers in lipophilic membranes. The tendency of the primary ammonium cation to complex with certain macrocyclic polyethers has been utilized in the optical resolution of amino ester salts and other amino compounds by means of chiral crown ether ligands [5, 7–9]. This has been extended to the synthesis of macrocyclic ligands with active sites capable of stereoselectively catalyzing reactions such as transacylations [10], thus mimicking enzymatic transformations found in nature. Another recent development concerns the synthesis of macrotricyclic ligands of the cryptand type which not only are capable of forming inclusion complexes with cations but also with anions such as Cl⁻, after protonation of the nitrogen atoms [11].

Crown ether and cryptand ligands have also been extensively used as anion

Journal of Polymer Science: Polymer Symposium 64, 267–280 (1978)
© 1978 John Wiley & Sons, Inc.   0360-8905/78/0064-0267$01.00

activators. The catalysis may result from an enhanced solubility of the complexed reagent, for example, when dealing with heterogeneous processes, or the ligand interaction with the cation may activate the anion by increasing the interionic ion pair distance or by breaking down less reactive aggregates into more reactive species. Examples in the area of polymerization can be found in the accelerating effects exerted by cryptands and crown ethers on the anionic polymerization of ethylene oxide, propylene sulfide, styrene, methylmethacrylate, isoprene, and butadiene [12, 13] as well as on that of lactones [14] and in the synthesis of polyesters from acrylic acid [15]. A review on the catalysis of organic reactions by crown ethers has recently been published [16]. Successful attempts have also been made to use macrocyclic ligands attached to an immobilized polymer matrix. The materials were found to be active as catalysts in displacement reactions of bromide by iodide and cyanide [17] and could also be utilized in resolving mixtures of enantiomers [18].

For some time now we have been investigating the behavior of polymers endowed with crown ether ligands as pendent groups, with the objective to compare their cation binding characteristics with those of their monomeric analogs [19]. The selectivity of cation binding is expected to be different for the polymers, especially when formation of stable complexes involves more than one ligand, or when cation desolvation requirements differ for the polymeric and monomeric ligands. Studies using extraction equilibria [19, 20], ion transport through liquid phases [20], optical spectroscopy [21] and conductance [22] demonstrate that significant differences exist between the two kinds of species.

Recent investigations on cation binding to poly(vinylbenzo-18-crown-6) (abbreviated as P18C6 and depicted below) in aqueous media,

have revealed that this polymer strongly adsorbs organic solutes by hydrophobic interactions [23, 24]. The association of anionic solutes to P18C6 can be further enhanced by addition of crown complexable cations. Thus, the polymer can be utilized to catalyze reactions that are sensitive to the polarity of the environment. In this paper, three aspects of the binding properties of poly(vinylbenzo-18-crown-6) will be stressed, viz., the association of organic solutes to P18C6, especially picrate salts; the catalytic decomposition of 6-nitrobenzisoxazole-3-carboxylate in water in the presence of P18C6, and, finally, the formation of polysalt complexes between P18C6 and carboxymethyl cellulose in aqueous electrolyte solutions.

EXPERIMENTAL

Picrate salts were prepared from picric acid and the appropriate base, and purified by recrystallization from ethanol. The synthesis of 6-nitrobenzisoxazole-3-carboxylate has been described in the literature [25]. The preparation of vinylbenzo-18-crown-6 [19, 26] and its polymerization by a radical initiator to P18C6 [19] has also been reported. The number average molecular weight of the poly(crown ether) used in most experiments was between 30,000 and 60,000.

The binding of picrate salts to P18C6 in water was measured spectrophotometrically in a Cary 15 spectrophotometer between 0 and 35°C. The transfer of the picrate anion from water to the polymer domain shifts the main absorption band maximum from 355 to 384 nm. The fractions of picrate ions in the two environments can be calculated from the known spectra of the salts in each domain separately. In several experiments electrolytes were added to determine the electrostatic binding of picrate to P18C6. Potentiometric measurements using an Orion 801 pH meter were carried out to determine the extent of binding of Na^+ and K^+ to P18C6 and 4'-methylbenzo-18-crown-6 [27, 28].

The decomposition of 6-nitrobenzisoxazole-3-carboxylate in water in the presence of P18C6 and KCl was followed in a Cary 15 spectrophotometer by monitoring the formation of the decarboxylation product, 2-cyano-5-nitrophenolate which has an absorption maximum in the 397–435 nm region. First order rate constants were calculated from the expression $k = (1/t)\ln[a_\infty/(a_\infty - a_t)]$, a_t and a_∞ being the observed optical densities at time t and $t = \infty$, respectively.

Polysalt complexes were formed by adding an aqueous solution of sodium carboxymethylcellulose (molecular weight, 900,000, degree of substitution, 0.875) to an aqueous solution of P18C6 containing a known quantity of potassium chloride. The precipitate was centrifuged off after the mixture was shaken overnight, and the supernatant liquid analyzed for P18C6 (spectrophotometrically) and CMC (using $CuSO_4$). Experimental details of these investigations as well as of the binding studies and catalysis by P18C6 have recently been published [28, 29, 47].

RESULTS AND DISCUSSION

Solute Binding to Poly(vinylbenzo-18-Crown-6)

The 355 nm absorption band of the picrate anion shifts to higher wavelength on addition of P18C6. At a $2 \times 10^{-4}M$ picrate concentration λ_m reaches its highest value of 384 nm at a crown to picrate ratio of about 50 (the polymer concentration is expressed in terms of crown monomer units). The final λ_m value is independent of the counterion. The appearance of an isosbestic point indicates the presence of two distinct picrate species. The 384 nm band is identical to that of a free picrate anion or of a loose picrate ion pair in an aprotic solvent such as acetone or tetrahydrofuran. The bathochromic shift results from a solubilization of the organic anion into the aprotic polymer core of P18C6, a process which resembles the solubilization of organic solutes into micelles. Binding of organic species to macromolecules in an aqueous medium has been observed for many synthetic and natural polymers, e.g., polyvinylpyrrolidone [30–33], bovine serum albumin [34, 35], crosslinked polylysine [36], sodium copoly(ethylacrylate-acrylic acid) [37] and many chemically modified polymers with hydrophobic side chains [38–40]. The interactions may be hydrophobic in nature, or caused by hydrogen bonding, charge transfer, or by electrostatic forces.

The P18C6 macromolecule has a hydrophobic polystyrene backbone to which 18-crown-6 moieties are attached. The crown molecules apparently interact quite effectively with water as shown by the relatively high water solubility of P18C6 ($\simeq 8$ g/100 cm^3, see ref. [19]) and its inverse temperature solubility, the cloud point being 37°C. The polymer in water resembles a soap micelle with the nonpolar polystyrene core surrounded by hydrophilic crown residues. The solubility of poly(crown ethers) can be considerably increased by adding crown complexable cations, as the neutral P18C6 is converted into a polycation. The formation of such polycations is demonstrated by the strong increase in the reduced viscosity on lowering the concentration of a mixture of P18C6 and KCl in mixtures of THF/CH$_3$OH [19], a behavior typical of that of polyelectrolytes.

Solute binding to polymers can often be conveniently described by using a rearranged form of the Langmuir isotherm [34], i.e.,

$$\frac{1}{r} = \frac{1}{n} + \frac{1}{nKa} \tag{1}$$

where in our system $1/r$ represents the ratio of total crown monomer units to bound picrate, a denotes the free picrate concentration, n is the number of binding sites per mole of crown units ($1/n = N$ represents the average number of crown units needed to bind one picrate anion, i.e., $1/n$ is the size of the binding site) and K is the intrinsic binding constant. The first binding constant, K_1, is equal to nK. A plot of $1/r$ versus $1/a$ for sodium picrate with P18C6 in water is depicted in Figure 1. While this plot is linear as predicted from eq. (1), a similar plot for potassium and cesium picrate shows a curvature, with binding increasing more rapidly at higher picrate concentration. This deviation is caused by the much higher binding constants of K$^+$ and Cs$^+$ to P18C6 (see below) and cations become bound to the polymer chain. Higher picrate concentrations will increase the charge density on the chain, and electrostatic attraction by the polycation causes more picrate to transfer to the polymer. To measure the binding caused by hydrophobic interactions it is necessary to use the sodium or lithium picrate salts, as these two cations have much lower binding constants to P18C6.

The binding of picrate was determined between 35 and 0°C, and the values of K and $1/n$ determined from the slopes and intercepts of the respective lines are listed in Table I. The first binding constants are also given. The log K versus $1/T$ plot is linear, ΔH being -11.9 kcal/mole and $\Delta S = -16.8$ e.u. The intrinsic binding constants appear to be considerably higher than found for other homopolymers. Although no values for picrates are reported, dialysis measurements

TABLE I

Binding of Sodium Picrate to Poly(vinylbenzo-18-crown-6) in Water

Temp °C	$K_1^a \times 10^{-3}$ M^{-1}	$K_1^b \times 10^{-5}$ g	$K^a \times 10^{-4}$ M^{-1}	$1/n$
35	1.17	3.47	5.1	44
25	3.19	9.45	13.4	42
17.5	4.85	14.3	17.0	35
0	20.9	61.9	67.3	32

[a] Values expressed in moles/liter of crown monomer units.

[b] Derived from r values calculated for 10^5 g of polymer. $K_1(10^5 \text{ g}) = K_1(M^{-1}) \times 10^5/338$.

reveal that under a set of identical conditions 70% picric acid binds to P18C6 to only 5% for polyvinylpyrrolidone. Binding of methylorange to P18C6 is only slightly less than for picrate, i.e., $K_1 \simeq 3 \times 10^5$ (10^5 g). This again is higher than the binding of this dye to polyvinylpyrrolidone ($K = 1.79 \times 10^4$; see ref. [33]) and even to serum bovine albumin ($K = 5.4 \times 10^4$; see ref. [36]). Binding to P18C6, therefore, compares favorably to this protein, and even to some of the chemically modified synthetic polymers reported by Takagishi et al. [39]. The modified polyethylenimines possessing lauryl side chains reported by Klotz [38] have higher binding constants.

The average number of crown monomer units needed to bind one picrate anion at 25°C is 42. Molyneux and Frank [30] reported a value 10 for polyvinylpyrrolidone, independent of solute. The value 42 is also obtained by determining $1/r$ at increasing picrate concentration, where, due to saturation, $1/r$ becomes constant and equal to $1/n$. The binding size appears to decrease on lowering the temperature. This may be due to a slight expansion of the polymer coil as indicated by a higher intrinsic viscosity of P18C6 at 0°C as compared to 25°C. It also appears that $1/n$ decreases as the polymer is converted into polycation by

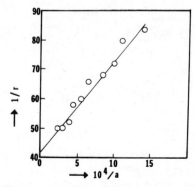

FIG 1. Plot of $1/r$ versus $1/a$ for binding of sodium picrate to P18C6 in water at 25°C.

addition of KCl. This again could be caused by polymer expansion, facilitating the accommodation of the large picrate anion in the polymer core.

The binding of sodium picrate to the neutral P18C6 polymer has been termed hydrophobic since no electrostatic forces appear to be involved. Addition of LiCl (up to $0.1M$) has no effect on the binding. On the other hand, the thermodynamic parameters, especially the low ΔS value, do not seem to be characteristic for what is usually observed in hydrophobic interactions, viz., a positive entropy. Even when corrected for the "cratic term" of $\Delta S = 7.98$ e.u. [30], the so-called unitary entropy [30] is still -8.8 e.u. Exothermic enthalpy values are not uncommon and have been found in the binding of solutes to polyvinylpyrrolidone [30] and bovine serum albumin [35]. Plots of log nK versus $1/T$ in some systems show a change from a negative to a positive enthalpy [33]. The rather pronounced exothermicity in our system could be the result of interactions between the π clouds of the picrate molecule and of the benzocrown ethers. The negative entropy may indicate that the usual loss of water structure around a polymer that exhibits hydrophobic associations and that leads to a gain in entropy, is not realized in our system. The picrate anion is probably located entirely in the nonpolar polymer core of the P18C6 molecule, and in such a case the transfer apparently would not greatly affect the water structure around the macromolecule. This seems to be confirmed by the observation (see next section) that in the catalysis of the decarboxylation of benzisoxazole-3-carboxylate the rate constant is not much different from that found in benzene.

It has already been pointed out that crown complexable cations enhance the binding of anionic organic solutes. This is demonstrated by the plot shown in Figure 2. Under the chosen conditions, approximately 45% picrate is bound in the absence of electrolytes ($[Pi] = 1 \times 10^{-5}M$, $[P18C6] = 5 \times 10^{-4}M$), different salts can be added and each time the increase in the fraction of bound picrate is measured as a function of electrolyte concentration. As Figure 2 shows,

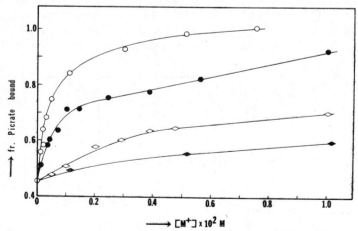

FIG. 2. Dependence of the fraction of sodium picrate bound to P18C6 in water as a function of salt concentration. [Sodium picrate] $= 1 \times 10^{-5}M$; [P18C6] $= 5 \times 10^{-4}M$; O, CsCl; ●, KCl; ◊, NH$_4$Cl; ■, NaCl; $T = 25°C$.

the electrostatic binding decreases in the order $Cs^+ > K^+ > NH_4^+ > Na^+$. Addition of LiCl has no effect. The order for the alkali ions should parallel their ability to bind to P18C6. Potentiometric measurements show that the complex formation constants for Na^+ and K^+ to P18C6 are $1.2M^{-1}$ and $35M^{-1}$, respectively (the values actually vary somewhat, and depend on the number of ions already complexed to P18C6).

The electrostatic binding of picrate anions to P18C6 can also be used to determine complex formation constants for cations that are not easy to measure potentiometrically. For example, CsCl or $AgNO_3$ may be added as the electrolyte, and the increase in the fraction of picrate bound determined spectrophotometrically. This is then compared with the binding of picrate by K^+ for which the binding constant can be measured potentiometrically. Hence, the number of cations bound to the chain can be calculated at different salt concentrations. For NaCl a binding constant of $1.4M^{-1}$ was obtained, which compares favorably with the potentiometric value $1.2M^{-1}$.

Binding of other solutes to P18C6 can be studied spectrophotometrically, such as the interaction of p-nitrophenol and its salts, phenolphtalein and chrysophenine. Compounds such as sodium tetraphenylboron and dodecylsulfate also bind to P18C6, as shown by the displacement of picrate from the polymer domain of P18C6 on addition of such solutes. Quantitative measurements are in progress to determine binding constants to P18C6 for various species. As expected, the hydrophobic interaction decreases on addition of alcohol, and no picrate binding is observed when the mixture water/alcohol contains about 40% ethanol by volume.

Poly(crown Ether) Catalyzed Decarboxylation of 6-Nitrobenzisoxazole-3-Carboxylate

The rate of decarboxylation of benzisoxazole-3-carboxylate and of its derivatives into the salicylonitrile salt is remarkably solvent dependent [41, 42].

The reaction, a concerted intermediateless E_2 elimination, is slow in water and alcohols, but is accelerated in aprotic solvents such as benzene and tetrahydrofuran, and is extremely fast in dimethylsulfoxide and hexamethylphosphoramide. Cationic micellar reagents such as cetyltrimethylammonium bromide catalyze the decomposition of the carboxylate in water by transferring the substrate to the micellar environment [43–45]. We briefly reported [23] that poly(vinylbenzo-18-crown-6) also catalyzes the decomposition of an aqueous solution of 6-nitrobenzisoxazole-3-carboxylate. The interaction of the substrate with P18C6 is much weaker than found for picrate anions, but it can be significantly enhanced by converting the neutral macromolecule into a polycation by adding K^+ or Cs^+ ions. Recently, Klotz et al. have shown that modified polyethylenimines endowed with apolar lauryl groups for promoting hydrophobic interactions and with quaternized amino nitrogen to provide electrostatic attraction forces, catalyze the decarboxylation of nitrobenzisoxazolecarboxylate and of cyanophenylacetate [40]. The interesting and unique feature of our system is that the reaction can be made cation specific by varying the structure of the crown moiety.

The reaction scheme for the P18C6 catalyzed decomposition of the carboxylate is similar to that of a micellar catalysis, i.e.,

$$\text{products} \xleftarrow[k_0]{} S + M \underset{K}{\leftrightarrows} SM \xrightarrow[k_m]{} \text{products} + M \qquad (2)$$

where S denotes the free carboxylate concentration and M in our system represents the concentration of free binding sites. This quantity is given by $[M] = [C]/N$, C being the concentration of free crown units and N the average number of monomer units required to form one binding site. It can be easily shown [46] that the observed rate constant k_ψ is given by

$$\frac{1}{k_\psi - k_0} = \frac{1}{k_m - k_0} + \frac{N}{(k_m - k_0)KC} \qquad (3)$$

the assumption being that N and k_m are independent of the number of bound substrate molecules. The first assumption appears justified on the basis of results obtained for picrate salts. If only a few binding sites are occupied, the concentration C may be replaced by C_0. The decarboxylation is pseudo first order as long as only a small fraction of substrate is bound. This is the case in our system when no crown complexable cations are added, but in the presence of a salt such as KCl, deviations from first order kinetics occur above 50% of substrate conversion. Another cause for this deviation is product binding which is neglected in the derivation of eq. (3). Evidence for this binding can be found in the bathochromic shifts observed in the absorption maximum of the 2-cyano-5-nitrophenolate, viz., from 397 nm in the absence of P18C6 to 433 nm at high P18C6 and KCl concentration. The k_ψ values calculated from our experimental data are all initial rate constants.

A plot of $1/(k_\psi - k_0)$ versus $1/c$ for the sodium salt is depicted in Figure 3A. A linear relationship is also found for the potassium salt, but the slope is significantly smaller as the binding of the substrate is enhanced due to complexing of K^+ to P18C6. Unfortunately, the intercepts are too small to yield an accurate k_m value, and K cannot be determined directly. However, the intrinsic binding constant can be determined by carrying out a second series of experiments, choosing conditions such that substrate is in excess, i.e., $S_0 \gg C_0/N$. In this case an expression for the initial rate constant k_ψ is obtained by applying the steady state for reaction 2 [40].

$$k_\psi - k_0 = \frac{k_m C_0}{(K^{-1} + S_0)N} \qquad (4)$$

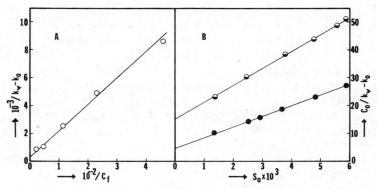

FIG. 3. Decomposition of sodium-6-nitrobenzisoxazole-3-carboxylate in water at 25°C in the presence of P18C6. (A) Dependence of the observed rate constant k_ψ on the P18C6 concentration (C = free crown ether concentration); (B) dependence of k_ψ on the initial concentration of carboxylate, S_0. O, no KCl present; ●, $10^{-3}M$ KCl present.

which can be rearranged to

$$\frac{C_0}{k_\psi - k_0} = \frac{N}{k_m K} + \frac{N}{k_m} S_0 \tag{5}$$

The binding constant, K, can be determined from the slope and intercept of plots derived from eq. (5), or by combining the slopes of the two plots obtained from eqs. (3) and (5). In the second set of experiments the [P18C6] is kept constant at $2.5 \times 10^{-3}M$ and the initial carboxylate concentration, S_0, is varied between $1 \times 10^{-3}M$ and $6 \times 10^{-3}M$ (to assure that no cations are bound to P18C6, we used the lithium carboxylate instead of the sodium salt.) For $N \simeq 40$ (see previous section) the ratio C_0/N is about $6 \times 10^{-5}M$, hence the condition $S_0 \gg C_0/N$ is fulfilled. The procedure can also be applied when experiments are carried out in the presence of a constant amount of an electrolyte such as KCl. Since the [P18C6] is also constant, the charge density on the polymer is not changed when S_0 is varied. This would, in principle, permit the measurement of a series of binding constants as a function of the charge density on the polymer.

Plots of $C_0/(k_\psi - k_0)$ versus S_0 in the absence and presence of KCl are depicted in Figure 3B. Values of N/k_m and K can be calculated from the straight lines. In the absence of KCl the intrinsic binding constant of carboxylate to P18C6 was found to be $400M^{-1}$ ($340M^{-1}$ when computed from the intercept of Fig. 3A and the slope of Fig. 3B plot), which is a factor 330 smaller than for picrate anions. Addition of $10^{-3}M$ KCl raises the binding constant to $K = 780M^{-1}$. At this salt concentration, about 10% of the crown moieties contain a K^+ cation. The charge density rapidly increases at higher KCl concentrations, so does the binding constant K as indicated from the much smaller intercepts. At $4 \times 10^{-3}M$ KCl the binding constant is approximately $4000M^{-1}$.

The decarboxylation rate constant in the polymer domain, k_m, can be reasonably estimated by assuming that the size of the binding site for the carboxylate is not too much different from that of the picrate anion. Taking $N = 42$ this yields $k_m = (7.0 \pm 1) \times 10^{-3}$ sec^{-1}. This value is not too different from that obtained by Kemp and Paul [42] in benzene (the value varies between 5×10^{-3} and 2×10^{-3} sec^{-1} when calculated from those given by the authors at 30°C, using $E = 24$ kcal/mole; see ref. [42]). This may indicate that the carboxylate molecule is almost completely surrounded by the benzene moieties of the poly-(crown ether) molecule.

In the presence of a sufficient quantity of KCl all substrate could conceivably transfer to the polymer domain, permitting a direct measurement of k_m. Results of measurements of k_ψ in solutions of increasing concentrations of K$^+$ ions are depicted in Figure 4. The ratio C_0/S_0 was kept constant at 50, hence, with the size of the binding site estimated to be about 40, no saturation could occur even when all substrate is bound. The rate constant rapidly increases with the concentration of cations on P18C6 (determined directly by potentiometry in a separate experiment with no substrate present) and seems to reach a maximum value of about 11×10^{-3} sec^{-1}. At C_0/S_0 ratios of 100 to 150 we have measured k_ψ values as high as 15×10^{-3} sec^{-1}, about twice the k_m value found when no cations are bound to P18C6. At very high KCl concentration k_ψ appears to decrease considerably. This was observed at different C_0/S_0 ratios. The counterion does not appear to affect k_ψ because addition of excess LiCl does not change

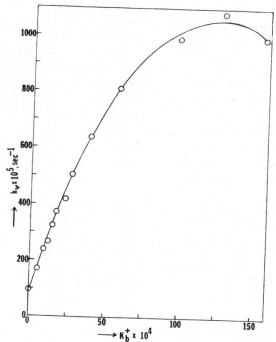

FIG. 4. Dependence of the observed rate constant, k_ψ, of decarboxylation of 6-nitrobenzisoxa-zole-3-carboxylate in a water–P18C6 mixture as a function of the concentration of K$^+$ charges bound to P18C6. $T = 25$°C.

the rate. The dependence of k_m on the charge density on P18C6 is in agreement with the observation made by Klotz et al., viz., that k_m in the laurylated polyethylenimines is enhanced by a factor 2 when the polymer is quaternized. The authors attribute this to stabilization of the anionic transition state by the positive ions on the polymer. A similar effect may be responsible in our system.

A k_ψ value of $15 \times 10^{-3} \sec^{-1}$ (we are not certain whether in this experiment all substrate was bound, hence, k_m may even be a little higher) represents an acceleration factor of 5000 as compared to the decarboxylation rate constant in water. The largest k_m value reported for the polyethylenimine systems is $4 \times 10^{-3} \sec^{-1}$ [40], while the highest rate enhancement for the detergent systems studied by Bunton et al. [45] was found to be 400, i.e, $k_m \simeq 1.2 \times 10^{-3} \sec^{-1}$. Undoubtedly the rate constant for the carboxylate decomposition will depend on the structure and polarity of the polymer phase in which the substrate is solubilized. It would be interesting to correlate k_m with the polarity of such a phase by binding to the polymer certain compounds whose fluorescent spectra are known to vary with the polarity of their micro environment. Further details on the P18C6-carboxylate system such as the effect of different electrolytes will be reported elsewhere [47].

Polysalt Formation

Addition of crown ether complexable cations to P18C6 converts the neutral polymer into a polycation. Hence, when a polyanion such as sodium carboxymethylcellulose (CMC) is added to a P18C6-salt solution, favorable conditions exist that may result in the formation of insoluble polysalt complexes. Both synthetic and natural polymers can form these type of complexes. Studies by several researchers [48–50] have demonstrated that variables such as charge equivalence, electrolytes, chain length, charge density, chain conformation, and chain flexibility all play an important role in the formation of the complex and in determining its ultimate structure and properties.

When sodium CMC is added to an aqueous solution of P18C6 containing potassium chloride, a precipitate is formed. The amount depends on the KCl concentration and the initial ratio of crown monomer units to carboxylate groups. At a 2:1 ratio of crown to carboxylate the solution must contain about $0.1M$ KCl to maximize the amount of P18C6 precipitated. The complex redissolves at a $[KCl] > 0.5M$, or by removing the supernatant electrolyte solution and adding water. The salt concentration at which the maximum quantity of insoluble polysalt complex is generated depends on the initial ratio of crown to carboxylate. This result is graphically depicted in Figure 5, where the fraction of precipitated P18C6 is plotted versus the CMC/P18C6 ratio at different KCl concentrations. Note that a decrease in the KCl concentration from $0.2M$ to $0.005M$ at low CMC/P18C6 ratio (e.g., 0.05) causes more P18C6 to precipitate, while the reverse is true at a CMC to P18C6 ratio in the vicinity of 0.5. The order in which reagents are added appears to have no effect on the final results.

Polysalt formation in the CMC-P18C6 system results from electrostatic interactions between the polyanion and the cation–crown complexes distributed

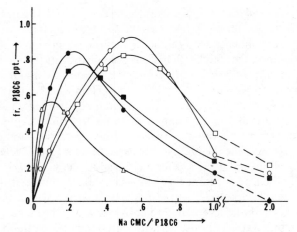

FIG. 5. Polysalt formation between sodium carboxymethylcellulose and P18C6 in the presence of KCl. Plot of fraction of P18C6 in the precipitated complex versus the ratio of sodium CMC to P18C6. KCl concentrations: △, 0.002M; ●, 0.005M; ■, 0.02M; ○, 0.1M; □, 0.2M. (Taken from ref. [28].)

along the P18C6 macromolecule. The charge density on the P18C6 chain depends on the salt concentration and the nature of the cation. Also, the mobility of these charges may play a role in determining the stability of the polysalt complex as steric restraints can be relaxed by readjusting the position of the positive charges.

In most systems that have been investigated, complex formation appears to be completed when charge equivalence is reached. To verify this in our system would require a knowledge of the complex formation constant of K^+ to P18C6 in water. However, the binding constant of K^+ to P18C6 varies with the charge density on P18C6, and only an approximate calculation can be made. Taking $K = 35M^{-1}$ (see first section) and assuming that the ratio complexed K^+ to total crown moieties does not exceed 0.5 [20, 22], one can calculate the concentration of bound K^+ in the precipitated complex. This is found to be in reasonable agreement with the carboxylate concentration in the precipitate for conditions where all CMC has precipitated, implying that also in our system charge equivalence is important in forming insoluble polysalt complexes.

Calculations, at least qualitatively, confirm that the average segment length of free crown ligands between points of charge interaction on the two macromolecules increases at lower KCl concentration. The conclusion can be drawn that at a high crown to carboxylate ratio (e.g., $\simeq 10$), maximum precipitation of P18C6 is accomplished at low KCl content. The polysalt complex thus formed has a gellike consistency as a result of the rather long segments of neutral crown moieties which interact with water and cause the precipitate to swell. A high KCl concentration is required to bring about maximum P18C6 precipitation when the initial P18C6/carboxylate ratio is low, e.g., $\simeq 2$, and the resulting polysalt complex is compact and easy to isolate.

Screening of the charges on the macromolecule by counterions reduces electrostatic interactions and causes the polysalt complex to dissolve at high salt concentration. This is also the reason for not observing a precipitate with NaCl. The binding constant of Na^+ to P18C6 is only $1.2M^{-1}$, and about $0.15M$ NaCl is needed to supply the necessary charges on P18C6 to effectuate its maximum precipitation at a crown to CMC ratio of 5. With KCl only $0.005M$ is needed. The already weak electrostatic interactions caused by the low concentration of charges is further reduced by the high ionic strength in the NaCl system. The result is that no precipitate is formed, while nearly all P18C6 precipitates with $0.005M$ KCl. If to the latter system $0.1M$ of the noncomplexable salt $N(CH_3)_4Cl$ is added, again no precipitation is observed. Cesium chloride is an effective polysalt complex former, the binding of Cs^+ to P18C6 being even stronger than that of K^+. The counterion also is of importance. At a constant CMC to P18C6 ratio and a potassium salt concentration of $0.1M$ the amount of poly(crown ether) in the precipitate decreases in the order $OH^- \sim Cl^- > Br^- > I^-$. A more detailed account of this work can be found elsewhere [28].

Recent work has revealed that polysalt complex formation with poly(crown ethers) is quite general. For example, polymers derived from crown esters of methacrylic acid form insoluble complexes with sodium carboxymethylcellulose in the presence of KCl, and polyanions such as poly(styrenesulfonates) also form precipitates with poly(crown ethers). Earlier reported work showed complex formation to exist between P18C6 and polynucleotides [51]. It may be recalled in this respect that P18C6 is known to inhibit the polyriboadenylate templated elongation of oligodeoxythymidine [51]. This inhibitory effect on reverse transcriptase activity is exhibited both in the absence and presence of KCl, indicating the formation of soluble complexes, possibly as a result of both electrostatic and hydrophobic interactions.

The authors gratefully acknowledge the financial support of this research by the National Science Foundation (CHE 7609805) and by the Donors of the Petroleum Research Fund, administered by the American Chemical Society.

REFERENCES

[1] C. J. Pedersen, *J. Amer. Chem. Soc., 89,* 7017 (1967); *J. Amer. Chem. Soc., 92,* 391 (1970); *J. Org. Chem., 36,* 254 (1971); C. J. Pedersen and H. K. Frensdorff, *Angew. Chem. Int. Edit., 11,* 16 (1972).

[2] J. J. Christensen, D. J. Eatough, and R. M. Izatt, *Chem. Rev., 74,* 35 (1974).

[3] C. Kappenstein, *Bull. Soc. Chim., 89* (1974).

[4] J. M. Lehn, *Structure and Bonding, 16,* 1 (1974).

[5] D. J. Cram and J. M. Cram, *Science, 103,* 183 (1973).

[6] D. N. Reinhoudt, R. T. Gray, C. J. Smit, and Ms. I. Veenstra, *Tetrahydron, 32,* 1161 (1976).

[7] R. C. Helgeson, J. M. Timco, P. Moreau, S. C. Peacock, J. M. Mayer, and D. J. Cram, *J. Amer. Chem. Soc., 96,* 7097 (1974).

[8] D. Curtis, D. A. Laidler, J. F. Stoddart, and G. H. Jones, *J.C.S. Chem. Comm.,* 835 (1975).

[9] J. M. Girodeau, J. M. Lehn, and J. P. Sauvage, *Angew Chem. Int. Edit.,* 14, 764 (1975).

[10] Y. Chao and D. J. Cram, *J. Amer. Chem. Soc., 98,* 1015 (1976).

[11] E. Graf and J. M. Lehn, *J. Amer. Chem. Soc., 98,* 6403 (1976).

[12] S. Boileau, B. Kaempf, J. M. Lehn and F. Schue, *J. Polym. Sci. Pol. Lett. Edit.*, *12*, 208 (1974);
     S. Boileau, P. Hemery, B. Kaempf, F. Schue, and M. Viguier, *J. Polym. Sci. Pol. Lett. Edit.*,
     12, 221 (1974).
[13] T. C. Cheng and A. F. Halasa, *J. Polym. Sci. Pol. Chem. Edit.*, *14*, 583 (1976).
[14] S. Slomkowski and S. Penczek, *Macromolecules, 9*, 367 (1976); A. Deffieux and S. Boileau,
     Macromolecules, 9, 369 (1976).
[15] B. Yamada, Y. Yasuda, T. Matsushita and T. Otsu, *J. Polym. Sci. Pol. Chem. Edit.*, *14*, 277
     (1976).
[16] G. Gokel and H. D. Durst, *Synthesis,* 168 (1976).
[17] M. Cinouini, S. Collona, M. Molinari, F. Montanari, and P. Tundo, *J.C.S. Chem. Commun.*,
     394 (1976).
[18] G. Dotsevi, Y. Sogah, and D. J. Cram, *J. Amer. Chem. Soc.*, *98*, 3038 (1976).
[19] S. Kopolow, T. E. Hoʒen Esch, and J. Smid, *Macromolecules, 6*, 133 (1973).
[20] K. H. Wong, K. Yagi, and J. Smid, *J. Memb. Biol.*, *18*, 379 (1974).
[21] S. Kopolow, Z. Machacek, U. Takaki, and J. Smid, *J. Macromol. Sci.*, *A7*, 1015 (1973).
[22] S. C. Shah, S. Kopolow, and J. Smid, *J. Polym. Sci. Pol. Chem. Edit.*, *14*, 2023 (1976).
[23] J. Smid, S. Shah, L. Wong, and J. Hurley, *J. Amer. Chem. Soc.*, *97*, 5932 (1975).
[24] J. Smid, *J. Pure Appl. Chem.*, *48*, 343 (1976).
[25] H. Lindemann and H. Cissee, *Justus Liebigs Ann. Chem.*, *469*, 44 (1929).
[26] J. Smid, B. El Haj, T. Majewicz, A. Nonni, and R. Sinta, *Org. Prep. Proc. Int.*, *8*, 193
     (1976).
[27] U. Takaki, T. E. Hogen Esch, and J. Smid, *J. Amer. Chem. Soc.*, *93*, 6760 (1971).
[28] A. J. Varma and J. Smid, *J. Polym. Sci. Pol. Chem. Edit. 15*, 1189 (1977).
[29] L. Wong and J. Smid, *J. Amer. Chem. Soc.*, *99*, 5637 (1977).
[30] P. Molyneux and H. P. Frank, *J. Amer. Chem. Soc.*, *83*, 3169 (1961).
[31] P. Bandyopadhyay and F. Rodriguez, *Polymer, 13*, 119 (1972).
[32] M. L. Fishman and F. R. Eirich, *J. Phys. Chem.*, *75*, 3135 (1971); *ibid., 79*, 2740 (1975).
[33] T. Takagishi and N. Kuroki, *J. Polym. Sci. Pol. Chem. Edit., 11*, 1889 (1973).
[34] I. M. Klotz, F. M. Walker, and R. B. Pivan, *J. Amer. Chem. Soc., 68*, 1486 (1946).
[35] T. Takagishi, K. Takami, and N. Kuroki, *J. Polym. Sci. Pol. Chem. Edit., 12*, 191 (1974).
[36] I. M. Klotz and J. H. Harris, *Biochemistry, 10*, 923 (1971).
[37] J. S. Tan and R. L. Schneider, *J. Phys. Chem., 79*, 1380 (1975).
[38] I. M. Klotz and A. R. Sloniewsky, *Biochem. Biophys. Res. Commun., 31*, 3 (1968).
[39] T. Takagishi, Y. Nakata and N. Kuroki, *J. Polym. Sci. Pol. Chem. Edit., 12*, 807 (1974).
[40] J. Suh, I. S. Scarpa, and I. M. Klotz, *J. Amer. Chem. Soc., 98*, 7060 (1976).
[41] D. S. Kemp and K. G. Paul, *J. Amer. Chem. Soc., 92*, 2553 (1970); *ibid., 97*, 7305 (1975).
[42] D. S. Kemp, D. D. Cox, and K. G. Paul, *J. Amer. Chem. Soc., 92*, 7312 (1975).
[43] C. A. Bunton and M. J. Minch, *Tetrahedron Lett., 44*, 3881 (1970).
[44] C. A. Bunton, M. Minch, and L. Sepulveda, *J. Phys. Chem., 75*, 2707 (1971).
[45] C. A. Bunton, M. J. Minch, J. Hidalgo, and L. Sepulveda, *J. Amer. Chem. Soc., 95*, 3262
     (1973).
[46] J. H. Fendler and E. J. Fendler, *Catalysis in Micellar and Macromolecular Systems,* Aca-
     demic, New York, 1975, p. 88.
[47] S. C. Shah and J. Smid, *J. Amer. Chem. Soc., 100*, 1426 (1978).
[48] A. Veis, in *Biological Polyelectrolytes,* A. Veis, Ed., Chap. 4, Marcel Dekker, New York,
     1970.
[49] A. S. Michaels, L. Mir, and N. S. Schneider, *J. Phys. Chem., 69*, 1447 (1965).
[50] H. Morawetz, *Macromolecules in Solution,* 2nd ed., Wiley, New York, 1975.
[51] J. Pitha and J. Smid, *Biochim. Biophys. Acta, 425*, 287 (1976).

ABIOTIC SYNTHESIS AND THE PROPERTIES OF SOME PROTOBIOCOPOLYMERS

C. I. SIMIONESCU, F. DÉNES, and I. NEGULESCU

"Petru Poni" Institute of Macromolecular Chemistry, Jassy, Romania

SYNOPSIS

The appearance and the properties of protobiocopolymers are critically discussed. Simulating the geologically relevant primeval Earth conditions, i.e., taking into account the early atmosphere, hydrosphere, and available energies, the possibility of simultaneous appearance of some protobiocopolymers is suggested. The self-assembling of these macromolecules in well-defined individual systems (microspheres), their special electrical properties, such as membrane potential and semiconductivity, and their stability in rather severe physicochemical conditions, e.g., pH, temperature, mechanical stress, and irradiation with UV light, bring forward the "protocell" which could arise on this track on our primitive planet.

INTRODUCTION

The results of the investigations on the origin of life carried out in the last two decades throw some light on a considerable number of fundamental problems connected both with the origin of biomolecules and with prebiotic evolution toward the protocell. Supported by relevant experiments simulating the primary conditions on the earth, different models have been proposed as possible ways of individual or simultaneous abiotic emergence of a large number of biologically significant chemical compounds. Starting from various gaseous mixtures thought as primitive atmospheres and using different energies supposed to characterize our planet $4-4.5 \times 10^9$ years ago, and taking into account the geological, hydrographic and climatic conditions of that time, many ways in which small organic molecules such as aldehydes, nitriles, amino acids, monosaccharides, purines, pyrimidines, porphyrins, and nucleotides might have abiotically arisen have now been demonstrated in a large number of laboratories interested in the study of the origin of life [1–62]. Much less are, however, the results reported on the abiotic synthesis of protobiopolymers. The appearance of the last compounds was of the first importance in developing protoenzymatic activities (e.g., hydrolysis, decarboxylation, amination, deamination, and peroxidation) or functions such as fermentation, the simplest, and the more complicated ones based on oxidation processes (e.g., photochemical reactions, photophosphorylation and photosynthesis). Szent-Györgyi divided the history of functions in

Journal of Polymer Science: Polymer Symposium 64, 281–304 (1978)
0360-8905/78/0064-0281$01.00

two widely different periods: the first, the α state, was characterized by a reducing atmosphere and the latter, the β state, by an oxidative atmosphere. What characterized the α state was the poverty of structure, fermentation and unbridled proliferation, while the β state was characterized by the added oxidative energy production and the building of new structures able to perform complex functions [63].

In most cases, however, the abiotic experiments did not overcome the stage of micromolecular biostructures. The existence of water since the early stages of our planet—the primitive ocean—and the requirement of water in any living system constitutes a thermodynamic barrier in formation of protobiopolymers by simple polycondensation reaction with elimination of water. For example, the formation of a peptide bond in aqueous media starting from amino acids requires a free energy as high as 4 kcal/mole:

$$
\overset{+}{H_3N}-\underset{\underset{R_1}{|}}{CH}-COO^- + \overset{+}{H_3N}-\underset{\underset{R_2}{|}}{CH}-COO^- + \overset{+}{H_3N}-\underset{\underset{R_3}{|}}{CH}-COO^- \text{ etc. } \rightleftharpoons
$$
$$
+nH_2O
$$

$$
\overset{+}{H_3N}-\underset{\underset{R_1}{|}}{CH}-CO-NH-\underset{\underset{R_2}{|}}{CH}-CO-NH-\underset{\underset{R_3}{|}}{CH}-CO- \text{ etc.}
$$

eq. 1 $\Delta G° = 2-4$ Kcal/mole   (1)

This thermodynamic barrier would be much greater for the formation of high molecular peptide polymers (proteins) in aqueous solutions. The simplest manner in which this impediment was visualized as surmountable by Fox et al. [64–70] late in the 1950s was by removal of water formed as a byproduct when peptide bonds were synthesized starting from a mixture of amino acids (eq. (1)). The proteinlike polymers of α-amino acids synthesized by Fox, and later by others, under simulated geological conditions, the so-called proteinoids, could contain all the amino acids common to contemporary proteins, have molecular weights of many thousands and possess an array of enzymelike activities [71–88]. On this thermal method of α-amino acids polycondensation, Fox's *thermal theory* of protobiopolymers formation is based. The use of inorganic templates (clays) to provide hypohydrous conditions for the aforesaid reaction has been advanced by A. Katchalsky [89–92] who proved experimentally (with amino acid adenylates) what Bernal [93] stated with regard to concentration of extreme diluted systems" . . . by absorption on very fine clays deposits in marine or fresh water." The hypohydrous conditions being provided by adsorption phenomena, Katchalsky was able to obtain long-chain amino acid polymers and suggested thus another way of the appearance of protobiopolymers (*the adsorption theory* of prebiotic synthesis of polymers). Protobiopolymers have been also synthesized using different chemical anhydrization agents [94–114] considered as primary derivatives, such as hydrogen cyanate (H—C≡NO), cyanogen (N≡C—C≡N), cyanoacetylene (H—C≡C—C≡N), a cyanovinylphosphate (N≡C—CH=CH—OPO$_3{}^{2-}$), and inorganic phosphates. However, the large number of chemical derivatives simultaneously co-existing on the primitive Earth and the high reactivity of anhydrization agents given by their labile electronic structure as well as the low frequence of collisions of three or more molecules are factors that decrease the appearance probability of protobiopolymers (of high molecular weight) in hypohydrous conditions provided by chemical agents. The polycondensation of amino acids in the presence of pyrosulfuric acid should be mentioned [115] in view of the existence of contemporary sulfuric acid pools, of the sulfuric acid-rich atmosphere of Venus, and the possibility of pyrosulfuric acid formation on primitive Earth by decomposition of sulfates.

The convergence of results of geological, cosmic, chemical, physical, and biological investigations marked out the limits of abiogene terrestrial conditions in which about 4 billion years ago the first self-replicable formations appeared. Hydrogen, methane, ammonia, and water are considered as the most probable main components of the primary atmosphere [116–119]. This oxygen-free atmosphere [120–127] gradually lost hydrogen due to the very low specific gravity of this element. The presence of CO and CO_2 in the primary atmosphere was also taken into consideration, but due to the reducing character of the main components, their concentration had to be very low. On the other hand, the assumption has been widely adopted that the following energies were available on the primitive earth: energy of electromagnetic radiations and of high energy radiations, caloric energy, electric energy, and mechanic energy. The average specific values of the most active energies available in the period of $4–4.5 \times 10^9$ years ago are given in Table I.

It should be remarked at this point that the energy types that could act upon different derivatives in the gaseous state (e.g., electric energy, energy of electromagnetic radiations, or high energy radiations) have a common characteristic: they could give rise to a large variety of active species, multifunctional free radicals especially [129–133]. Taking into account the fact that the multi-

TABLE I

Energies Available from Various Sources on the Earth, 4×10^9–4.5×10^9 Years Ago—Hypothetical Values [128]

Type of energy	Total $(cal\ y^{-1})$	Projected per cm^2 surface $(cal\ cm^{-2}\ y^{-1})$
Total optical solar radiation	8.5×10^{23}	170,000
Solar radiation below 2000 Å	1.5×10^{20}	30
High-energy radiation (from the crust, 35 Km)	2.4×10^{20}	47
Electrical discharges	2.0×10^{19}	4
Heat from volcanic emissions (rocks and lava)	7.5×10^{17}	0.15

molecular recombinations are especially peculiar to heterogeneous phases through the agency of adsorption processes, which are favored by low temperatures, the existence on the early earth of optimum conditions for such reactions (e.g., the permanence of cold poles, the sequence of glacial periods and the greenhouse effect) could have been of major importance in formation of certain macromolecules.

Focusing our attention upon the role of low temperature surfaces in the recombination of active species generated in the gaseous phase, we obtained in a large number of experiments carried out in simulated prebiotic conditions (various mixtures of "primary gases," i.e., methane, ammonia, and water vapors, and radiofrequency electric discharges as energy source) different polymeric structures (e.g., polypeptidelike, polysaccharidelike and lipidlike structures), visualizing thus a new way of appearance of protobiopolymers. Our results would try to answer some unsolved problems, for example, the question as to whether the amino acids preceded the proteins or vice versa, or how can one explain the simultaneous formation in similar conditions of the aforementioned polymeric structures.

EXPERIMENTAL

The experiments were carried out in the apparatus shown in Figure 1. Known amounts of methane, ammonia, and water vapors were introduced through needle valves (1–3) in the mixing chamber (4). The obtained mixture was continuously admitted into the spherical part (6-liter Pyrex flask) of the apparatus (7), where the electric discharge was provided by a high frequency generator (5 kW and 13.6 MHz) through the agency of two external silver plated electrodes. The vacuum (2–4 mm Hg) controlled by vacuum gauge (5) was assured by pump (12) protected by a trap (13). The active species formed in cold plasma are adsorbed on the ice previously deposited on the wall and at the bottom of the cylindrical flask (8), cooled −60 to −40°C with ethanol (mantle (9)), and recirculated through a cooling·system (14). The obtained products were characterized by conventional methods, as discussed below.

RESULTS AND DISCUSSION

The simulated abiotic syntheses performed in our laboratory show without doubt the possibility of simultaneous formation on the primitive Earth of different protobiopolymers, with the prevalence of a certain species as a function of the quantitative composition of the same gaseous mixture.

Starting from a gaseous mixture with a high content of $NH_3:CH_4,H_2O$, the raw material (i.e., the material obtained after lyophilization of the melted reaction milieu) is mainly constituted from polypeptidelike structures and linear and cyclic polymers and copolymers of hydrogen cyanide and nitriles [134–136]. The components identified subsequent to the hydolysis of the raw material were amino acids, purines, and pyrimidines, accompanied by neutral and basic ninhydrin positive derivatives of unknown structure. Some analytical data regarding

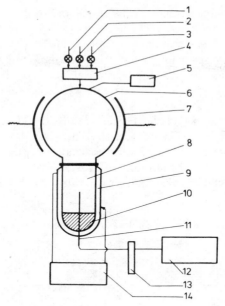

FIG. 1. Diagram of radiofrequency (RF) cold plasma apparatus: (1, 2, 3) vacuum glasscock for feed components; (4) gas mixture chamber; (5) vacuum gauge; (6) spherical upper part of the reactor; (7) external spherical RF electrodes; (8) cylindrical part of the reactor; (9) thermostated mantle; (10) ice; (11) connecting vacuum line; (12) vacuum installation; (13) low temperature (dry ice) trap; (14) low temperature thermostat.

the raw material and its hydrolyzate are given in Tables II and III and in Figures 2–7.

Fractionation of the reaction product (Figs. 2–5) yielded polymeric fractions of molecular weight up to tens of thousands. The determination of polypeptidelike structure (Lowry method, [137]) and the quantitative analysis of free and bonded amino acids (Table II) gave a total content of peptidelike compounds of 11–13 mg/ml of unlyophilized raw material and indicated the prevalence of four amino acids, i.e., glycine, alanine, glutamic and aspartic acids. It should be pointed out that only negligible amounts of free amino acids as compared with the total amount of amino acids (free and bound in peptidelike bonds) were formed. The IR spectrum of the raw material exhibits the main bands characteristic of the CO—NH peptide bond: 3300 and 3080 cm^{-1} (NH and NH. . .H), 1630 and 1240 (CO) and 1575 and 1540 cm^{-1} (NH—CO). The presence of a very intense peak at 3130 cm^{-1} ($+NH_3^+$) and the disappearance of the peaks from 1575 and 1540 cm^{-1} in the ir spectrum of the hydrolyzed material is an additional proof for the existence of peptidelike bonds in the raw material. Right after the thawing of the reaction product (cylindrical flask (8), Fig. 1) an insoluble dark material, designated as the black polymer, starts to precipitate. The IR spectrum of this insoluble fraction is quite different from that of the soluble fraction discussed above. The intense band of 2200 cm^{-1} (C≡N), the general attenuation of the spectrum and the maxima from 3200 cm^{-1} ($+NH_2$) and 1620 cm^{-1} (—C≡N—) suggest that this black polymer is derived from hydrogen cyanide and related derivatives. The basic pH (8.0–9.5) of solutions obtained after the melting of the reaction milieu is favorable for the formation of HCN polymers. The NMR spectrum of the black polymer (Fig. 6) exhibits a very intense signal centered at 3.5 ppm. In presence of NaOD or HD the signal shifted downfield, indicating that the protons are of NH rather than OH nature (Fig. 7).

Four bases of nucleic acids have been identified (one dimensional paper chromatography for preliminary detection combined with UV analysis of spot extracts) in the hydrolyzate of the raw material: adenine, guanine, hypoxanthine, and cytosine (Table III). The tetramer of hydrogen cyanide, diaminomaleodinitrile, could be considered as the starting derivative for closing of the imidazole

TABLE II
Amino Acid Spectrum of Raw and Fractionated Products[a]

Analyzed Product	Concentration, mg/ml								
	Free amino acids				Proteid bound amino acids				
	Serine	Glycine	Alanine	NH$_3$	Aspartic Acid	Serine	Glutamic Acid	Glycine	Alanine
Raw product	0.052	0.540	0.012	2.280	0.070	0.053	0.052	8.000	0.148
Fraction, M$_w$:									
100,000–150,000	–	–	–	0.088	–	–	–	0.007	Trace
70,000–100,000	–	–	–	0.501	0.014	Trace	0.007	1.426	0.027
20,000–40,000	–	–	–	0.060	–	–	–	0.059	–
less than 4,000	–	0.004	Trace	0.372	–	–	–	0.033	–
Black polymer[b]								(0.031)[c]	

[a] Hydrolyzation conditions: HCl 6N, 24 hr in vacuum condition closed neutral glass tube. Experimental conditions: buffer solutions, 5.28 pH sodium citrate 0.35N; 4.25 pH sodium citrate 0.2N; 3.25 pH sodium citrate 0.2N; 1 ml solution of raw product after the separation of insoluble dark-colored compound contains 10.1 mg proteinlike compound.

[b] C. I. Simionescu, I. I. Negulescu, M. I. Totolin and G. Bloos, in *Protein Structure and Evolution*, J. L. Fox, Z. Deyl, and A. Blažej (Eds.), Marcel Dekker, New York and Basel, 1976, Chap. 10, p. 185. Fractionation diagram, Fig. 7.

[c] Concentration in mg/mg of lyophilized raw product.

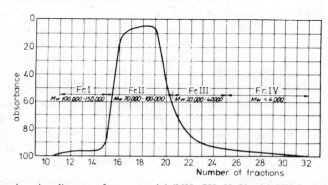

FIG. 2. Fractionation diagram of raw material (NH$_3$:CH$_4$,H$_2$O) with UV detection. Column packing, Sephadex G-100. Column dimensions: height, 520 mm; diameter, 20 mm. UV detection, 280 nm.

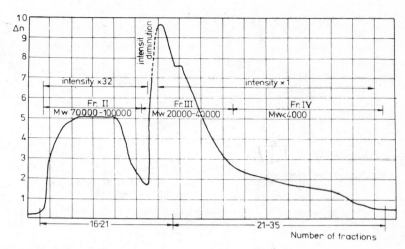

FIG. 3. Fractionation diagram of raw material ($NH_3:CH_4,H_2O$) with interferometric detection.

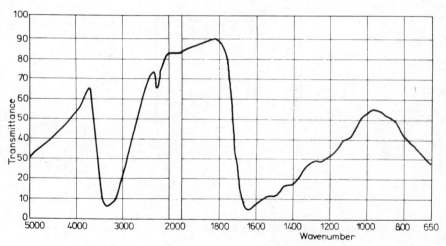

FIG. 4. Infrared spectrum of the black polymer. KBr pellet.

cycle (adenine is the cyclic pentamer of HCN) and for synthesis of purines [138, 139] while cyanoacetylene as a precursor of cytosine [140].

The very low concentration of free amino acids (Table II) should be noted. Free bases of purines and pyrimidines were not identified by paper chromatography in the raw material. At the same time one should note the high degree of transformation (60%) of the raw material into amino acids, suggesting the major peptidelike structure of this material. It is established that the polymers of hydrogen cyanide allow after hydrolysis a recovery of only 5–8% amino acids [141]. The prevalent existence of the aforementioned four amino acids in the hydrolyzates of other products of simulated prebiotic syntheses as well as in

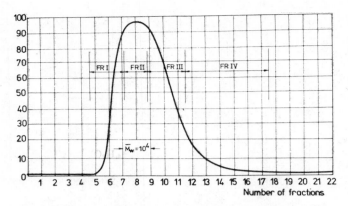

FIG. 5. Fractionation diagram of black polymer-rich raw material (NH_3:CH_4,H_2O) after separation of insoluble material. Column dimensions: height, 476 mm; diameter, 22 mm. Column packing, Sephadex G-100.

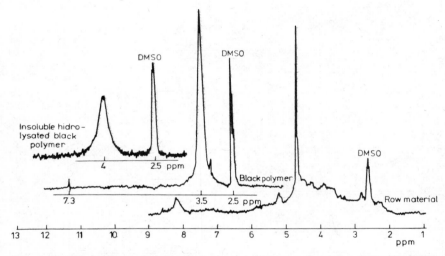

FIG. 6. 60 MHz NMR spectra of products separated from reaction milieu (NH_3:CH_4,H_2O). Solvent, DMSO-d_6; room temperature.

hydrolyzates of organic matter of lunar or meteoric origin (Table IV) points to the formation of protopeptides through the agency of certain statistic structure, regardless of the nature of the energy source.

The appearance of peptidelike structures played a decisive role in the development of the first protoenzymatic activities. The structural and conformational variety of these protobiopolymers assured their adaptation to the environment by first selection processes. It should be pointed out that both the protobiopolymers and the contemporary biopolymers are actually co- and terpolymers. Fox's thermal proteinoids, the contemporary proteins, nucleic acids, etc. are composed of different monomeric units which are released by hydrolysis. Our protobiopolymers are also made up of different units consisting of bonded amino

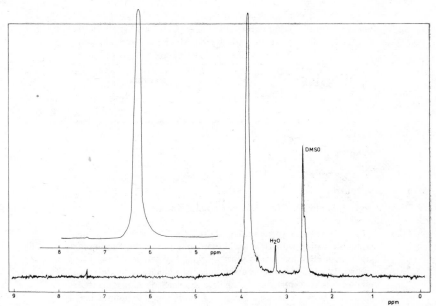

FIG. 7. 60 MHz NMR spectra of black polymer: basic medium (NaOD), bottom; acid medium (DCl), left. Solvent, DMSO-d_6; room temperature.

acids, bases of nucleic acids or saccharides, as will be seen later. However, for the sake of simplicity we will continue to refer to these copolymeric structures as protobiopolymers.

In evolutionary thought, it is well established that the higher organisms evolved from the simpler ones as a result of selection and diversification processes. The investigation of the contemporary simplest unicellular organisms could lead by extrapolation to the understanding of the first living systems. The diversification of the primary functions and the compartmentalization of primitive cells took place in a long evolutive way. The first functions had to be cumulative and not specific. The subsequent specialization of functions should be related to the appearance of certain chemical and morphological structures. At the same time, the first energetic processes (energy being a "sine qua non" condition for the life to start), occurring in primitive cells, had to be based also on electron transfer involving membrane processes. The wall of primitive cells appeared also in the course of adaptation to the environment and perhaps in the early stages the wall and membrane functions were performed by the same structure. The majority of microbial cell walls are fairly robust structures. The wall must possess sufficient tensile strength to protect the cell against osmotic explosion. It seems nature has utilized the polysaccharides as the principal type of structural polymer. It is evident from studies of the chemistry of cell walls that although the walls of a number of microorganisms are predominantly polysaccharide they contain protein and lipid constituents also. Consequently, the identification of polysaccharide structures in the "primary mixture" is important for the understanding of functions and for survival of "membrane-walls." In order to establish

TABLE III

Content of Bases of the Purines and Pyrimidines in the Hydrolyzate of the Reaction Product Obtained from Methane, Ammonia, and Water in Electrical Discharge[a]

Identified base[b]	Concentration in the raw material	
	liophylized mg/mg	unliophylized mmole/ml
Adenine	1.4×10^{-3}	3.0×10^{-5}
Guanine	2.5×10^{-2}	4.8×10^{-4}
Hypoxanthine	1.3×10^{-3}	3.0×10^{-5}
Cytosine	1.0×10^{-3}	7.5×10^{-4}

[a] C. I. Simionescu, M. Totolin, F. Dénes, and I. Negulescu, unpublished data.

[b] Experimental: Hydrolysis, $HClO_4$ 70% for 2 hr at 100°C, followed by neutralization with $10N$ KOH. Chromatography on Whatman 1 paper, double development by descending method with n-butanol:acetic acid:water, 4:1:5, followed by spraying with alcoholic eosine-$HgCl_2$ mixture. The concentration of each base was determined reading the uv absorption between 255 and 265 nm of extraction of determined area (3×4 cm, $0.1N$ HCl for 48 hr) from paper chromatograms of the sample under investigation centered at the R_f value corresponding to the colored spot from preliminary identification.

TABLE IV

Amino Acid Spectra of Hydrolyzed Amino Acid Precursors Obtained in Various Energy Conditions[a]

Electrical discharge[b]	Meteorites (Chondrites)	Lunar sample (Apollo missions)	Terrestrial lava (Exterior of sample heated)	Chemical synthesis (from $CH_2O + NH_3$)
Glycine	Glycine	Glycine	Glycine	Glycine
Alanine	Alanine	Alanine	Alanine	Alanine
Glutamic acid	Glutamic acid	Glutamic acid	Glutamic acid	Glutamic acid
Aspartic acid	Aspartic acid	Aspartic acid	Aspartic acid	Aspartic acid
Serine	Valine	Serine	Serine	Serine
	Proline	Threonine	Threonine	Valine
			Isoleucine	Proline
			Leucine	
			Valine	

[a] S. W. Fox, *Sci. Public Affairs, XXIX* (10), 46 (1973).
[b] Present paper.

the conditions in which polysaccharides might be formed prebiotically, the influence of water concentration in the feed composition (i.e., the ratio H_2O: CH_4, NH_3) upon the composition of the reaction product has been investigated. Consequently, in comparison with the former experiments, larger amounts of water were introduced in the plasma reactor. Therefore the partial pressure of the water vapor in the discharge field was determined by the temperature of the deposited ice plus the flow rate of the additional water vapor (16 ml/hr; reaction time, 15 h) which passed through the discharge during the reaction period. The yellow–brown solution (the final raw material) resulted in a dark colored solid

TABLE V

Elemental Analysis of Raw and Fractionated Products Obtained
with a High Ratio of $H_2O:CH_4,NH_3$

Analyzed sample	N (%)	C (%)	H (%)	O (%)[a]
Raw material	32.33	36.44	5.95	25.28
Fraction 1	0.00	42.12	6.70	51.18
Fraction 2	10.60	30.70	5.31	53.39
Fraction 3	29.10	36.71	5.64	28.59
Fraction 4	14.40	31.00	5.47	49.13
Fraction 5	16.50	30.50	5.60	47.40
Fraction 6	5.03	24.07	4.80	66.10
Fraction 7	0.82	30.90	5.15	63.13
Fraction 8	0.00	b	b	b

[a] Calculated at $100 - (H + C + N)$.
[b] Insufficient amount for analysis.

by vacuum concentration (25°C). Gel filtration chromatography on Sephadex
G-100 of the raw material, together with UV and interferometric detection,
resulted in eight different fractions (experimental conditions: column dimensions,
20×520 mm; eluant, bidistilled water; flow rate, 2 ml/min; fraction volume,
10 ml; UV detection system, UV Absorptiometer LKB 8300 Uvicord II; interferometric detection system, Waters R-4 Detector) (Fig. 8.) A black insoluble
material separated on the top of the column during the fractionation. The elemental analysis data of the two white (1 and 8) and the six yellow (2–7) colored
fractions are presented in Table V. One can notice a different nitrogen content
in the case of the yellow compounds and a total absence of the same element for
fractions 1 and 8. The high nitrogen content of the raw material was due to the
black insoluble product, separated on the top of the gel, identified as HCN
polymer previously (the black polymer). The IR data of the raw material, of the
individual fractions and of their hydrolyzates, as well as the amino acid analysis
of nitrogen-containing fractions, and thin layer and gas chromatography of
nitrogen-free fractions complete the array of methods used for the elucidation

TABLE VI

Amino Acid Spectrum of Raw and Fractionated Products. High Initial Ratio $H_2O:CH_4,NH_3$

Analyzed products[a]	Free amino acids mg/mg raw mater. Glycine	Bound amino acids, mg/mg raw material Aspartic acid	Glutamic acid	Glycine	Alanine
Raw material	4.1×10^{-3}	3.7×10^{-3}	3.5×10^{-3}	2.9×10^{-1}	6.7×10^{-3}
Fraction 2	–	1.9×10^{-4}	5.0×10^{-4}	3.4×10^{-2}	2.8×10^{-4}
Fraction 3	–	1.3×10^{-4}	2.2×10^{-4}	1.6×10^{-2}	1.5×10^{-4}
Fraction 4	–	8.5×10^{-5}	2.0×10^{-3}	3.7×10^{-3}	5.1×10^{-5}
Fraction 5	–	7.6×10^{-5}	8.7×10^{-5}	3.2×10^{-3}	Trace
Fraction 6	–	Trace	4.7×10^{-4}	1.6×10^{-3}	0,0
Fraction 7	7.4×10^{-5}	Trace	Trace	1.0×10^{-3}	0,0

[a] For experimental conditions, see Table II.

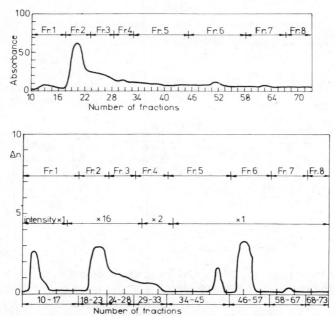

FIG. 8. Fractionation diagram of raw material ($H_2O:CH_4,NH_3$): with UV detection, top; with interferometric detection, bottom.

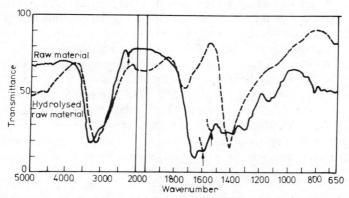

FIG. 9. Infrared spectrum of the raw material ($H_2O:CH_4,NH_3$) and of its hydrolyzate. KBr pellet.

of the "mosaic structure" of the reaction product. The Lowry method of investigation of the raw material resulted a 0.2 mg/mg total proteinlike compound content. The free and bound amino acid composition is presented in Table VI. Glycine, alanine, glutamic and aspartic acids are the main units in the nitrogen-containing fractions. One can notice at the same time a very low content of free amino acids. IR analysis of the samples points to the presence of peptidelike linkages for the nitrogen-containing fractions as well as the raw material and the absence of these bonds in the case of fractions 1 and 8. Thus, the IR spectrum of the raw material shows all of the peptide absorption discussed in the case of the major ratio NH_3. CH_4,H_2O, i.e., 3300, 3080, 1630, 1575, 1540, and 1240 cm^{-1} (amide absorptions). Vibrations at 1013, 1128, 1162, and 2220 cm^{-1} are present also. The last one, peculiar to the C≡N group, is absent in the hydrolyzate. Again the appearance of a strong peak at 3130 cm^{-1} (NH_3^+) and the absence, of 1575 and 1540 cm^{-1} vibrations (CO—NH) in the hydrolyzate spectrum supports the presence of the peptide bonds in the raw material (Fig. 9). The IR spectrum of the nitrogen-free fractions (1 and 8) presents the characteristics of mono and polysaccharidelike compounds (Fig. 10): 3430 cm^{-1} (OH), 1162 cm^{-1} (C—O—C), 1120 cm^{-1} (pyranose ring vibration) and 1013 (C—O from C—O—H). Absorptions characteristic of polypeptide structures are also present in the ir spectra of nitrogen-containing fractions. Because of the similar spectrum patterns only the diagrams of the 2–4 samples are presented (Fig. 11). Thin layer chromatography data on monosaccharide content of the hydrolyzates of fractions 1 and 8 are shown in Figure 12. In each case an unidentified low R_f value spot has been noticed, probably corresponding to unhy-

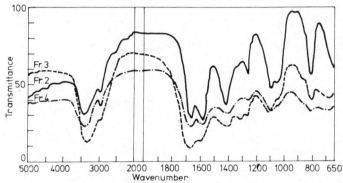

FIG. 10. Infrared spectra of fractions 2–4 separated from (H_2O:CH_4,NH_3) raw material. KBr pellet.

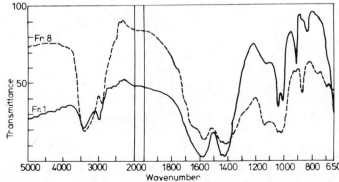

FIG. 11. Infrared spectra of fractions 1 and 8 separated from (H_2O:CH_4,NH_3) raw material. KBr pellet.

drolyzed polymeric material. The GC data of the same hydrolyzates are in good agreement with the thin layer chromatography results (Fig. 13). Two saccharides were evidenced, namely galactose and xylose. This experiment confirms that working in simulated primeval Earth conditions, saccharidelike compounds can be synthesized [141]. The water content in the feed composition is essential and required in large quantity for formation of saccharide structure. An equally important observation is that these compounds were formed simultaneously with the polypeptidelike structures by the reaction of active species, generated in gas phase cold plasma conditions, on the low temperature ice surface. The present finding provides support for the view that both the functional and structural protobiopolymers could appear at the very beginnings under similar geological conditions. The variety of protobiopolymers and their copolymer nature assured

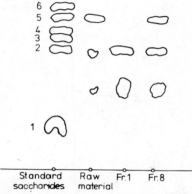

FIG. 12. Thin layer chromatography of the raw material ($H_2O:CH_4,NH_3$) and nitrogen-free fractions: (1) glucuronic acid; (2) galactose; (3) glucose; (4) arabinose; (5) xylose; (6) ribose.

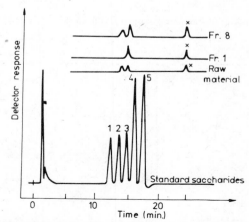

FIG. 13. Gas chromatograms of silylated standard saccharides, silylated raw material (H_2O: CH_4,NH_3) and fractions 1 and 8. Column packing, 3% SE 30; column dimensions: length, 2500 mm; diameter, 3 mm. Column temperature, 150°C. Flow rate, 11.75 ml/min. Detection with ionization flame. (1 and 2) xylose; (3) gamma galactose; (4) alpha galactose; (5) beta galactose.

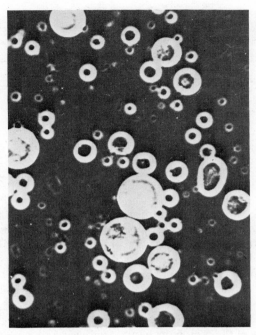

FIG. 14. Photomicrograph of microspheres formed by self-assembly of hydrophobic lipidlike structure-rich raw material ($CH_4:NH_3H_2O$); magnification, ×600.

the polyfunctionality required by the "adaptation to the environment" and consequently their accumulation. On the other hand, the random distribution of biologically interesting micro- and macromolecular compounds in "primary soup" could not satisfy the requirements for the energy accumulation and transfer. The necessity of a "minimal order" accomplished by the rearrangement of functional products in well-defined individual units in order to set the energy flux (mechanic, electric) marked one of the most important events of the evolution toward the living matter: self-assembly of protobiopolymers in individual super-structures.

The first results in this domain have been reported by S. W. Fox early in the 1960s [143–147]. The self-assembly of polypeptidelike structures, i.e., Fox's thermal poly(α-amino acids) obtained by heating anhydrous mixtures of α-amino acids, paved the way for the understanding of the protocell appearance, showing at the same time the possibility of creation of individual units with functions characteristic of cellular systems. It should be pointed out that the aspartic and glutamic acid-rich proteinoids form microspheres (2–8 μ) which are stable in acidic media and exhibit membranelike properties, e.g., selectivity in retention of certain chemical compounds, enzymelike activities [127, 148–151]. Furthermore, excitabilities were observed in microspheric membranes assembled from synthetic thermal proteinoids, especially a leucine-rich variety, and lecithin [152].

However the thermal theory does not explain the appearance of hydrophobic

TABLE VII

Elementary Analysis of the Raw Material Obtained in Electrical
Discharge from Various Methane, Ammonia, and Water Content in the
Initial Gaseous Mixture

Main component in the initial gaseous mixture	N (%)	C (%)	H (%)	O (%)[a]
NH_3 ($NH_3:CH_4,H_2O$)	32.91	44.96	7.02	15.11
H_2O ($H_2O:CH_4,NH_3$)	32.33	36.44	5.95	25.28
CH_4 ($CH_4:NH_3,H_2O$)	13.89	53.33	5.39	27.39

[a] Calculated as $100 - (H + C + N)$.

lipid structures as well as structural protobiopolymers. At the same time the stability of these microsystems (i.e., thermal proteinoid microspheres) in acidic media only, comes in conflict with the proposed slight alkalinity (pH 8–9) of the primitive ocean [153]. The formation of certain derivatives with hydrocarbon backbones of considerable length having polar ends is of particular interest in the appearance of the protomembrane, and consequently in the initiation of the first osmotic and enzymatic processes.

The spontaneous arising of a microsphere population (10–50 μ) as a result of self-assembly of the protobiopolymers synthesized in our laboratory, as it is shown in Figure 14, as well as the increase of microsphere number with feed ratio $CH_4:NH_3,H_2O$, are factors that plead for a possible appearance of certain well-defined individual structures in this way on the primitive Earth.

The comparative analysis of the raw material obtained at various feed compositions, i.e., various ratios of methane, ammonia, and water, shows that the amount of (hydrophobic) hydrocarbon components in the reaction milieu increased with the increase of the methane concentration in the initial gaseous mixture. The partial solubility of the (water-insoluble) raw material obtained at a high $CH_4:NH_3,H_2O$ ratio in the solvents of lipid structures, e.g., chloroform and detergent solution, and the analyses (elementary analysis, IR, and NMR) of the chloroform-soluble part emphasize the existence of some hydrocarbon structures decisive for hydrophobicity which is in turn required for "protomembrane" formation.

Elementary analysis of the raw material synthesized at various feed compositions, where ammonia, methane, and water were alternately the main component, is listed in Table VII.

Polypeptidelike (N:O ratio larger than unity in the material) and polysaccharidelike (O:N ratio larger than unity in the raw material) materials have been preferentially formed when ammonia and water were the main components of the feed composition successively, versus predominant hydrophobic lipidlike structures (high carbon content raw material) obtained in the case of the methane-rich feed composition.

The IR spectra of these last structures exhibit the absorptions characteristic of CH, CH_2 and CH_3 (2800–2920 cm^{-1}) groups, having at the same time the bands of OH (3800 cm^{-1}), CN (2200 cm^{-1}), CO (1720 cm^{-1}) and C—O—C (1000–1100 cm^{-1}) groups, as seen in Figure 15. NMR spectrum of the chloroform-soluble part (Fig. 16), in good agreement with ir data, contains between 0.5 and 2 ppm the signals of alkyl protons and between 2 and 3 ppm the signals of protons from CH$_3$, CH$_2$ and perhaps CH groups bound to —CHO and —COOH, the amino (imino) protons (3.5–4.5 ppm) being much less represented in comparison with NMR spectrum of amino acid precursor-rich raw material (Fig. 6). The formation of the lipidlike structure is important for the development of hydrophobicity, an essential property in the view of preparing the conditions for differential accumulation of the energy, its transformation, transmission, and cyclic reaccumulation through the agency of an abiotic well defined individual system, the primitive model of which is the microsphere [135].

The microspheres formed by self-assembly of our protobiopolymers are not affected in a large pH range (6–9) by rather severe thermal (up to 75°C) and mechanical (the microspheres can be separated by centrifugation at 15,000 rpm) stress and though they disappear during UV illumination, they reassemble after cessation of the irradiation (Fig. 17).

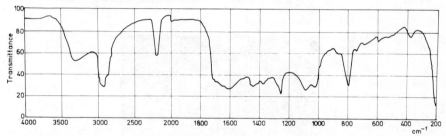

FIG. 15. Infrared spectrum of the chloroform-soluble raw material fraction (CH$_4$:NH$_3$,H$_2$O). KBr pellet.

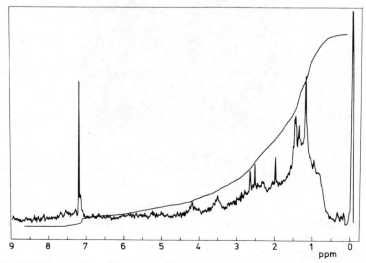

FIG. 16. 60 MHz NMR spectrum of the chloroform-soluble raw material fraction (CH$_4$: NH$_3$,H$_2$O). Solvent, CDCl$_3$; room temperature.

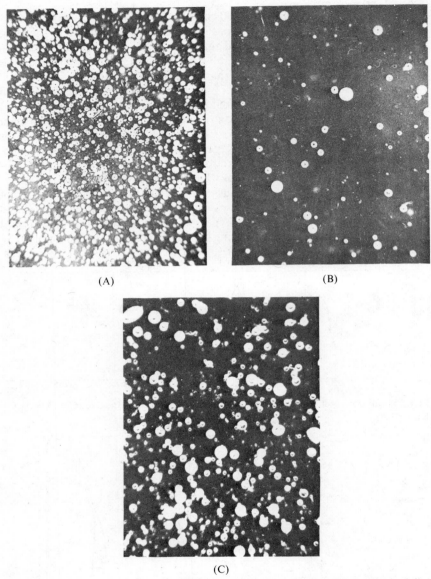

(A)   (B)

(C)

FIG. 17. Photomicrographs of self-assembled protobiopolymers (microspheres obtained from methane-rich feed mixture): (A) before UV irradiation; (B) after 2 min of UV irradiation (200 W mercury lamp placed at 10 cm from the sample); (C) 15 min after cessation of UV irradiation (×150).

In the early 1940s Szent-Györgyi [154] suggested that electron transport between the insoluble oxidation enzymes occurred through an energy band, as in semiconductors (eq. (2)), passing

$$\sigma = \sigma_0 e^{(-\Delta E / 2kT)} \text{ ohm}^{-1} \text{ cm}^{-1} \qquad (2)$$

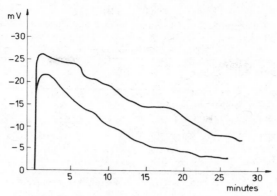

FIG. 18. Membrane potential of microspheres.

through the fibrous proteins of the supporting membranes. Later on, Evans and Gergely [155], using a simple LCAO-MO theory, showed that the extended π-electron system of the transverse C=O...NH hydrogen bonds in the protein structure would give rise to three narrow energy bands, with a gap of 3.2 eV between the second filled band and the upper third unfilled band. The wall structure of the "protocell" model (i.e., microsphere) synthesized in our laboratory, the special electrical properties manifested as shown in Figure 18 by an electrical potential (measured as shown in Fig. 19) between the two phases, i.e., in and out the microsphere, separated by "protomembrane," and the value of the electrical conductivity, σ, and of activation energy, ΔE (Fig. 20, Table VIII) suggest the possibility of an electron transfer process through semiconduction, even at an early evolutionary stage, bridging consequently the two stages, α and β, proposed by Szent-Györgyi [63]. On the other hand, the stability of microspheres suggests some highly organized aggregates could "survive" on the primitive Earth [156].

These properties of self-assembled protobiocopolymers prefigure the coexistence of various energy transfers, and consequently of performing various

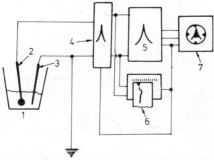

FIG. 19. Diagram of the installation used for measurement of the membrane potential of microspheres: (1) perpex chamber; (2) glass microelectrode (3 M KCl, 15–25 Mohm); (3) reference electrode (AgCl); (4) cathodic follower with MF filter; (5) dc preamplifier; (6) potentiometric recorder; (7) oscilloscope with camera.

TABLE VIII

Comparative Data Regarding the Electrical Conductivity at
318 K and the Activation Energy of the Electrical Conductivity
of Some Protobiopolymers

Protobiopolymer	σ, ohm^{-1}cm^{-1}	ΔE, eV
Protobiopolymer synthesized at a high NH_3:CH_4,H_2O ratio	1.95×10^{-8}	1.09
Thermal proteinoid[a]	9.54×10^{-13}	0.79

[a] Proteinoid synthesized from α-amino acids in presence of fuming H_2SO_4 [115].

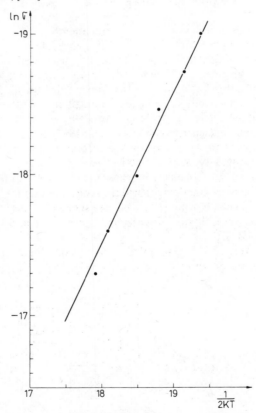

FIG. 20. Dependence of electrical conductivity of raw material (NH_3: CH_4,H_2O) versus temperature (metallized pellet).

work (mechanical, electrical and osmotic), required in all living systems, representing thus a new step in the understanding of the evolution toward the biomatter.

REFERENCES

[1] S. L. Miller, *Science, 117*, 528 (1953).

[2] S. L. Miller, *J. Am. Chem. Soc., 77*, 2351 (1955).

[3] S. L. Miller, *Ann. N.Y. Acad. Sci., 69*, 260 (1957).

[4] P. H. Abelson, *Science, 124*, 935 (1956).

[5] P. H. Abelson, *Ann. N.Y. Sci., 69*, 274 (1957).

[6] J. Oró, *Nature, 197*, 971 (1963).

[7] G. Ponnamperuma and J. Flores, Abstr. 152nd Nat. Meeting Am. Chem. Soc., N.Y. (1966).

[8] K. Harada and S. W. Fox, *Nature, 201*, 335 (1964).

[9] W. Groth, *Angew. Chem., 69*, 681 (1957).

[10] W. Groth and H. V. Weyssenhoff, *Naturwissenschaften, 44*, 510 (1957).

[11] K. Dose and B. Rajewsky, *Biochim. Biophys. Acta, 25*, 225 (1957).

[12] M. Calvin, *Perspectives Biol. Med., 5*, 147 (1962).

[13] C. Ponnamperuma and R. Mack, *Science, 148*, 1221 (1965).

[14] M. Taube, St. Z. Zdrojewski, K. Samochocka, and K. Jezierska, *Angew. Chem., 79*, 239 (1967).

[15] N. Friedman, W. J. Haverland and S. L. Miller, in *Chemical Evolution and the Origin of Life*, R. Buvet and C. Ponnamperuma, Eds., North Holland, Amsterdam, 1971, p. 123.

[16] B. Franck, *Chem. Ber., 93*, 446 (1960).

[17] K. Bahadur, *Nature, 173*, 1141 (1954).

[18] G. Steinman, A. E. Smith, and J. J. Silver, *Science, 159*, 1108 (1968).

[19] A. R. Deschreider, *Nature, 182*, 528 (1958).

[20] T. Hasselstrom, M. C. Henry, and B. Murr, *Science, 125*, 350 (1957).

[21] K. Dose and K. Ettre, *Z. Naturforsch., 136*, 784 (1958).

[22] R. Paschke, R. W. H. Chang, and D. Young, *Science, 125*, 881 (1957).

[23] K. Dose and C. Ponnamperuma, *Radiation Res., 31*, 650 (1967).

[24] S. W. Fox, J. E. Johnson, and M. Middlebrook, *J. Am. Chem. Soc., 77*, 1048 (1955).

[25] J. Oro and S. W. Kamat, *Nature, 190*, 442, (1961).

[26] S. W. Fox and C. R. Windsor, *Science, 170*, 984 (1970).

[27] C. Sagan and B. N. Khare, *Nature, 232*, 577 (1971).

[28] A. Bar-Nun, N. Bar-Nun, S. H. Bauer, and C. Sagan, *Science, 168*, 470 (1970).

[29] C. U. Lowe, M. W. Rees, and R. Markham, *Nature, 199*, 219 (1963).

[30] C. N. Mathews and R. E. Moser, *Nature, 215*, 1230 (1967).

[31] R. A. Sanchez, J. P. Ferris, and L. E. Orgel, *Science, 154*, 784 (1966).

[32] N. Friedmann and S. L. Miller, *Science, 166*, 766 (1969).

[33] J. E. Van Trump and S. L. Miller, *Science, 178*, 859 (1972).

[34] C. I. Simionescu, N. Asandei, and F. Dénes, *Compt. Rend. Acad. Sci. Paris, 213*, 599 (1971).

[35] C. I. Simionescu, F. Dénes, and M. Macoveanu, *Dokl. Akad. Nauk, 207*, 469 (1972).

[36] C. I. Simionescu, F. Dénes, and M. Macoveanu, *Biopolymers, 12*, 237 (1973).

[37] C. I. Simionescu, F. Dénes, and E. Schneider, *Compt. Rend. Acad. Sci. Paris, 275*, 703 (1972).

[38] W. V. Allen and C. Ponnamperuma, *Currents in Mod. Biol., 1*, 24 (1967).

[39] W. M. Garrison, D. C. Morrison, J. G. Hamilton, A. A. Benson, and M. Calvin, *Science, 114*, 416 (1951).

[40] K. Dose and S. Risi, *Z. Naturforsch., 23b*, 981 (1968).

[41] J. Oró and J. Han, *Science, 153*, 1393 (1966).

[42] J. Oró, *Ann. N.Y. Acad. Sci., 108*, 464 (1963).

[43] R. A. Sanchez, J. P. Ferris, and L. E. Orgel, *J. Mol. Biol., 38*, 11 (1968).

[44] J. P. Ferris, R. A. Sanchez, and L. E. Orgel, *J. Mol. Biol., 33*, 693 (1968).

[45] J. Oró and A. P. Kimball, *Arch. Biochim. Biophys., 94*, 217 (1961).

[46] J. Oró and A. P. Kimball, *Arch. Biochim. Biophys., 96*, 293 (1962).

[47] M. Calvin, *Chem. Eng. News, 39*, 96 (1961).

[48] R. M. Kliss and C. N. Matthews, *Proc. Nat. Acad. Sci., 48*, 1300 (1962).

[49] S. W. Fox and K. Harada, *Science, 133*, 1923 (1961).
[50] C. Ponnamperuma, R. Mariner, and C. Sagan, *Nature, 198*, 1199 (1963).
[51] C. Ponnamperuma, C. Sagan, and R. Mariner, *Nature, 199*, 222 (1963).
[52] R. Mayer, K. Runge, and K. Dreschsel, *Z. für Chemie, 3*, 134 (1963).
[53] N. W. Gabel and C. Ponnamperuma, *Nature, 216*, 453 (1967).
[54] C. Reid and L. E. Orgel, *Nature, 216*, 455 (1967).
[55] C. Ponnamperuma, in *The Origins of Prebiological Systems*, S. W. Fox, Ed., Academic, New York, 1965.
[56] J. Oró and A. C. Cox, *Fed. Proc., 21*, 80 (1962).
[57] G. W. Hodgson and B. L. Baker, *Nature, 216*, 29 (1967).
[58] G. W. Hodgson and C. Ponnamperuma, *Proc. Nat. Acad. Sci., 59*, 22 (1968).
[59] A. Szutka, *Nature, 212*, 401 (1966).
[60] R. A. Sanchez and L. E. Orgel, *J. Mol. Biol., 47*, 531 (1970).
[61] W. D. Fuller, R. A. Sanchez, and L. E. Orgel, *J. Mol. Biol., 67*, 25 (1972).
[62] W. D. Fuller, R. A. Sanchez, and L. E. Orgel, *J. Mol. Evolution, 1*, 249 (1972).
[63] A. Szent-Györgyi, *Life Sciences, 15*, 863 (1974).
[64] S. W. Fox, *Am. Sci., 44*, 347 (1956).
[65] S. W. Fox, A. Vegotsky, K. Harada, and P. D. Hoogland, *Ann. N.Y. Acad. Sci., 69*, 328 (1957).
[66] S. W. Fox and K. Harada, *J. Am. Chem. Soc., 82*, 3745 (1960).
[67] S. W. Fox and K. Harada, *Arch. Biochem. Biophys., 86*, 281 (1960).
[68] S. W. Fox, K. Harada, K. R. Woods, and C. R. Windsor, *Arch. Biochem. Biophys., 102*, 439 (1963).
[69] S. W. Fox and T. Nakashima, *Biochim. Biophys. Acta, 140*, 155 (1967).
[70] S. W. Fox and T. V. Waehneldt, *Biochim. Biophys. Acta, 160*, 246 (1968).
[71] J. E. Germain, P. A. Finot, and G. Biserte, *Bull. Soc. Chim. Biol., 45*, 40 (1963).
[72] T. Oshima, *Arch. Biochem. Biophys., 126*, 478 (1968).
[73] H. G. Hardbeck, G. Krampitz, and L. Wulf, *Arch. Biochem. Biophys., 132*, 72 (1968).
[74] F. Hare, *J. Chim. Phys., 68*, 330 (1971).
[75] G. Hennon, R. Plaquet, M. Doutrenaux, and G. Biserte, *Biochemie, 53*, 215 (1971).
[76] M. A. Dixon and E. C. Webb, *Enzymes,* Academic, New York, 1958, p. 666.
[77] K. Dose and L. Zaki, *Z. Naturforsch., 26 b*, 144 (1971).
[78] S. W. Fox and K. Harada, *J. Am. Chem. Soc., 82*, 3745 (1960).
[79] S. W. Fox and D. Joseph, in *The Origin of Prebiological Systems*, S. W. Fox, Ed., Academic, New York, 1965, p. 371.
[80] S. W. Fox and C. T. Wang, *Science, 160*, 547 (1968).
[81] G. Krampitz, D. S. Diehl, and T. Nakashima, *Naturwissenschaften, 54*, 516 (1967).
[82] G. Krampitz and S. W. Fox, *Proc. Nat. Acad. Sci., 62*, 399 (1969).
[83] G. Krampitz and H. Hardbeck, *Naturwissenschaften, 53*, 81 (1966).
[84] D. L. Rohlfing, *Nature, 216*, 657 (1967).
[85] D. L. Rohlfing, *Arch. Biochim. Biophys., 118*, 468 (1967).
[86] D. L. Rohlfing, *Science, 169*, 998 (1970).
[87] V. R. Usdin, M. A. Mitz, and P. J. Killos, *Arch. Biochim. Biophys., 122*, 258 (1967).
[88] D. L. Rohlfing and S. W. Fox, *Advan. Catalysis, 20*, 373 (1969).
[89] M. Paecht-Horowitz, J. Berger, and A. Katchalsky, *Nature, 228*, 636 (1970).
[90] M. Paecht-Horowitz and A. Katchalsky, *Biochim. Biophys. Acta, 140*, 14 (1967).
[91] A. Katchalsky and G. Ailam, *Biochim. Biophys. Acta, 140*, 1 (1967).
[92] A. Katchalsky, *Naturwissenschaften, 60*, 215 (1973).
[93] J. D. Bernal, in *The Physical Basis of Life*, J. D. Bernal, Ed., Routlege and Kegan Paul, London, 1951.
[94] C. Ponnamperuma and E. Peterson, *Science, 147*, 1572 (1965).
[95] G. D. Steinman, R. M. Lemmon, and M. Calvin, *Proc. Nat. Acad. Sci., 52*, 27 (1964).
[96] J. D. Ibanez, A. P. Kimball, and J. Oró, *Science, 173*, 444 (1971).
[97] R. Lohrmann and L. E. Orgel, *Science, 161*, 64 (1968).
[98] G. D. Steinman, D. H. Kenyon, and M. Calvin, *Nature, 206*, 707 (1965).
[99] H. Halman, R. A. Sanchez, and L. E. Orgel, *J. Org. Chem. 34*, 3702 (1969).

[100] G. D. Steinmann, R. M. Lemmon, and M. Calvin, *Biochim. Biophys. Acta, 124*, 339 (1966).
[101] G. D. Steinmann, R. M. Lemmon, and M. Calvin, *Science, 147*, 1574 (1965).
[102] S. L. Miller and M. Paris, *Nature, 204*, 1248 (1964).
[103] J. P. Ferris, *Science, 161*, 53 (1968).
[104] T. V. Waenheldt and S. W. Fox, *Biochim. Biophys. Acta, 134*, 1 (1967).
[105] A. Schwartz and S. W. Fox, *Biochim. Biophys. Acta, 87*, 696 (1964).
[106] A. Schwartz and S. W. Fox, *Biochim. Biophys. Acta, 134*, 9 (1967).
[107] J. P. Ferris, G. Goldstein, and D. J. Beaulieu, *J. Am. Chem. Soc., 92*, 6598 (1970).
[108] S. Chang, J. Flores, and C. Ponnamperuma, *Proc. Nat. Acad. Sci., 64*, 1011 (1969).
[109] J. Rabinowitz, S. Chang, and C. Ponnamperuma, *Nature, 218*, 442 (1968).
[110] A. Schwartz and C. Ponnamperuma, *Nature, 218*, 443 (1968).
[111] W. Feldmann, *Z. für Chem., 9*, 154 (1969).
[112] N. Chung, R. Lohrmann, L. E. Orgel, and J. Rabinowitz, *Tetrahedron, 27*, 1205 (1971).
[113] J. Rabinowitz, *Helv. Chim. Acta, 53*, 1350 (1970).
[114] J. Rabinowitz, J. Flores, R. Kerbsbach, and G. Rogers, *Nature, 224*, 795 (1969).
[115] F. Dénes and S. W. Fox, *Biosystems, 8*, 83 (1976); S. W. Fox, F. Dènes, J. Pollack, R. Pethig and T. R. Rodewald, *Science*, in press.
[116] H. N. Russel, *Science, 81*, 1 (1935).
[117] A. I. Oparin, in *The Origin of Life*, A. I. Oparin, Ed., Macmillan, New York, 1938.
[118] H. C. Urey, in *The Planets*, H. C. Urey, Ed., Yale University Press, New Haven, 1952.
[119] S. L. Miller and H. C. Urey, *Science, 130*, 245 (1959).
[120] D. H. Holland, *Proc. Nat. Acad. Sci., 53*, 1173 (1965).
[121] D. H. Holland, *Geochim. Cosmochim. Acta, 36*, 637 (1972).
[122] H. Lepp and S. S. Goldich, *Geol. Soc. Amer. Bull., 70*, 1637 (1959).
[123] P. E. Cloud, Jr., *Science, 160*, 729 (1968).
[124] M. G. Rutten, *Space Life Science, 1*, 1 (1970).
[125] L. Van Valen, *Science, 171*, 439 (1971).
[126] S. L. Miller and L. E. Orgel, in *The Origin of Life on the Earth*, S. L. Miller and L. E. Orgel, Eds., Englewood Cliffs, Prentice-Hall Inc., 1974.
[127] S. W. Fox and K. Dose, in *Molecular Evolutuion and the Origin of Life*, Fox and Dose, Eds., W. H. Freeman, San Francisco, 1972.
[128] F. K. McTaggart, in *Plasma Chemistry in Electrical Discharges*, Taggart, Ed., Elsevier, Amsterdam, 1969.
[129] S. S. Tseng and S. Chang, *Origins of Life, 6*, 61 (1975).
[130] J. R. Hollahan and A. T. Bell, in *Techniques and Applications of Plasma Chemistry*, Hollahan and Bell, Eds., Wiley, New York, 1974.
[131] M. Venugopalan, in *Reactions under Plasma Conditions, Vols. 1 and 2*, Venugopalan, Ed., Wiley-Interscience, New York, 1971.
[132] B. D. Blaustein, in *Chemical Reactions in Electrical Discharges*, Blaustein, Ed., ACS Publication, Washington D.C., 1969.
[133] D. K. Lam and R. F. Baddour, *J. Macromol. Sci. Chem., A 10 (3)*, 383 (1976).
[134] C. I. Simionescu, F. Dénes, D. Onac, and G. Bloos, *Biopolymers, 13*, 943 (1974).
[135] C. I. Simionescu, F. Dénes, and M. Dragnea, *Comt. Rend. Acad. Sci. Paris, 278*, 29 (1974).
[136] C. I. Simionescu, I. Negulescu, M. Totlin, and G. Bloos, in *Protein Structure and Evolution*, J. L. Fox, Z. Deyl and A. Blažej, Eds., Marcel Dekker Inc., New York, 1976, p. 185.
[137] O. H. Lowry, N. F. Rosenbrough, A. L. Farr, and R. F. Randall, *J. Biol. Chem., 193*, 265 (1951).
[138] J. P. Ferris and L. E. Orgel, *J. Am. Chem. Soc., 88*, 1074 (1966).
[139] R. A. Sanchez, J. P. Ferris, and L. E. Orgel, *Science, 147*, 149 (1966).
[140] R. A. Sanchez, J. P. Ferris, and L. E. Orgel, *Science, 153*, 72 (1966).
[141] S. W. Fox, in *Protein Structure and Evolution*, J. L. Fox, Z. Deyl and A. Blažěj, Eds., Marcel Dekker Inc., New York, 1976, p. 198.
[142] C. I. Simionescu, M. Totlin, and F. Dénes, *Biosystems, 8*, 153 (1976).
[143] S. W. Fox, K. Harada, and J. Kendrich, *Science, 129*, 1221 (1959).

[144] S. W. Fox, *Science, 132*, 200 (1960).
[145] S. W. Fox and S. Yuyama, *Comp. Biochem. Physiol., 11*, 317 (1964).
[146] S. W. Fox, R. J. McCouley, and A. Wood, *Comp. Biochem. Physiol., 20*, 773 (1967).
[147] S. W. Fox and S. Yuyama, *Ann. N.Y. Acad. Sci., 108*, 487 (1963).
[148] R. S. Young, in *The Origins of Prebiological Systems,* S. W. Fox, Ed., Academic, New York, 1965, p. 347.
[149] S. W. Fox, R. J. McCauley, P. O'B. Montgomery, T. Fukushima, K. Harada, and C. R. Windsor, in *Physical Principles of Biological Membranes,* F. Snell, J. Wolken, G. Iverson and J. Lam, Eds., Gordon and Breach, New York, 1969, p. 417.
[150] L. Hsu, S. Brooke, and S. W. Fox, *Currents Mod. Biol., 4*, 12 (1971).
[151] S. W. Fox, *Nature, 205*, 328 (1965).
[152] Y. Ishima and S. W. Fox, in *Abstracts—Neuroscience Meeting*, San Diego, California, 1973.
[153] P. H. Abelson, *Proc. Nat. Acad. Sci., 55*, 1365 (1966).
[154] A. Szent-Györgyi, *Science, 93*, 609 (1941).
[155] M. G. Evans and J. Gergely, *Biochim. Biophys. Acta, 3*, 188 (1949).
[156] C. I. Simionescu, F. Dénes, I. Negulescu, and M. Totolin, in *Abstracts—International Symposium on Macromolecules*, Stockholm, 1976, Chap. V, p. 1.

CONSIDERATIONS OF SOME COPOLYMERS OBTAINED BY RADICAL MECHANISMS

CRISTOFOR I. SIMIONESCU

Polytechnic Institute of Jassy, Romania

SYNOPSIS

The modification of the chemical character of certain macromolecular products by radical copolymerization and grafting reactions with the object of improving some properties, constitutes a field of continuous and current debate about the correlation between synthesis, outcoming structure, and resulting properties of the final products. In this respect some results have been briefly presented, referring to this dependence in the case of multicomponent copolymers (comprising three to four components) and of grafted copolymers; in the same respect the behavior of some biocopolymers (of the alginate type) has been correlated with their structure deduced by the application of certain principles of classical copolymerization to the natural products under study.

MULTICOMPONENT COPOLYMERS. ELEMENTS OF CHEMICAL STRUCTURE DETERMINING THE BEHAVIOR OF THE FINAL PRODUCTS

The worldwide development and diversification of plastic materials and synthetic fiber production contribute an important part to multicomponent copolymerization. The fact that systems consisting of many monomers, chosen according to well established criteria, are likely to ensure the envisaged set of properties by means of the introduced functional groups accounts for the above mentioned part played by multicomponent copolymers.

The need for the practical achievement of certain copolymers having the properties required by the specific field of use called, as a matter of course, for the concentration of research on the possibilities of knowing and perfectly handling the initial conditions apt to result in the realization of certain directed structures, having a direct influence upon the behavior of the final product.

The Composition of Multicomponent Polymers

The possibility of predicting by calculation the composition of copolymers represents an essential theoretical problem with important practical applications. Its solution leads to the achievement of copolymers with the envisaged chemical composition and forecasts the possibility of determining the initial monomer composition according to criteria of chemical homogeneity, assuring a strict check on the quality of the copolymers.

Journal of Polymer Science: Polymer Symposium 64, 305-327 (1978)
© 1978 John Wiley & Sons, Inc.
0360-8905/78/0064-0305$01.00

Concerning composition and its variation with conversion, there are to be found in the literature many data regarding binary copolymerization, but fewer data referring to the ternary type, and only some publications concerned with systems comprising more than three components. This situation is accounted for by the experimental and also computational difficulties which increase in parallel with the number of monomers within the system.

One of the objects of our research has been a simple computation scheme likely to permit the numerical integration of differential composition equations for multi-component systems, capable of rapid solution, a condition imposed by laboratory and industrial practice.

Using the simplified forms of the differential compositional equations for. multicomponent systems [1, 2], the following type of equation has been suggested [3]:

$$[m_l] = [M_1] \frac{r_{l1}}{r_{1l}} \sum_{i=1}^{n} [M_i] r_{li}^{-1} \bigg/ \sum_{i=1}^{n} [M_j] \frac{r_{ji}}{r_{ij}} ([M_i] r_{ji}^{-1}) \qquad (1)$$

where M's represent the feed composition, m is the instantaneous composition of the copolymers, r's are the reactivity ratios of the comonomers, and $l = 1,2 \ldots n$.

These equations are the basis of the calculation program, the problem being solved by means of an electronic computer, applying the step-by-step method of integration. The unreacted comonomer composition for "t" calculation step is determined by the equation

$$[M_l]_t = [M_l]_0 - [m_l]_t \, \alpha \, t/(1 - \alpha_t) \qquad (2)$$

where $[M_1]_0$ represents the initial concentration (mole fraction) of 1-component and α_t is the molar conversion at "t" calculation step, $[M_1]_t$ being the molar fraction of the same component in comonomer mixture at "t" calculation step.

The mean molar fraction of 1-monomer at t calculation step (the mean composition of the copolymer) is defined also in the usual manner as

$$[m_l^*]_t = \sum_{t=1}^{t} [m_l]_t$$

where $l = 1,2, \ldots, n$.

A case which has not been sufficiently dealt with in the literature is the particular system of monomers (comprising more than three components) from which one or more have a reduced capacity of homopolymerization. Taking into account the large participation of those monomers in the plastics industry, the problem of finding a mathematical model apt to describe most accurately the composition of the respective copolymers, regardless of the number of components, has constituted another goal of our research [3], resulting in equations of the following type:

$$d[M_1]:d[M_2]: \ldots d[M_n] = [M_1] \sum_{i=1}^{n} [M_i] \, R_{1pi} : [M_2] \, R_{1/2} \sum_{i=1}^{n} [M_i] \frac{r_{21}}{rpi}$$

$$[M_{j2}] \, R_{1pj_2} \sum_{i=1}^{n} [M_i] \, R_{j_2pi}: \tag{3}$$

$$[M_{jk}] \, R_{1pj_k} \sum_{i=1}^{n} [M_i] \, R_{j_kpi}:$$

$$[M_h] \, R_{1ph} \sum_{i=1}^{n} [M_i] \, \frac{r_{hl}}{r_{hi}}:$$

$$M_n \, R_{1pn} \sum_{i=1}^{n} [M_i] \, \frac{r_{nl}}{r_{ni}}$$

where $R_{hji} = r_{hj}/r_{hi} = K_{hi}/K_{hj}$ is the constant of the ternary copolymerization.

The application of these equations in the ternary copolymerization systems in which one or two monomers do not homopolymerize but copolymerize among themselves and with the third one, results in the equations of composition suggested by Ham [4].

The azeotropic copolymerization of n-component systems has also had little notice in specialized publications. Applying matrix calculus in solving the systems of n-linear equations with n-unknown quantities, we can get a simple equation of a general character which predicts the composition of the azeotropic copolymer irrespective of the number of components in the system and the presence of nonhomopolymerizable monomers [3].

The Compositional Intermolecular Nonhomogeneity of Multicomponent Polymers

The study of the chemical heterogeneity of copolymers has become an essential problem for research, taking into account the direct dependence of the quality of plastic products on this parameter.

The chemical nonhomogeneity of monomer systems is quite advanced. Generally in respect to the homogeneity of reaction products, the behavior during copolymerization of a certain mixture of monomers may be known *a priori* by following their step-by-step vanishing from the substratum at different times [3, 5]. From Figure 1 it may be seen in the case of the quaternary acrylonitrile (AN)-methacrylic acid (MA)-styrene(S)-vinyl chloride (VCl) system, for instance, at the theoretical total conversion of 100%, the reaction product is practically a physical mixture made up from a quaternary polymer, a terpolymer, and a binary copolymer, respectively.

For the quantitative evaluation of the chemical inhomogeneity in multicomponent copolymers, there are no distribution functions yet. We have made some attempts to evaluate the chemical inhomogeneity, just as in the case of binary copolymers using differential and integral theoretical curves of compositional distribution on the basis of data obtained from the integration of composition differential equations (Fig. 2).

This estimation of the compositional inhomogeneity of the copolymers only by the width of the distribution curves, is not sufficient as it can, obviously, lead only to some general information [6]. A correct appraisal can be made only by knowing the two values of the instantaneous copolymer-composition and of conversion $(d \, [(1 - n)n_0^{-1}]/dm$.

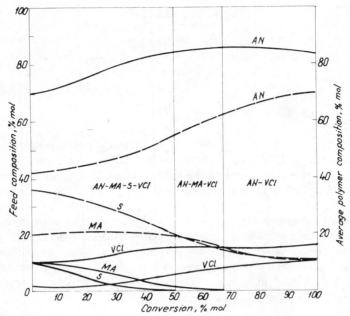

FIG. 1. The variations of feed and copolymer composition as functions of conversion for the acrylonitrile (AN)–methacrylic acid (MA)–styrene (S)–vinylidene chloride (VCl) polymer [3].– –, monomer mixture;—, copolymer.

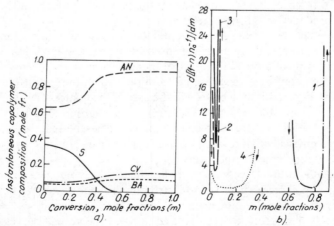

FIG. 2. The integral (a) and differential (b) theoretical curves of compositional distribution in the acrylonitrile (AN)–methacrylic acid (MA)–styrene (S)–vinylidene chloride (VCl) polymer [3, 6].

Intramolecular (Sequential) Inhomogeneity Multicomponent Polymer Composition

Until 1970, in the study of sequence distribution with more than two components, only two structural details were available by calculation, namely, the fraction of sequences containing n-units of monomer [7] and the clustering tendency of monomers [8], all the other characteristics being left undefined.

Tosi [9] supplemented this study in the subject of terpolymers, suggesting mathematical expressions for the main characteristic values for the sequence distribution in them. Our researches [5, 10] have also led to the elaboration of some mathematical relationships for the evaluation of homo- and hetero-bond fractions in terpolymers having the form:

a) $i = j$:

$$\frac{1}{F_{ii}} = 1 + 2\left(\frac{M_{ki}}{r_{ij}} + \frac{M_{ki}}{r_{ik}}\right) + M_{ji}^2\frac{r_{ji}}{r_{ij}} + M_{ki}^2\frac{r_{ki}}{r_{ik}} + 2M_{ji}M_{ki}\frac{r_{ki}}{r_{ij}r_{kj}} \qquad (4)$$

where

$$M_{ji} = [M_i][M_i] \qquad \text{and} \qquad M_{ki} = [M_k][M_i]$$

b) $i \neq j$:

$$\frac{1}{F_{ij}} = 2 + M_{ij}r_{ij} + M_{ji}r_{ji} + \frac{r_{ij}}{r_{ik}}(2M_{kj} + M_{ki})\frac{r_{ki}}{r_{kj}}$$

$$+ M_{ki}\frac{r_{ki}}{r_{ik}}\left(\frac{r_{ij}}{r_{kj}} + M_{kj}r_{ij}\right) \qquad (5)$$

where

$$M_{ij} = \frac{[M_i]}{[M_j]}, \qquad M_{ki} = \frac{[M_k]}{[M_i]}, \text{ etc.}$$

r's being the reactivity ratios.

These equations, based on the steady-state assumption and on the general expression for the bond fractions, differ from those suggested by Tosi in that they use only the initial composition of the monomers mixture and the reactivities of the components, respectively. By means of eqs. (4) and (5) it is possible to evaluate the initial conditions corresponding to the maximum alternation of the monomer units in terpolymers using eq. (6) (steady state is also assumed).

$$\Sigma F_{ij} = F_{ij} + F_{jk} + F_{ki} \qquad (6)$$

For instance, in the case of the acrylonitrile (M_1)–methacrylic acid (M_2)–methyl methacrylate (M_3) terpolymer, the calculations by means of a simple FORTRAN IV program have led to the data listed in Table I.

As can be seen, in binary copolymers, under the conditions of maximum alternation, both components are present in equimolar quantities, while terpolymers comprise different proportions of the three components. The fact that the mean sequence length of each component is the same both in binary copolymers and in terpolymer constitutes a common feature.

The $\bar{n}_1 = \bar{n}_2 = \bar{n}_3 = 1$ value for the terpolymer shows that at initial composition corresponding to the maximum degree of alternation, no radical present in the system adds its own monomer; thus, a perfect intra- and intermolecular homogeneity for the respective terpolymers is ensured.

TABLE I
The Initial Monomer Composition Corresponding to the Maximum Alternation Degree in Copolymers

F	F_{max}	Initial monomer composition % mol			Copolymer composition % mol			Mean sequence length		
		M_i	M_j	M_k	m_i	m_j	m_k	$\bar{n}_1$	$\bar{n}_2$	$\bar{n}_3$
F_{12}	0.329	0.829	0.171	0.000	0.5	0.5	0.0	1	1	–
F_{23}	0.287	0.000	0.447	0.553	0.0	0.5	0.5	–	2	2
F_{31}	0.356	0.729*	0.000	0.271	0.5	0.0	0.5	1	–	1
Σ Fij	0.4141	0.687	0.124	0.189	0.405	0.394	0.201	1	1	1

For indifferent initial conditions using our reference [5] and the literature data [9], the conversional variation of the main characteristic parameters within 0–100 conversion range has been determined by applying the same "step by step" integration method. (Figs. 3 and 4.)

It is obvious that, for certain degrees of transformation, interesting for practice, products with known, predetermined parameters can be obtained, through the knowledge of the dependence between initial conditions and final structure, a dependence which constitutes one of the decisive factors in controlled syntheses.

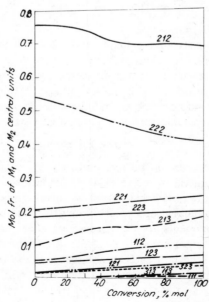

FIG. 3. The variation of mean concentration of the compositional triads in the acrylonitrile (M_1)–methacrylic acid (M_2)–methyl methacrylate (M_3) terpolymer ($M_1:M_2:M_3 = 0.3:0.6:0.1$ moles) as a function of conversion [3].

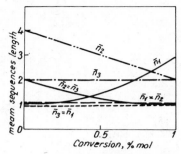

FIG. 4. The variation of the mean sequences length as a function of conversion in the acrylonitrile (M_1)–methacrylic acid (M_2)–methyl methacrylate (M_3) terpolymer $(M_1:M_2:M_3 = 0.3:0.6:0.1$ moles) [5].

The Correlation between Chemical Structure and Properties

At present only the dependence of the copolymer properties on its composition is satisfactorily studied. Making use of all the data referring to the chemical structure (the content of monomers, the distribution of the units, the mean length of the sequences) we illustrated the dependence of the thermal behavior of the acrylonitrile–methacrylic acid–methyl methacrylate terpolymer on the above mentioned factors, under the conditions of thermal oxidative degradation (Table II) [11, 12].

At different comonomer proportions in the copolymer, the data from Table II show that by including comonomers an effect of increasing thermostability is obtained in the terpolymers with predominant content of methacrylic acid and methyl methacrylate, respectively, as compared with corresponding homopolymers, and an effect of decreasing it in comparison with the homopolymer in the case of the product with prevalent acrylonitrile. In both situations, the reason is the disturbance of the chemical structure peculiar to homopolymers, that is, the interruption of homosequences from the predominant monomer by the addition of the other two comonomers. In the copolymer with more acrylonitrile units, for example, (terpolymer I), as a result of the preferential addition of foreign monomers to the acrylonitrile radicals, the long sequences of acrylonitrile are missing (the mean sequence length is shortened from $\bar{n}_1$ to 1.3); in consequence, the interaction of —CN adjacent groups is inhibited and the cyclizations which lead to the known thermal stability of polyacrylonitrile do not take place any longer. In the copolymers with higher contents of methacrylic acid (terpolymer II) and methyl methacrylate (terpolymer III), the shortening of the mean sequence length of the basic comonomer reduces the depropagation tendency and has, therefore, a favorable effect on the thermal stability.

In conclusion, if the behavior of the final product were under the control of structural factors, these depend upon the initial conditions, that is, the comonomer ratio, the reactivities, and the applied method of synthesis, respectively. The directing and checking of final properties are being, theoretically, made possible by the suitable determination and preservation of these conditions.

It is a matter of course that the criteria of chemical structure, commented

TABLE II

Chemical Structure–Thermal Behavior Correlation for the Acrylonitrile–Methacrylic Acid–Methyl Methacrylate Terpolymer of Different Compositions

Polymer	Initial chemical structure of polymers											Degradation step and corresp. temp. °C	Weight losses, W %	Chemical structure resulted by heating	Remarks on the structural effects
	Average composition, % mol			Main sequence	Concn. of seq. % mol	$\bar{n}_i$			N_i						
	M_1	M_2	M_3			$\bar{n}_1$	$\bar{n}_2$	$\bar{n}_3$	N_1	N_2	N_3				
PAN	100	0	0	$-(M_1)\bar{n}_1-$	100	$\bar{n}_1$	–	–	–	–	–	I 100–268 II 268–400 III 400–720	8 11 81		Thermal stability up to 600° due to cyclization
Terpolymer I	60	35.6	4.4	$-M_2-M_1-M_2-$ $-M_2-M_1-M_3-$ $-M_1-N_1-M_2-$	60 24	1.5	1	1	1	2	18	I 130–278 II 278–465 III 465–600	18.2 57.9 24		Initially degradation in inhibited; over 130° the thermostability decreases as result of the shortening of consecutive sequences of AN
PMA	0	100	0	$-(M_2)\bar{n}_2-$	100		$\bar{n}_2$	–	–	–	–	I 90–327 II 327–463 III 463–570	57 38 5		Rapide depolymerization in the first step; over 200° stabilization in some extent appears
Terpolymer II	27	67	6	$-M_2-M_2-M_2-$ $-M_2-M_2-M_2-$	30 27 22	1	2.5	1	1	3.5	1 8	I 80–322 II 322–458 III 458–552	14.30 63 22.4	H bonds	Depolymerization probability is reduced due to the shortening of consecutive sequence of MA; stability up to 200°
PMMA	0	0	100	$-(M_3)n_3-$	100			$\bar{n}_3$	–	–	–	I 100–220 II 220–332 III 332–570	18 54 28	unstable at 220°	Rapide de polymerization up to 270° (50 %)
Terpolymer III	20	13	67	$-M_1-M_3-M_3-$ $-M_3-M_3-M_1-$ $-M_3-M_3-M_2-$	37 37.5 10	1	1	3	1	3.7	6 1.5	I 100–260 II 260–430 III 430–563	5.8 64 13		Monomer production is inhibited by unsaturated structures. At high temp. terpolymer degradation becomes efficient by the AN units

[a] M_1 = AN; M_2 = MA; M_3 = MMA.

[b] $\bar{n}_i$ = mean homosequences length; N_i = mean number of m_j and m_k units between two successive m_i units.

upon briefly in this paper with regard to the possibilities of deliberately acting upon the copolymerization processes, do not cover the whole range of methods for the solving of practical needs in view of the variety and quality of final properties. Undoubtedly, the modifications of physical structure by means of the technological synthesis and processing parameters have to supplement the solution of this outstandingly important problem.

BIOCOPOLYMERS. STRUCTURE STUDY (BASED ON CLASSICAL COPOLYMERIZATION THEORY) AND ITS IMPLICATIONS ON PROPERTIES

If, for synthetic polymers, the establishment of chemical structure (component ratio, distribution of structural units in macromolecular chain, etc.) is not a problem today, this information is generally obtained in a different way in the case of natural polymers.

A first attempt to elucidate the structure of such a natural polymer—alginic acid separated from brown seaweed, *Laminaria digitata*—was made by Haug and co-workers [13-15] who by translating the copolymerization theory to these natural polymers established that alginic acid is a linear block-copolymer characterized by three sequences types: long blocks of D-mannuronic acid (M), L-guluronic acid (G), and also long heteropolymeric sequences of predominantly alternating structure of these two uronic acids. Since in numerous studies the dependence of chemical composition on this polysaccharide was established under biosynthesis conditions (brown seaweed species, the anatomo-morphological element, the place, and the harvesting season) it should be supposed that the same factors would also influence comonomer distribution in the macromolecular chain.

Our studies [16-19] had confirmed these suppositions for the alginic acid separated from brown seaweed *Cystoseira barbata*, establishing a different structure as compared with the same product extracted from *Laminaria digitata*.

The experimental data based on partial acid hydrolysis of polyuronides allowed the separation of three distinct fragments from the point of view of chemical composition (conventionally noted with A, B, and C).

Fragment A (M/G = 63/37). The experimental results regarding the supplementary hydrolysis of fragments A, B, and C, the fractionation, and the product characterization were interpreted in terms of a kinetic model, according to the classical theory of the "ultimate unit" in copolymerization reactions.

Supposing that the kinetic and statistical treatments which are valid for very long chains could also be applied for lower degrees of polymerization and that the terminal kinetic model satisfactorily describes the behavior of comonomers in bio-synthesis process, fragment A may be described in statistical terms (Table III [16].

The calculations aimed at a general characterization of fragment A and at a quantitative evidence of the alternating tendency using the three alternating indices from synthetic polymer chemistry, i.e., the mean sequence length (ho-

TABLE III

Structural Characteristics of Fragment A Statistically Calculated with the Assumption of the Terminal Kinetic Model (M/G = 1.63[a]; $\overline{P}_n$ = 24[b] transition probabilities[c]: P_{MM} = 0.432; P_{MG} = 0.568; P_{GG} = 0.074; P_{GM} = 0.926)

Mean sequence length		Diad concentration[d]				Triad concentration[e]				
Homo-sequence[f]	Alternating sequence[g]	P(MM)	P(MG)	P(GG)	P(GM)	P(MMM)	2P(MMG)	P(GMG)	P(GGG)	2P(GGM) P(MGM)
$\overline{n}_M$ = 2 $\overline{n}_G$ = 1	$\overline{n}_{MG,GM}$ = 4	0.268	0.352	0.028	0.352	0.115	0.305	0.2	0.002	0.052 0.326

[a] Electrophoresis and NMR spectra. [b] From iodine number. [c] $P_{MM} = a/(1 + d)$; $P_{MG} = 1 - P_{MM}$; $P_{GG} = b/(1 + b)$; $P_{GM} = 1 - P_{GG}$. [d] $P(MM) = P(M).P_{MM}$; $P(MG) = P(M).P_{MG}$, etc. [e] $P(MMM) = P(M).P_{MM}^2$; $P(MMG) = 2.P(M).P_M.P_{MG}$; $P(GMG) = P_{MG}^2$, etc. [f] $\overline{n}_M = 1/P_{MG}$ or $\overline{n}_M = 1$ a. [g] $\overline{n}_{MG,GM} = 2/(P_{MM} + P_{GG})$.

mosequences and alternating parts), the numerical fraction of alternating diads ($F_{MG,GM}$) and the numerical fraction of MGM and GMG triads ($F_{MGM,GMG}$) in copolymer. The value $F_{MG,GM}$ = 70 suppose that about 70% of all possible bonds in fragment A are of M–G and G–M type, respectively. The length of strictly alternating parts—MGMGMGM—(^{1}MG.GM = 4) attests that it contains only 33% of all MG(GM) bonds of the entire fragment. It seems that the difference of 37% of MG(GM) bonds are quite uniformly distributed along the chain, being separated by units of M type. This affirmation is supported by the high enough value of P_{MM} diad (0.268) and by the mean length of continuous mannuronic acid sequences ($\overline{n}_M$ = 2), respectively.

In order to make complete the general picture of fragment A it was considered necessary to study the oligomers with the aim of confirming the calculations regarding those structures and implicitly of the initial fragment (Table IV).

The distribution of "mers" in oligomers suggest a probable structure of fragment A, as follows:

—MMG—MMG—MMG—GM—GM—GM—GM—

MMG—MMG—MMG—

The central alternating portion characterized by an "alternating period" constituted from —MG— is limited in both ends by alternating fragments of a longer period of —MMG— type.

Fragment B (M/G = 97/3) is considered a homopolymer of D-mannuronic acid.

Fragment C (M/G = 15/85) may be concerned as a copolymer with a prevalent content of L-guluronic acid.

TABLE IV

Distribution of Units of D-Mannuronic Acid and L-Guluronic Acid in Mono- and Oligomers Isolated from Fragment A

Fragment	M, %	G, %
Tetramer	27.50	11.60
Trimer	45.00	24.40
Dimer Monomer	27.50	64.00

Regarding its structure, the three structural models may be assumed, corresponding to the behavior of this fragment in different conditions, as follows [17]:

(a) an almost ordered distribution of uronic acid, where G sequences of certain length—m— are interrupted by isolated M units

$$-M-(G)_m-M-(G)_m-$$

(b) a less uniform structure characterized by the presence of a central part constituted from continuously bonded—n— units of guluronic acid, separated from some isolated units from other blocks of —p—or—r— length smaller than —n

$$-(G)_p-M-(G)_n-M-(G)_r-$$

(c) the possibility of the following type of monomer clustering cannot be excluded

$$-(M)_m-(G)_n-; \qquad m \ll n$$

There is no doubt that the existence of the agglomeration of guluronic acid units irrespective of the supposed structural model, is on the one hand due to the prevalent content of this component, and on the other hand, to the specific behavior of fragment C as a homopolymer in conditions of its treatment with potassium chloride solution of different normalities (Fig. 5).

By acid hydrolysis, a fairly large amount of guluronic acid goes into solution. Taking into account the order of hydrolysis rates ($k_{MM} > k_{MG(GM)} > k_{GG}$) this fact cannot be explained only if stochastic existence of small marginal blocks of guluronic acid units is admitted.

The variation of the average degree of polymerization and the composition of fragment C for different times of hydrolysis gives an idea about the average length of the central block of guluronic acid units (Fig. 6).

The calculated data used and in this case the "reactivity indices" (given by variation of the inverse of the polymerization degree ($1/\bar{P}_n$) versus the scission degree (α) indicate for the average length of mannuronic acid sequences a value of $\bar{n}_M = 1 + a \simeq 1$ and for average length of guluronic acid sequences a value of $\bar{n}_G = 1 + b \simeq 8$.

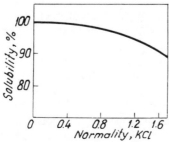

FIG. 5. Solubility of the fragment C versus the normality of potassium chloride solution [17].

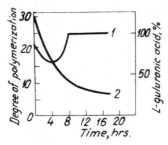

FIG. 6. Variation of polymerization degree (- -) and of L-guluronic acid content (—) as a function of time, in residue resulting from acid hydrolysis of fragment C [17].

The correlation of this information with the M/G ratio and the polymerization degree enable us to assume that about 42% of all G units are contained in sequences of a length $\bar{n}_G \simeq 10$, while the others (58%) are statistically distributed in sequences of —p— length varying under this value. The experimental and theoretical results data indicate as a more probable model this model (b) in which the so-called central "core" of an average length of at least 10 units of guluronic acid is separated from other shorter blocks of the same type, by isolated units of mannuronic acid.

Representation of Alginic Acid Macromolecule. By the comparative study of polysaccharides extracted from brown seaweeds *L. digitata* and *C. barbata* the following facts have been established (Table V): (1) the macromolecules of polyuronides from the two species are constituted from the three types of chemically distinct blocks; (2) the comonomer distribution, presented comparatively for the three fragments (see Table III), is a characteristic of seaweed species.

Dependence of Some Properties by Comonomer Ratio and Their Distribution. In the analyzed conditions it is to be expected that the properties of alginates extracted from different seaweed species are to be influenced by chemical composition and comonomer distribution in case of the "synthesis–structure–properties" interrelation.

This assertion is supported, for example, on the one hand by the rheological behavior of sodium alginate solutions (Fig. 7), and on the other hand by the ability of the polysaccharide to form complexes with iodine in the presence of some electrolytes [18].

Regarding the ability of complex formation with halogen (UV = 587 nm) one keeps attention on chemical composition homogeneity in complexing, and complex formation is possible only in the case of fragment B (homopolymer of D-mannuronic acid).

In these conditions, when fragments A, B and C complex differently with iodine, it can be considered that as a whole, the macromolecular chain will form a deficient helix (helicoidal portions—evidenced by complexes—alternating with disordered portions).

Based on data obtained in the study of "polymer–iodine" complexes (five to six structural units/iodine atom) a model of helicoidal structure is proposed [18], in the form of a "solenoid" (Fig. 8).

TABLE V
Comparative Representation of Alginate Macromolecules from Different Seaweed Species

Seaweed type		Laminaria digitata		Cystoseira barbata
Fragment	M/G	Assumed structure (15)	M/G	Assumed structure (17)
A	54.5/45.5	Predominant alternating copolymer	63/37	Uniform copolymer with two alternation "steps"

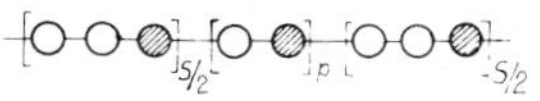

p ∼ 4; s ∼ 6

| B | 92/8 | Homopolymer based on D-mannuronic acid  | 97/3 | Homopolymer based on D-mannuronic acid 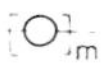 |

| C | 13/87 | Homopolymer based L-guluronic acid | 15/85 | Copolymer with different G blocks lengths separated by isolated M units |

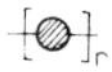

n ≃ 10   p, r < 10

─○─ D- mannuronic acid

─◉─ L- guluronic acid

These considerations, already done in this field, had the aim of underlining the possibility of extension of the synthetic copolymer theory to biocopolymers whose structure, correlated with complex conditions of biosynthesis, present unsolved aspects up to now.

GRAFTED COPOLYMERS. THE ACTIVATED POLYMER (UNDER THE EFFECT OF PHYSICAL AND CHEMICAL AGENTS)—A "MATRIX" FOR THE STRUCTURE OF THE COPOLYMER

The grafted copolymers are a matter of interest in the modification of chemical characteristics of fibers, with the object of making up the balance between "natural products–synthetic products." However, there is still a long way to go from ideas, the stock of grafted fibers in laboratories, to a rational taking into account of techniques.

At this level, earlier concepts as well as other explorations are to be dwelt upon, in order to answer the following question: "Which are the most effective fiber grafting methods, by means of which the structure and the envisaged properties can be ensured?"

Our contribution to solving this complex problem is marked by some conclusions drawn from our studies undertaken within the framework of mutual

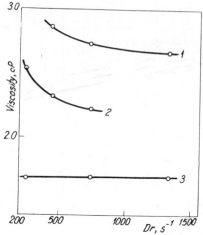

FIG. 7. Variation of viscosity of the alginate solutions extracted from different seaweed species as a function of shear rate [17].

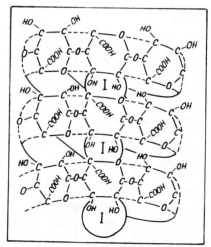

FIG. 8. Proposed model for the helicoidal structure of the alginic acid [18].

relations between "synthesis–structure–properties" which has found a particular expression in the "method of grafting–support polymer–grafted copolymer."

These conclusions show the following: (1) Regardless of the way of achieving grafts by means of (a) initiating reactions or/and (b) termination reactions, the activation of the support polymer through a physical or a chemical method is needed.

In this respect, there should be a selection of the activating agent which at the limits of allowed fiber modifications is likely to provide, along with the preservation of the material, a suitable grafting degree. Consequently, a process of method selection is imposed by the above mentioned mutual relations.

(2) The activation of fibers through physical and chemical methods points

to the existence of an intermediate form—*the activated support-polymer*—(as it has been characterized by deactivation) which exhibits (on its surface or/and in its bulk):

active species (e.g., macromolecular initiator, macromolecular disrupting agents, or transfer) apt to realize grafting reactions, and

structural modifications, more or less significant, depending on the nature of the initial support-polymer and the specific method used (containing the activating agent, technique, and experimental conditions).

According to this viewpoint, the transition from the initial fiber (Spi) to the grafted copolymer (Gc) takes place by an intermediate stage (Sp*) as shown below:

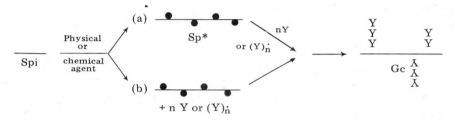

This holds true both for preactivating (a) and for simultaneous activation and grafting (b) with a given monomer (y).

The analysis of the activated support-polymer (in the absence of the monomer, but with some experimental conditions of activation and grafting conditions being observed) permits a more reliable description of the matrix on which the grafts and the homopolymers are to be set up.

In the case of the mentioned principles, the activated support-polymer as well as the initial polymer may be regarded as an element to be referred to, for checking of the structural modifications induced by grafting.

These two premises will be accounted for, in examples to be reviewed briefly further on.

Some Remarks on the Support-Activated Polymer under the Influence of Physical Agents

Among physical methods under study, attention should be given to low energy irradiation discharges in a radiofrequency field (cold plasma), less used in grafting, besides electric discharges effluvia along with UV radiations, used for the first time in grafting.

It may be stated that electric radiofrequency discharges (cold plasma) produce, as a consequence of the electronic bombardment, free radicals on the surface of the fibers. The concentration of these active species depends mainly upon the power and the pressure within the reactor. The value obtained for cellulose fibers is of the same order as the one determined in the case of γ-irradiation (10^{15}–10^{17} spins/g; RES and DPPH).

Regarding the behavior of fibers at the cold plasma activation, mention should be made of a series of structural modifications depending upon the nature of the support-polymer and the experimental conditions (mainly the applied power): (a) all types of fibers after being activated exhibit weight losses (in relation to Spi). In the case of cotton, for instance, decreases of 5–14% have been recorded, as well as the appearance of ketonic groups (at 263–270 mμ in UV), change in color, reduction of the hydroxylic group content, along with the increase of free radicals, the reduction of the polymerization degree by 8–47%, and the appearance of crosslinks (less significant though, owing to the absence of the monomer from the reaction system);

(b) the support-polymer is activated on the surface (Fig. 9) and the synthesized product is deposited as a thin film (the thickness of which varies in accordance with the power and the time of grafting (Figs. 9 and 10c) [21, 22]. This method ensures a sufficiently homogeneous grafting, a real packing of the fibrous materials, and permits the extension of the range of monomers to practically nonpolymerizable products (e.g., benzene);

(c) the grafting of cotton fibers, viscose, polyester with acrylonitrile, acrylamide, or benzene has led to some materials with improved surface properties. For instance, the polyester changes its hydrophilic properties after grafting with acrylamide (PAA 22.93%) from 0.88% (Spi) and 0.50% (Sp*) to 9.60% (Gc); UR 95% [21].

The mechanical strength of grafted is greater than the same property of the activated support-polymer, and similar to the strength of the initial fiber.

Under the influence of the effluvia along with UV radiations, free radicals are formed as a consequence of the photolytic scission of the same chemical bonds. In the case of wool, the active species were identified by ESR of the cystyl and tyrosyl type, with an average life of 20 hr [22].

Regarding certain structural modifications of wool by activation through this method, as compared with the initial and grafted fiber, we mention the following facts: (a) the modifications of the chemical composition of macromolecular chains by activation, and the transformation of certain amino acids (especially by oxidation) only at the cuticular level; the appearance of lysino-alanine and

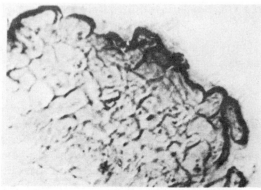

FIG. 9. Microphotograph of cotton yarn grafted with acrylonitrile (1200 W); cross-section (magnification 400 X) [20].

(a)

(b)

FIG. 10. Electron micrograph. Polystyrene film replica; C–Pd shadowed; magnification 4200 ×; (a) viscose rayon witness (Spi); (b) viscose rayon after the action of radiofrequence field (2750 W) in argon (Sp⁺); (c) viscose rayon grafted with benzene (520 W) [21].

methionine sulfoxide being also significant. In similar experimental conditions, the latter product increases its amount under UV radiation.

By grafting with styrene, no important variations of the contact in amino acids were determined, except for an "unknown compound + threonine";

    (b) the transformations occur only on the surface of the fiber. The parameters calculated from the X-ray diffraction spectra (at small and large angles) are not noticeably modified by activating or grafting. The nonuniform clusters of synthetic polymer are placed preferentially at the limit of the scales (Fig. 11a and b);

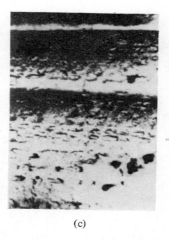

(c)

FIG. 10. (*continued from previous page*)

(c) the variation of some physicomechanical characteristics renders evident a stiffening of fibers during the activation process (an increase of the value of elasticity modulus and breaking load as compared with Spi) which is accounted for by the appearance of crosslinks on the surface (which may appear even in the presence of monomer). By grafting in the same mild activation conditions (1–5 min), the stiffening decreases, while the mechanical strength increases. Grafted wool (Gc) has improved textile characteristics, and a small capacity of felting (as compared to Spi and Sp).

Under identical experimental conditions, the grafting of wool with vinylidene chloride or styrene reaches a content of synthetic polymer of 1.6–6.2% (by preirradiation) and of 4.2–8.2% (by simultaneous activation and grafting); the content is noticeably larger than in the case of activation exclusively with UV radiations (0–2.8%).

Some Remarks Concerning the Activated Support-Polymer under the Influence of Chemical Agents

With respect to the chemical methods, reference is made to diazotization and redox system of the reducer-oxidizer type "macromolecular compound-metal with varying valency."

It may be expected that the diazotization method (introduction of $-NH_2$ groups by means of reactive dyes) should lead to the formation of "diazo" macromolecular products on viscose fibers, calcium alginate fibers, and wool; the diazo compound has been made evident by coupling with naphtholes.

For the first time, we have used the direct diazotization of wool [23, 24] with a view to being grafted, taking into consideration certain groups of the chemical structure. The subsequent thermal decomposition of macromolecular diazonium compounds ensures the appearance of free radicals on the chain.

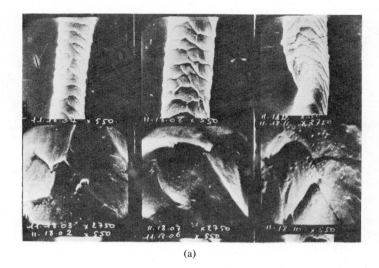

(a)

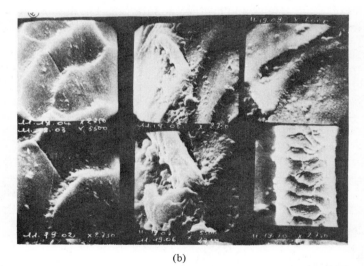

(b)

FIG. 11. Electronic ballayage microscopy: (a) wool witness; (b) grafted wool with vinylidene chloride by electric discharges effluvia along with UV radiation. [22].

The support-polymer undergoes the diazotization method (fibers dyed by reactive dyes) which brings about important structural modifications along with the successive stages in order to be activated. From the experimental data, we may infer that the most significant influence in the structural transformation of wool is during the proper diazotization stage.

Further on we shall refer to direct diazotization of wool and to some structural

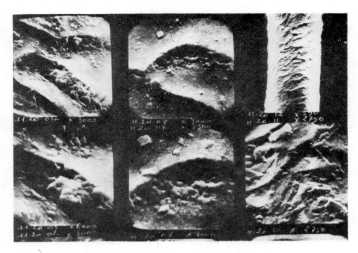

FIG. 12. Electric ballayage microscopy. Grafted wool with acrylonitrile by diazotization [22].

modifications (as a consequence of activation): (a) the amino acids from the cuticular layer are subjected to chemical transformations, from which we should mention the oxidation of cystine, the increase of cysteic acid content, the appearance of lysino-alanine and methionine sulfoxide. The dosing of the amino acid after grafting wool with acrylonitrile shows the appearance of three unidentified compounds (isolated after the separation of isoleucine, phenylalanine, and cysteic acid) because of the influence exerted by the new component in the structure;

(b) both the diazotization and the grafting process under the studied conditions occur on the surface of the fibers (assertion based on the X-ray spectra and electronic ballayage microscopy, with nonuniform deposits of synthetic polymer (Fig. 12), splittings and flattening of the scales;

TABLE VI
Some Physicomechanical Indices of Grafted Alginate Fibers

Activation method	Polymer	Activation and grafting conditions	Composition % Alg(COO)$_2$Ca/ modified^{a}Alg	Physico-mechanical indices		
				$\bar{d}_M$	$\frac{T}{gf}$/fiber	Al %
Redox and complexing reactions	Calcium alginate (Sp1)	–	94.97/ 5.03	19.62	3.23	5.00
	Calcium alginate activated Sp$^+$	30 min.	58.30/41.70	18.40	2.28	6.00
	Calcium alginate-Co-g-polyacrylo-nitrile (PAN – –13.720 %) Gc	AN 100 l/h 20° ; 60 min	–	18.80	3.23	8.33
Forming and eliminating of diazo-groups	Calcium alginate with amino-groups	–	69.11/30.89	19.71	3.20	12.61
	Calcium alginate activated	NaNO$_2$ + HCl	60.95/39.05	16.40	1.26	15.00
	Calcium alginate –Co-g-polyacrylo-nitrile (PAN - 1.19)	AN 78° 60 min	–	17.40	3.88	12.74

^{a} $\bar{d}_M$, mean fiber diameter; T, fiber tenacity; Al, fiber elongation.

TABLE VII

Some Roentgenographic Characteristics of Grafted Alginate Fibers, Depending on the Chemical Method of Activation

Chemically activation method	Polymer	Roentgenografic characteristics				
		Maxima number	d_{hkl} $\overset{o}{A}$	Degrees		
				$\alpha/2$	$\beta/2$	
Witness	Calcium alginate	1	2.01	43	0	
		2	1.57			
		3	1.15			
Redox and complexing reactions	Calcium alginate treated in solution 0.01 MCe $(SO_4)2$ 0.1 M H_2SO_4 (30 min)	1	5.21			
		2	4.03	0	15	
		3	1.43			
		4	1.15			
	Calcium alginate-CO-g- -polyacrylo- nitrile (PAN 20%)	1	3.87			
		2	1.82	0	52.30	
Introduction and elimination reactions of	Calcium alginate with amino-groups	1	2.01			
reactive groups (diazotization)	Calcium alginate -Co-g-polyacrylo- nitrile (PAN 1.92%)	1	2.14			
		2	1.57	7.30	35	
		3	1.15			

[a] d_{hkl}, interplanar spacing; $\alpha/2$, $\beta/2$, angular length in degrees of the measured interference arc at half maximum intensity, perpendicular and parallel with the fiber axis, respectively.

(c) the direct diazotization of wool as a varient of the classic method is superior to the method applied for the dyed fibers with reactive dyes (with subsequent reduction). This concerns the preservation of certain physicomechanical properties of economic efficiency.

The strength of the grafted fibers (VCl_2 = 4.2%) is higher than that of the activated material (Sp*) and may be compared with the initial polymer (Spi).

The redox reactions in the calcium alginate–Ce^{4+} system, lead to the appearance of free radicals used in the grafting of fibers with vinyl monomers (3–20% PAN) [25].

The case we have studied represents an example of structural modifications undergone by the support-polymer (Spi) by activation (Sp*) and grafting (Gc) and are partly recorded in Table VI.

It should be observed that for the same fiber and monomer, depending upon the chemical method used (redox or diazotization reactions), there result products with different röentgenographic characteristics (Table VII). If the activated support-polymer stage is not taken into consideration (Sp^+), the decrease of orientation of the grafted copolymer ($\beta/2$ = 52°30′) would by wholly attributed to grafts. The influence of the synthetic polymer (in Gc), merely

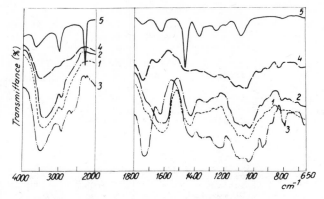

FIG. 13. The infrared spectra of PAN-homopolymer synthesized by every initiation system studied [25]: (1) in aqueous acid solution of cerium salt ($t = 20°C$); (2) in aqueous solution of benzene diazonium salt ($t = 80°C$); (3) with metallic sodium in liquid ammonia ($t = 78°C$); (4) with sodate malonic ester in DMF; (5) with sodium ethoxide in DMF ($t = 40°C$).

emphasizes the orientation in the longitudinal direction, which has been recorded in previous stages of activation (Sp*).

The insolubility in diluted alkaline solutions after structural transformations of calcium alginate leads to the following order:

$$Gc \gg Sp* > Spi$$

At the close of these considerations, regarding some structural modifications of the support-polymer in the activation stage, another aspect, ignored in the literature, has been underlined. This refers to the synthesized grafts, namely, under specific synthesis conditions the *grafting method leads to certain structural peculiarities* (depending on the nature of the monomer) for *the new macromolecular chains*. This assertion is based upon the difference of chemical structure in the polyacrylonitrile grafts synthesized in different reaction conditions (Fig. 13).

Comparison of the grafting methods under the action of physical and chemical agents shows—by the facts emphasized above—that the choice of the direction of fiber modification is determined by the nature of the support-polymer and the envisaged properties, ascertaining the mutual relation "synthesis–structure–properties."

Regarding the nature of the initiating mechanisms, despite their radical character, the complex factors which have an influence on the grafting reaction in the heterogeneous phase (including the nature of the active species) tend to blur the conventional barriers existing among these mechanisms.

In a more general respect, a unitary view of the range of problems connected with grafting is to be reached nowadays in order to do away with the crisis within the theory and practice of this field.

Will successful techniques for the industrial grafting of fibers be found? Will all the technical and economical shortcomings be surpassed? Will the physical methods prevail in this contest, ensuring a simultaneous improvement of a large number of properties?

REFERENCES

[1] A. Valvassori and G. Sartori, *Chim. Ind.* (Milan), *44*, 1091 (1962).

[2] G. E. Ham, *J. Polym. Sci. A-2*, 169 (1964).

[3] M. Tomescu, Thesis, Polytechnic Institute of Jassy, 1973.

[4] G. E. Ham, *J. Macromol. Sci. Chem., 1*, 93 (1967).

[5] A. Liga, Thesis, Polytechnic Institute of Jassy, 1973.

[6] M. Tomescu and Cr. Simionescu, *Die Makromol. Chem.,* in press.

[7] T. Alfrey and G. Goldfinger, *J. Chem. Phys., 12*, 205 (1944).

[8] J. A. Seiner, *J. Poly. Sci., 2*, 985 (1964).

[9] C. Tosi, *E. Polym. J., 2*, 161 (1970).

[10] A. Liga and I. Negulescu, *Bull. Inst. Polit. Iasi, 25*, 75 (1975).

[11] N. Asandei, A. Liga, and A. Biro, *Eur. Polym. J., 7*, 317 (1971).

[12] Cr. Simionescu, A. Liga, and N. Asandei, *Rev. Roum. de Chim.,* in press.

[13] A. Haug, B. Larsen, and D. Smidsrød, *Acta Chem. Scand., 23*, 2955 (1969).

[14] B. Larsen, O. Smidsrød, A. Haug, and T. Painter, *Acta Chem. Scand., 23*, 2375 (1969).

[15] B. Larsen, T. Painter, A. Haug, and O. Smidsrød, *Acta Chem. Scand., 23*, 355 (1969).

[16] Cr. Simionescu, V. Popa, V. Rusan, and A. Liga, *Cell. Chem. Technol., 9*, 213 (1975).

[17] Cr. Simionescu, V. Popa, A. Liga, and V. Rusan, *Cell. Chem. Technol., 9*, 547 (1975).

[18] Cr. Simionescu, V. Popa, and V. Rusan, *Cell. Chem. Technol., 9*, 641 (1975).

[19] V. Popa, Thesis, Polytechnic Institute of Jassy, 1976.

[20] Cr. Simionescu, N. Asandei, and F. Denes, *Cell. Chem. Technol., 3*, 165 (1969).

[21] Cr. Simionescu, M. Macoveanu, and N. Olaru, unpublished data.

[22] Cr. Simionescu and O. Milcomete, unpublished data.

[23] I. Rusznak and F. Peter, *Bull. Inst. Textile France,* 267 (1961).

[24] M. Leveau, M. Caillet, *Bull. Inst. Textile France, 97,* 7–23 (1960); *93,* 75–79 (1961).

[25] V. Rusan, Thesis, Polytechnic Institute of Jassy, 1973.

COPOLYMERIZATION OF VINYL CHLORIDE WITH BIS(BETA-CHLOROETHYL) VINYLPHOSPHONATE

R. GALLAGHER
Stauffer Chemical Company, Eastern Research Center, Dobbs Ferry, New York 10522

J. C. H. HWA
Stauffer Chemical Company, Westport, Connecticut 06880

SYNOPSIS

Bis(beta-chloroethyl) vinylphosphonate and vinyl chloride were copolymerized free-radically in water suspension and in toluene solution at 55°C. From compositional analysis of the low-conversion copolymers, the reactivity ratios of bis(beta-chloroethyl) vinylphosphonate (r_1) and vinyl chloride (r_2) were found to be 0.37 ± 0.1 and 0.26 ± 0.05, respectively, indicating that the two monomers have a marked tendency to alternate. From the Q and e values of vinyl chloride, the corresponding Q,e values for the vinylphosphonate are calculated to be 0.23 and 1.73, respectively. Interestingly, the rate of copolymerization is greater than that of homopolymerization, and is also greater when water is present. These copolymerization characteristics are deemed favorable in industrial productions. The copolymers have lower viscosities and glass temperatures than poly(vinyl chloride). From patent literature, the copolymers also show useful flame retardant properties.

INTRODUCTION

The copolymerization of vinyl chloride with other vinyl monomers has been extensively reviewed [1, 2]. Until recently, relatively little has been published on the copolymerization of bis(beta-chloroethyl) vinylphosphonate with vinyl halides. The vinylphosphonate monomer is now available on a commercial scale as FYROL Bis-Beta (Stauffer Chemical Company, abbreviated in this paper as Bis-Beta) and it has shown utility because of the favorable flame retardant properties that it imparts to copolymers. The free radical copolymerization of Bis-Beta with vinyl chloride has briefly been mentioned in earlier publications of the Stauffer Chemical Company [3, 4]. This paper describes in greater detail the copolymerization characteristics involving these two monomers, particularly the determination of the reactivity ratios.

Journal of Polymer Science: Polymer Symposium 64, 329–337 (1978)
© 1978 John Wiley & Sons, Inc.   0360-8905/78/0064-0329$01.00

EXPERIMENTAL

In a typical free-radical, suspension copolymerization, a 32 oz (600 ml) soda bottle containing 16.0 g Bis-Beta (commercial grade material washed with dilute sodium hydroxide and then with water, and fractionally distilled; monomer was 99.6% pure), 0.10 g azoisobutyionitile (AIBN, Vazo 64, DuPont), 0.2 g Methocel (hydroxypropyl methylcellulose K-35, Dow) as a 1% solution, and 350 ml distilled water was placed in a freezer at $-20°C$ for at least 3 hr. The bottle with frozen contents was placed on a large-capacity, sensitive balance and liquid vinyl chloride (commercial grade, B. F. Goodrich), in excess of the desired 84.0 g, was added from an inverted tank. When the excess was boiled off, thus flushing the atmosphere, the bottle was immediately capped (cap lined with cork). The bottle was placed in a protective metal container and secured in a water bath preheated to 35–40°C. The bath was heated to 55°C (in 15 min) and the bottle was rotated end-over-end at 20 rpm. After 2.6 hr at 55°C, the bath was quickly cooled (in 10 min). The bottle was removed and the cap was punctured, allowing unreacted vinyl chloride monomer to vaporize. The solid polymer was separated by filtration, dried, and reprecipitated three times (5–8 g polymer) from tetrahydrofuran (25 ml) solution in ether (800 ml). The polymer was dried under infrared lamp for 24 hr. In runs where the Bis-Beta/vinyl chloride ratio was high, the copolymer was isolated at low conversions by first separating the oil phase (unreacted Bis-Beta and dissolved copolymer), and then adding the oil phase to ether (15-fold excess). The gummy polymer was further purified by precipitation from tetrahydrofuran solution, as in other cases. Polymerizations in toluene solutions followed the general procedure, except that in the work-up, toluene was first evaporated from the contents by purging with nitrogen in the hood at room temperature. Copolymer compositions were determined by phosphorus analysis in duplicate and glass transition temperature (T_g) by differential thermal analysis. Viscosity measurements were made in 1.0% cyclohexanone solution at 25°C.

RESULTS AND DISCUSSION

Duplicate free-radical copolymerizations of vinyl chloride and Bis-Beta were run in aqueous suspension and in toluene solution at 55°C. The compositions of the copolymers, which were obtained at low conversions, were determined by phosphorus analysis. The results are shown in Tables I and II.

From the results, a plot of monomer–copolymer composition (Fig. 1) shows a marked tendency toward alternation in the copolymerization, with an azerotropic composition (copolymer and feed compositions being identical) at about 55 mole % of Bis-Beta. Monomer reactivity ratios r_1 (Bis-Beta) and r_2 (vinyl chloride) were determined using the method of Fineman and Ross [5]. As shown in Figures 2 and 3 (expanded plot), a plot of $(f-1)/F$ as ordinate versus f/F^2 as abscissa leads to a straight line whose slope is $-r_2$ and intercept is r_1, where F and f are mole ratios of Bis-Beta/vinyl chloride in feed and in copolymer, respectively. The r_1 and r_2 values thus determined were 0.37 ± 0.1 and $0.26 \pm$

TABLE I

Copolymerization of Vinyl Chloride/Bis-Beta in Water at 55°C

Monomer Composition						Con-	Copolymer Composition		
VC g.	BB g.	H_2O ml.	AIBN g.	F^a	Time hr.	version %	P %	f^a	RV^b
92.0	8.0	350^c	0.10	0.0231	2.6	5.2	3.28,3.20	0.087	1.79
92.0	8.0	350^c	0.10	0.0231	2.6	5.4	3.14,3.20	0.084	1.78
84.0	16.0	350^c	0.10	0.0506	2.6	5.5	5.04,5.25	0.169	1.60
84.0	16.0	350^c	0.10	0.0506	2.6	5.3	5.68,5.54	0.196	1.63
76.0	24.0	350^c	0.10	0.085	2.6	11.8	6.62,6.64	0.267	1.52
76.0	24.0	350^c	0.10	0.085	2.6	10.2	6.86,6.88	0.286	1.52
44.4	55.6	150	0.10	0.336	1.0	12.2	9.4, 9.4	0.644	1.28
44.4	55.6	150	0.10	0.336	1.0	12.5	9.4, 9.5	0.660	1.28
21.0	79.0	55	0.10	1.01	1.25	15.0	10.8,10.8	1.16	1.16
21.0	79.0	55	0.10	1.01	1.25	13.0	10.6,10.5	1.03	1.17
20.0	360^d	160	0.60	4.71	4.3	5.5	11.9,11.9	2.28	1.06
15.0	500^d	160	1.20	8.73	4.3	2.7	12.1,12.1	2.71	1.05

^{a} Mole ratio BB/VC = (wt % BB/233)/(wt % VC/62.5).
^{b} Relative viscosity.
^{c} Containing 0.2 g Methocel.
^{d} Containing 2.5% inert solvent.

0.05, respectively, and were not dependent on the medium of copolymerization (i.e., water versus toluene).

From the reactivity ratios and using the Price–Alfrey equation [6], the Q and e values for Bis-Beta were calculated to be 0.23 and 1.73, respectively, using 0.044 and 0.20 as the corresponding values for vinyl chloride [7].

It is of interest to compare Q and e of Bis-Beta with those obtained by other workers who used styrene [8–10] or vinyl acetate [8] instead of vinyl chloride. The results (Table III) show that the Q,e values of Bis-Beta differ, depending on the comonomer used in its copolymerization and on the copolymerization conditions used for measuring the reactivity ratios. The high e value for Bis-Beta (1.73), obtained in this work, corresponds more closely to the e value based on vinyl acetate (1.59) than those based on styrene (0.10 to 0.80) as the comonomer. This discrepancy may be attributed to the approximate nature of the Alfrey–Price equation [11] from which Q and e are derived. Notwithstanding the theoretical limitation and using the Q,e values obtained in this work as the basis, the reactivity ratios of Bis-Beta and other common vinyl monomers were calculated. The results, shown in Table IV, could serve as a guide to workers contemplating such copolymerizations.

Interestingly, the alternating tendency of vinyl chloride/Bis-Beta copolymerization was also manifested in the rates of copolymerization, which were

TABLE II
Copolymerization of Vinyl Chloride/Bis-Beta in Toluene at 55°C

VC g.	BB g.	Tol. ml.	AIBN g.	F[a]	Time hr.	Conversion %	P %	f[a]	RV
90.0	10.0	140	0.20	0.0296	6.5	9.0	3.30,3.30	0.0878	1.25
90.0	10.0	140	0.20	0.0296	6.5	9.0	3.24,3.25	0.0870	1.26
80.0	20.0	140	0.20	0.0670	6.5	12.0	5.70,5.70	0.202	1.20
80.0	20.0	140	0.20	0.0670	6.5	10.0	5.50,5.60	0.199	1.20
50.0	50.0	120	0.20	0.268	6.5	13.0	9.00,8.95	0.557	1.12
50.0	50.0	120	0.20	0.268	6.5	16.0	9.00,8.96	0.560	1.11
20.0	80.0	140	1.0	1.073	6.5	14.0	11.2,11.1	1.408	1.05
20.0	80.0	140	1.0	1.073	6.5	12.0	10.7,10.6	1.087	1.04
30.9	269[b]	165	2.7	2.28	9.0	6.4	11.5,11.4	1.66	1.05
20.0	360[b]	165	4.0	4.71	9.0	4.5	11.9,12.0	2.37	1.04
15.0	500[b]	150	5.0	8.73	9.0	3.5	12.4,12.0	2.97	1.04

Monomer Composition / Copolymer Composition

[a] Mole ratio BB/VC = (wt % BB/233)/(wt % VC/62.5).
[b] Containing 2.5% inert solvent.

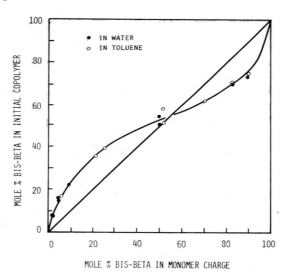

Fig. 1 Monomer copolymer composition diagram.

determined by measuring the amount of polymer formed at different feed compositions, but under identical polymerization conditions. The results (Fig. 4) show unexpectedly that the rates of copolymerization were higher than those of homopolymerization of either monomer. The maximum rate occurred at about

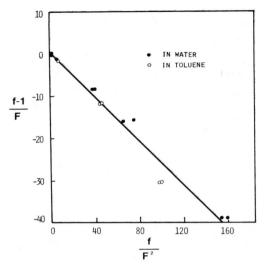

Fig. 2. Fineman–Ross plot for determination of monomer reactivity ratios.

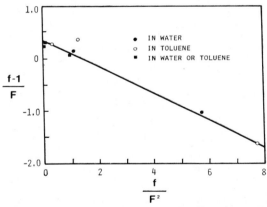

Fig. 3. Expanded Fineman–Ross plot for determination of monomer reactivity ratios.

50 wt % or 20 mole % Bis-Beta. Also, it was observed that the rate of copolymerization was greater in the presence of water (Table V). No similar acceleration by water or methanol took place when Bis-Beta was homopolymerized. These observations, along with the copolymer/feed composition data (see Fig. 1), suggest the possibility that the two monomers form, to a limited extent, a donor/acceptor charge transfer complex. The copolymerization may involve both free monomers and the complex in a manner described by Yoshimura et al. [12] and Seiner and Litt [13], producing the observed alternation and acceleration effects. The presence of water probably increased the polarity of the monomer phase since water was soluble in Bis-Beta in about 16% at room temperature. Increased polarity could enhance charge transfer complex formation. Additional supportive data, however, are required to substantiate these speculations.

TABLE III
Comparison of Q,e Values of Bis-Beta

Comonomer	Q	e	Reference
Vinyl chloride	0.23	1.73	This work
Vinyl acetate	0.40	1.59	8
Styrene	0.20	0.10	8
Styrene	0.11	0.80	9
Styrene	0.195	0.20	10

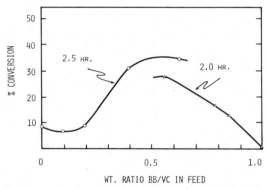

Fig. 4. Rate of copolymerization of BB/VC in water suspension (55°C, 0.1% AIBN).

We have also measured the viscosity and glass temperatures of the vinyl chloride/Bis-Beta copolymers. It was found that for both low-conversion (Table I) and high-conversion copolymers (Table VI), increasing amounts of Bis-Beta in the feed reduced somewhat the relative viscosities of the copolymers. This is probably due to chain transfer of the growing chain with Bis-Beta, either as a monomer or in the polymer chain, since homopolymer of Bis-Beta was generally obtained only as a low molecular weight viscous syrup or semi-solid [9, 14]. The glass temperatures of high-conversion vinyl chloride/Bis-Beta copolymers decreased appreciably with increasing Bis-Beta content (Table VI). This decrease is caused mostly by the softening effect of Bis-Beta. Bis-Beta homopolymer has a considerably lower T_g than that of poly(vinyl chloride).

We wish also to comment on some of the practical aspects of vinyl chloride/Bis-Beta copolymers. For most potential applications, one probably would want to incorporate a minor (5–30%) amount of the more expensive Bis-Beta from the view of modifying the properties of poly(vinyl chloride). Within this concentration range, the rate of one-shot copolymerization is similar to that of homopolymerization of vinyl chloride. Because of the alternation effect, there is no residual Bis-Beta in the polymer at the 80–90% conversion level which is commonly practiced. Complete consumption of the higher boiling component without excessive compositional drift is regarded as beneficial in the industrial

TABLE IV
Calculated Reactivity Ratios of Bis-Beta (M_1) and Other Vinyl
Monomers (M_2)

M_2	Q_2	e_2	r_1	r_2
Bis-Beta	0.23	1.73	1.00	1.00
Styrene	-0.8	1	0.002	0.57
Vinyl Acetate	-0.22	0.026	0.30	0.074
Vinyl Chloride	0.2	0.044	0.37	0.26
Methyl Methacrylate	0.4	0.74	0.031	5.48
Methyl Acrylate	0.6	0.42	0.78	3.60
Methacrylic Acid	0.65	2.34	0.015	20.5
Acrylic Acid	0.77	1.15	0.037	10.5
Methacrylonitrile	0.81	1.12	0.041	10.3
Acrylonitrile	1.2	0.6	0.15	4.9
Vinyl Fluoride	1.28	0.012	8.8	0.09
Maleic Anhydride	2.25	0.23	2.46	0.31
Acrylamide	1.3	1.18	0.093	8.9
Ethylene	-0.2	0.015	0.54	0.044
Isobutylene	-0.96	0.033	0.066	0.011
Vinyl Bromide	-0.25	0.047	0.159	0.124
Vinylidine Chloride	0.36	0.22	0.098	1.57
Butadiene	-1.05	2.39	0.0007	0.56
Isoprene	-1.22	3.33	0.0004	0.40
Diallyl Phthalate	0.36	0.044	0.49	0.31
Vinylpyrrolidone	-1.14	0.14	0.011	0.023

TABLE V
Effect of Medium on Rate of Copolymerization[a]

BB g.	VC g.	Medium H_2O g.	CH_3OH g.	Time hrs.	Yield g.
100	0	0	0	2.0	0.1
100	0	15	0	2.0	0.5
100	0	0	14	2.0	0.1
85	15	0	0	1.5	1.5
85	15	0	0	3.0	10
85	15	170	0	1.5	52
85	15	60	0	3.0	79

[a] 55°C, 0.6 g AIBN.

TABLE VI
Viscosity and T_g of High Conversion VC/BB
Copolymers[a]

%BB in Charge	Rel. Visc.	Tg, °C
0	2.10	84
10	2.03	79
15	2.00	67
20	1.93	62

[a] Aqueous suspension, 0.1% AIBN, 55°C, 12 hr, >95% conversion.

production of successive batches of a polymer. The monomer recovery and compositional drift problems encountered in the copolymerization of vinyl chloride with other monomers such as vinyl acetate, acrylonitrile, and vinylidene chloride, though surmountable, are much simplified in the case of Bis-Beta. The lowering of the T_g of poly(vinyl chloride) by incorporating Bis-Beta can also be beneficial in some applications. Perhaps the most outstanding characteristic of vinyl chloride/Bis-Beta copolymers and terpolymers is the flame retardant property. A plasticized vinyl chloride/Bis-Beta (92.5/7.5) copolymer showed significantly reduced flame spread and smoke generation, when compared to poly(vinyl chloride) [15]. A vinyl chloride/Bis-Beta copolymer (75/25) gave a clear blend with a methyl methacrylate/ethyl acrylate (80/20) copolymer (at 44/56 blend ratio). The blend was found to be nonburning by ASTM D-635 test method for flammability [16]. Also, a plasticized vinyl chloride/Bis-Beta/vinyl acetate (87/6/7) terpolymer could be blown chemically to a fine-cell, uniform foam at low foaming temperatures [17].

The helpful discussions with E. D. Weil and R. Pease are gratefully acknowledged.

REFERENCES

[1] C. A. Brighton, in *Encyclopedia of Polymer Science and Technology*, Interscience, New York, 1971, Vol. 14, p. 345-358.
[2] I. B, Kollyar and E. N. Zilberman, in *Adv. in Polym. Sci.,* Z. A. Rogovin, Ed., Wiley, New York, 1974, p. 240.
[3] E. D. Weil and A. N. Aaronson, lecture presented at the University of Detroit Polymer Conference on Recent Advances in Combustion and Smoke Retardance of Polymers, May 25–27, 1976.
[4] E. D. Weil, *JFF/Fire Retardant Chem.*, 1, 125 (1974).
[5] M. Fineman and S. D. Ross, *J. Polym. Sci., 5,* 259 (1950).
[6] T. Alfrey, Jr. and C. C. Price, *J. Polym. Sci., 2,* 101 (1947).
[7] J. Brandrup and E. H. Immergut, *Polymer Handbook,* Interscience, New York, 1966, p. II–352.
[8] G. B. Fridman, Ya. A. Levin, and B. E. Ivanov, *Sb. Nekot. Probl. Org. Khim. Matr. Nauk Sess., Inst. Org. Fiz. Khim., Akad. Nauk SSSR*, 89–93 (1972).
[9] S. Konya and M. Yokoyama, *Kogyo Kagaku Zasshi, 68*, (No. 6), 1080 (1965).
[10] S. Fujii, *J. Science Hiroshima University*, Ser. A-11, *32* (no. 2), 89 (1967).
[11] T. Alfrey, Jr. and L. J. Young, in *Copolymerization*, G. E. Ham, Ed., Interscience, New York, 1964, p. 67.

[12] M. Yoshimura, Y. Shirota and H. Mikawa, paper presented to Division of Polymer Chemistry, 172nd National Meeting, American Chemical Society, August, 1976, San Francisco, Calif.; *Polym Preprints, 17* (no. 2), 590 (1976).
[13] J. A. Seiner and M. Litt, *Macromolecules, 4,* 308 (1971).
[14] R. G. Beaman, U.S. Patent 2,854,434 (1958).
[15] P. Kraft and L. Smalheiser (Stauffer Chemical Company), Br. Patent 1,382, 625 (1975).
[16] P. Kraft and S. Altscher (Stauffer Chemical Company), U.S. Patent 3,725,509 (1973).
[17] J. C. Goswami (Stauffer Chemical Company), Bel. Patent 832,287 (1976).

CONTRIBUTIONS TO VINYL CHLORIDE SUSPENSION POLYMERIZATION WITH CONSTANT RATE

D. FELDMAN and A. MACOVEANU

Polytechnic Institute of Jassy, "Petru Poni," Jassy, Romania

SYNOPSIS

In the present paper the authors study the possibility of conducting vinyl chloride suspension polymerization at constant rate using a temperature program, as well as some physical and morphological characteristics of the obtained polymer. The elaboration of the temperature program necessary for conducting vinyl chloride suspension polymerization with constant rate was realized by two different means: empirical, on the basis of experimental data, and theoretical, on the basis of a mathematical model, which describes the kinetics of suspension radical vinyl chloride polymerization. In performing the polymerization with constant rate, the design of caloric exchange surfaces becomes much easier, an optimal using of the reaction space being in the same time ensured during the whole time of the process. The interpretation of experimental data permits us to affirm that by polymerization with constant rate can be obtained a poly(vinyl chloride) similar to the polymer with the same average molecular weight synthesized at a constant temperature with slightly weaker volumetric properties but some better thermal stability.

INTRODUCTION

The synthesis and processing technologies of the high polymers can be undoubtedly considered among the newest chemical technologies.

The first studies regarding polymerization reaction engineering appeared after the Second World War, their aim being the transition from batch polymerization processes to continuous ones. This tendency is also preserved nowadays.

In a polymerization plant the main equipment is the reactor. A rational design of such a reactor destined to realize high polymers with preestablished properties, implies a succession of complex problems.

One of those—which may be a main one—is the agitation and homogenization of great amounts of reaction mass with complex rheological properties at suitable price costs.

For big reactors, building in the scale-up process, the most delicate problem is the design of the caloric exchange surfaces, the polymerization heat being in most cases very high. From this viewpoint the nonstationary processes as vinyl chloride suspension polymerization raise very difficult problems. At the suspension polymerization of vinyl chloride at constant temperature and at conversions around 60–70%, a marked decrease of the pressure into the reactor is

Journal of Polymer Science: Polymer Symposium 64, 339–350 (1978)
© 1978 John Wiley & Sons, Inc.   0360-8905/78/0064-0339$01.00

noted; also the temperature is maintaining itself constant. In the same time the amount of evolved heat is strongly increasing, decreasing afterwards rapidly to a low value. The rate of the heat amount variation is so high that the mass reaction can no longer be maintained constant [1].

It is desirable that such nonstationary polyreactions developing with regard to the time variation of the polymerization rate can be led with constant rate during the time; this makes much easier the design of polymerization reactors and the optimal use of the reaction space during all the development time of the process.

In view of attenuating the nonstationary working conditions in vinyl chloride suspension polymerization, one used blends of slow and rapid initiators or one introduced some weak inhibitors into the reaction medium.

In the present paper the possibility of conducting vinyl chloride suspension polymerization with constant rate is presented by means of a temperature program.

ESTABLISHMENT OF THE TEMPERATURE PROGRAM FOR VINYL CHLORIDE POLYMERIZATION WITH CONSTANT RATE

The elaboration of the temperature program necessary for conducting vinyl chloride suspension polymerization with constant rate was carried out both empirically, on the basis of some experimental data [2], and theoretically, on the basis of a mathematical model which describes the kinetics of vinyl chloride radical, suspension polymerization [3].

The empirical method begins by experimentally determining at least three conversion/time curves, for three different polymerization temperatures. On the basis of these curves, by interpolation the conversion/time curves corresponding to each polymerization temperature are determined in the domain in which one wishes to perform the reaction (Fig. 1).

For interpolation, the curves seen in Figure 2 are used, which represent the variation of the reaction time necessary to realize a certain conversion at different polymerization temperatures.

For the purpose of establishing the temperature program necessary for performing vinyl chloride suspension polymerization with constant rate a priori imposed by requirements (e.g., in Figure 1 with a rate of 33% conversion/hour), the following graphic proceeding was used:

In the fascicle of conversion/time curves from Figure 1, one looks for segments which can be estimated as straight lines with the same slope as that of the line after which the conversion in time is varying when conducting polymerization with constant rate (33% conversion/hour in Fig. 1). With that end in view one translated the line which describes the conversion variation at polymerization with constant rate up to the meeting of a conversion-time curve which has a segment that can be estimated with a straight line and which has the same slope as the given straight line. The temperature corresponding to this conversion-time curve will be chosen for the performance of polymerization with constant rate; this temperature will be maintained during all the time the slope of the considered segment remains equal with the slope of the given straight line.

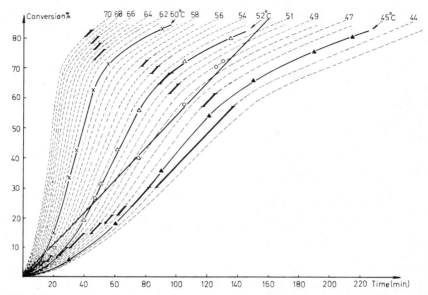

FIG. 1. The conversion/time curves obtained at vinyl chloride suspension polymerization at 45, 52, and 60°C, respectively. — curves experimentally obtained; - - - curves obtained by interpolation.

By this method the temperature program represented in Figure 3 can be obtained.

For the experimental verification of this technique, the vinyl chloride suspension polymerization was performed by using the temperature program shown in Figure 3.

The variation of the conversion has been gravimetrically followed. One observes that the obtained values (marked by little circles) are disposed on the straight line (Fig. 1); hence the polymerization occurs with constant rate as one wished.

The described empirical method is simple, rapid, and certain because it starts from experimental conversion/time curves and therefore takes into account the real conditions in which the polymerization is occurring in the reactor used. Among its drawbacks are to be mentioned the error appearing at the curve segments linearization and at the estimate of the slope of these linearized segments, as well as the fact that the method does not permit the use of a computer.

Consequently, the tendency of doing industrial processes by means of computers being more and more pronounced, the possibility of elaborating the temperature program using a computer on the basis of a mathematical model which describes the kinetics of the reaction was studied.

Till nowadays, the theoretical elaboration of a kinetic model correctly describing the process on its whole duration, irrespective of the reaction temperature, has not been possible because of the complexity of the vinyl chloride polymerization reaction. The models proposed by Talamini [4] and Hamielec

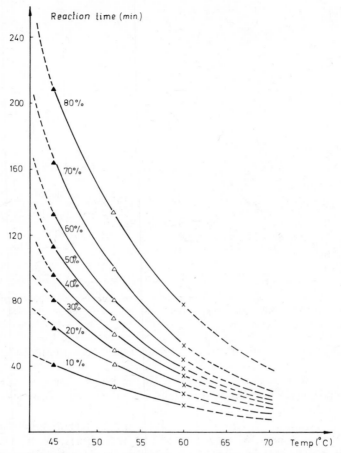

FIG. 2. The variation of the reaction time necessary to realize a certain conversion at different polymerization temperature.

[5] give good results but some of the parameters used need experimental studies.

For the elaboration by computer of the temperature program, the mathematical model proposed by Hamielec and co-workers was used [5], which is:

$$\frac{dx}{dt} = \frac{1 + Qx}{\sqrt{1 - Bx}} k_1 I_0^{1/2} \exp\left(-\frac{k_d t}{2}\right) \tag{1}$$

At conversion $x > x_f$, the following equation is valuable:

$$\frac{dx}{dt} = \frac{Pk_1}{1 - x_f} I_0^{1/2} \frac{(1 - x)^2}{\sqrt{1 - Bx}} \exp\left(-\frac{k_d t}{2}\right) \tag{2}$$

The application of this model was studied by Hamielec and co-workers [6] for vinyl chloride polymerization at 50°C in the presence of more currently used radical initiators.

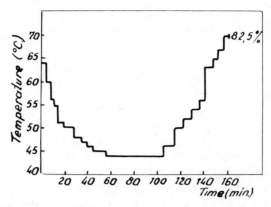

FIG. 3. Temperature program empirically established in order to obtain a constant polymerization rate.

In Figure 4 the conversion/time curves are presented theoretically obtained by means of Hamielec's model and the other one experimentally obtained at vinyl chloride suspension polymerization at 60°C in the presence of isopropylperoxidicarbonate (IPP). The experimental data are those used for the empirical establishment of the temperature program.

The parameters used in the model for 60°C have the following values: $I_0 = 0.001$ moles IPP/mole vinyl chloride; $k_d = 0.2257$ (h^{-1}); $k_1 = 0.3611$ (h^{-1}); $B = 0.393$; $Q = 5.47$; and $x_f = 0.76$. The equation which describes the dependence on temperature of the dissociation constant (k_d) of IPP is [6]:

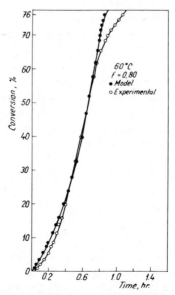

FIG. 4. Suspension vinyl chloride polymerization at 60°C. Comparison between the conversion/time curve obtained experimentally [2] and the one obtained by means of the mathematical model proposed by Hamielec and co-workers [5].

$$k_d = 1.786 \ 10^{18} \exp(-14500/T) \ (h^{-1}) \tag{3}$$

The constant k_1 was calculated as a function of the dissociation constant:

$$k_1 = 2f k_d \tag{4}$$

where f represents the efficiency of initiation. Taking into account the conditions in which the reaction was carried out, we admitted $f = 0.80$ [7]. Considering $\rho_{monomer} = 0.85 \ g/cm^3$ and $\rho_{polymer} = 1.40 \ g/cm^3$, $B = 0.393$. Rigorously exact establishment of the parameters Q and x_f requires the correlation of the data obtained from the model with the experimental conversion-time curves for each temperature from the considered domain.

At the elaboration of the temperature program on the computer, the mathematical relationship which describes the dependence on temperature of Q and x_f was established by the method of least squares applied to the data from the literature [5]:

$$Q = -181.9210 + 1.1455T - 0.0018T^2 \tag{5}$$

$$x_f = -1.1922 + 0.0142T - 0.00003T^2 \tag{6}$$

At the elaboration of the temperature program on the computer, the performance of vinyl chloride polymerization with a constant rate of 33%/hr was taken into account.

The principle of the working program of the computer, its logical schema being presented in Figure 5, is the following:

The computer crosses the admitted domain of temperature variation until it finds the temperature at which the value series of parameters (depending on temperature), which intervene in the mathematical model for the considered conversion, will lead to a rate of polymerization value (%/hr) identical with the imposed rate. By this way, as in the case of the graphic method, one looks for linearized curve segments of the same slope as the imposed slope corresponding to the constant polymerization rate. Thus are retained the necessary temperatures for different additives considered times, so that the polymerization rate may be maintained on the whole duration of the process at the imposed value.

The temperature program established on the computer according to the above described technique is presented in Figure 6.

If one compares both temperature programs, the empirical established one and the other established on the computer, one finds that the variation domain of the temperature is similar. The differences which appear between the two programs are due, on the one hand, to the small deviations of the model confronted by the experimental curves and, on the other hand, to the inherently subjective appreciation regarding the interpolations and linearization imposed by the use of the empirical method.

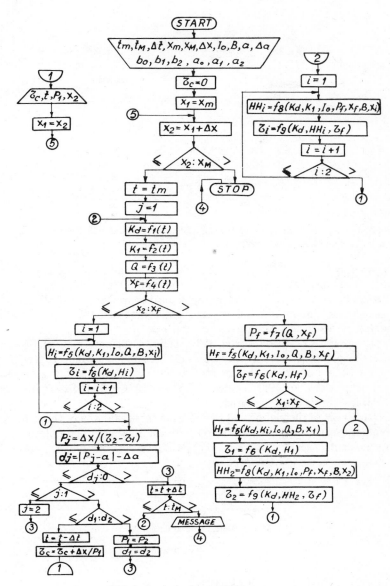

FIG. 5. Logical schema of the program.

CHARACTERIZATION OF THE POLYMER OBTAINED WITH THE TEMPERATURE PROGRAM

In the processing process, the characteristics of the PVC particles determined by the mode of performing the reaction synthesis have a particular importance. The results regarding the characterization of PVC powder obtained by polymerization with constant rate were interpreted by comparing with the data

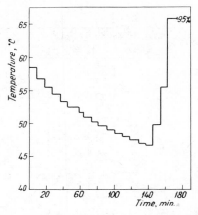

FIG. 6. The temperature program required for leading vinyl chloride polymerization with constant rate of 33%/hr established by a computer.

corresponding to the PVC samples obtained in the same polymerization conditions, but at constant temperature of 45, 52, and 60°C, respectively.

Table I lists some physical and morphological characteristics of PVC obtained by suspension polymerization at constant temperature and at variable in time/temperature according to the established program [2, 8].

As the analysis of the data from Table I shows, the poly(vinyl chloride) synthesized by polymerization with constant rate has an identical average molecular weight with that of a PVC obtained by suspension polymerization at a constant temperature of about 49°C (between 45 and 52°C). The distribution of the molecular weights is also identical.

On the basis of the elaborated temperature program and of the dependence on temperature of the molecular weight, for establishing by calculation the average molecular weight of poly(vinyl chloride) obtained with constant rate, the following relationship was proposed:

$$\overline{M}_x = \frac{\sum_i \overline{M}_{x_i} \cdot m_i}{\sum_i m_i} \tag{7}$$

TABLE I
Physical and Morphological Characteristics of PVC-S Powder

| Characteristic | Polymerization Temperature | | | Program |
	45°C	52°C	60°C	
$\overline{M}_w$ (g/mole)	115,000	90,000	71,000	96,000
$\overline{M}_n$ (g/mole)	50,000	51,000	45,000	52,000
$\overline{M}_w/\overline{M}_n$	2.1	1.8	1.6	1.8
Unsagged density, g/cm^3	0.450	0.360	0.320	0.295
Sagged density, g/cm^3	0.520	0.410	0.385	0.330
Relative compression, %	16.0	13.9	11.9	11.9
Natural slope angle, degrees	61	59	53	57
Pycnometric density (CH$_3$OH), g/cm^3	1.180	1.265	1.280	1.260
Rate of flow, g/sec	5.1	3.1	3.6	3.7

where $x = w$ or n and m is the mass expressed in grams of PVC obtained in the stage "i" of temperature. Using eq. (7) and the temperature program shown in Figure 3, the values $\overline{M}_w = 100,000$ and $\overline{M}_n = 53,500$ were obtained which are in good agreement with the $\overline{M}_n$ experimentally determined (Table I).

In PVC processing, the unsagged and sagged densities have an increasing importance as a result of the development of the continuous and of high capacity proceedings, in most cases the feeding of the equipment being volumetric. It seems that because of the higher polymerization temperature from the beginning of the process, the values of the unsagged and sagged densities are a little lower in the case of PVC obtained with the temperature program.

The synthesis temperature, which determines the value M, influences the structure of the microparticles component of the PVC particles in the following way [9]: with the increase of the temperature the number of the microparticles which form the PVC particles decreases; on the other hand, their interpenetration degree decreases too; thus will be realized a "spongier" structure of the macromolecules which form the microparticles.

Thus is explained the fact that PVC obtained with constant rate corresponds in average molecular weight to a PVC obtained at about 49°C, but in the case of the unsagged and sagged densities it is comparable with PVC obtained at a constant temperature of 60–65°C, corresponding to the initial phase in the polymerization with the program.

From Table II one sees that the PVC sample obtained by polymerization with the temperature program corresponds in porosity and plastifier sorption to the polymer synthesized at constant temperature (in the present case at about 49°C) which leads to the same value of the average molecular weight [8]. The thermal stability of the PVC samples was tested by several methods [8].

In Table III are the results obtained by using the method with Congo red, as well as the values of Brabender thermal stability. The best value of the Brabender thermal stability is that obtained in the case of PVC synthesized with the temperature program. As is known, the method of Brabender thermal stability was verified as the most significant of those used for the PVC processing evaluation.

If the weight loss (%) is examined in Table IV, it can be noted that PVC obtained with constant rate (conversion, $C = 80\%$) has a better thermal stability than the other samples.

TABLE II
Characteristics Regarding the Porosity and the Plastifier Sorption

| | Polymerization Temperature, °C | | | |
Characteristic	45	52	60	Program
Sorption time at 75°C, min	120	33	17	28
Irreversible plastifier adsorption at cold, %	52.6	48.8	37.5	50.4
Absorption rate p DOP at low p PVC/min	0.1	0.5	1.1	1.3
Specific surface, m²/g	1.8	2.8	2.8	2.6

TABLE III
Data Regarding the Thermal Stability of PVC Samples

| Method | Polymerization Temperature, °C | | | Program |
	45	52	60	
Congo Red	38 sec	42 sec	45 sec	43
Total stability, min	14.75	15.50	13.50	19.00
Alligation power, kgf	1.44	1.28	1.10	1.50
Duration constant couple, min	12.0	13.0	11.0	14.5

TABLE IV
The Behavior at Thermal Destruction of PVC Samples

| Sample Name | ATG | | | Loss in Weight (% at T_m) | Order of Destruction Reaction, n | Activation Energy (kcal/mole) |
	$T_i°C$	$T_m°C$	$T_f°C$			
PVC obtained at 45°C ($C = 80\%$)	220	282	370	29.6	0.9	44.43
PVC obtained at 52°C ($C = 80\%$)	200	282	370	34.0	0.8	38.35
PVC obtained at 60°C ($C = 80\%$)	198	275	368	33.0	1.3	44.73
PVC obtained with Temperature Program ($C = 9\%$)	210	282	370	30.0	0.6	44.92
PVC obtained with Temperature Program ($C = 30\%$)	210	275	380	35.0	0.5	43.26
PVC obtained with Temperature Program ($C = 55\%$)	200	275	370	34.3	0.7	35.38
PVC obtained with Temperature Program ($C = 80\%$)	210	280	375	24.7	0.0	40.46

Studies performed on inferior compounds used as models demonstrated that the thermal stability of PVC is determined by the amount of labile groups, by the number of units bonded in syndiotactic chains, as by the position of labile groups in respect to the syndiotactic segments from the macromolecules [10]. The thermal stability of PVC [11] is influenced also by the size of the particles.

The stereoregularity of PVC obtained with the temperature program was determined by NMR spectrography and compared with the stereoregularity of PVC obtained at constant temperature of 45, 52, and 60°C, respectively [12]. The obtained data are presented in Table V.

The interpretation of the data by using dispersional analysis led to the conclusion that one can affirm with 95% probability that PVC obtained with the temperature program presents the same stereoregularity as the PVC samples

TABLE V

Data Regarding the Stereoregularity of PVC Obtained by Means of NMR Spectrography

Polymeriza-tion Tempera-ture (°C)	Conversion (%)	Triads			Diads	
		S (%)	I (%)	H (%)	S (%)	I (%)
45	30	40.0	38.0	22.0	51.0	49.0
	55	34.0	35.0	31.0	50.5	49.5
	80	37.0	48.0	15.0	44.5	55.5
52	30	41.0	35.0	24.0	53.0	47.0
	55	42.3	41.5	16.5	50.5	49.7
	80	40.0	37.0	23.0	51.5	48.5
60	30	34.0	30.4	35.4	51.7	48.10
	55	35.4	34.0	31.8	51.3	49.0
	80	32.0	37.0	31.0	47.5	52.5
With program	30	21.5	20.5	58.0	50.5	49.5
	55	42.4	32.9	25.6	55.2	45.7
	80	40.0	30.0	30.0	55.0	45.0

obtained at constant temperatures. The fact that the variable temperature used for polymerization with constant rate has no significant influence on the syndiotacticity was confirmed also by X-ray crystallinity determination. PVC samples, all having the same conversion of 80%, obtained at a temperature of 45, 52, and 60°C, and, with the temperature program, all presented a crystallinity degree of around 5%. According to the literature data [13], one can affirm that in all studied samples the syndiotactic sequences with four to five monomer units are present in very little amount.

We also established the content of labile groups (Table VI) in the samples obtained by polymerization with the temperature program and at constant temperature [8].

Studying the data presented in Table VI, one finds that PVC sample obtained by polymerization with the temperature program has the lowest content of labile chlorine which is identical with that of the PVC sample obtained at a constant temperature of 45°C.

According to the data found by Caraculacu and co-workers [14], there is an inversely proportional dependence between the average molecular weight and the labile chlorine content. Because the average molecular weight of the PVC sample obtained with the temperature program is lower than that of the sample obtained at 45°C, the former ought to have the highest amount of labile chlorine, but in fact it is not so. Consequently, the amount of labile chlorine could be considered a factor which is determining improvement of the thermal stability

TABLE VI

The Amount of Labile Chlorine in PVC Samples

Polymerization Temperature (°C)	45	52	Program
Cl_L (%)	0.107	0.125	0.101

of PVC obtained with constant rate against the polymer with the same average molecular weight but obtained by polymerization with constant temperature.

In relation with the size of PVC particles is established that the sample obtained with the temperature program contains 76% higher particles than 315 μ, in comparison with the content of 71.5% of the sample obtained at 45°C [8]. According to the data of Guyot and Bert [11] this difference between the sizes of PVC particles can also be an improvement factor of the thermal stability of PVC obtained with constant rate.

CONCLUSIONS

(a) Suspension polymerization with constant rate of vinyl chloride can be realized by using a temperature program which can be experimentally established or by starting from a mathematical model which describes the kinetics of the process.

(b) Conducting the polymerization with constant rate, the design of caloric exchange surfaces becomes much easier, an optimal using of the reaction space being in the same time ensured during the whole time length of the process.

(c) According to the temperature program, a certain average molecular weight of the polymer can be preestablished, and consequently the plastification and gelation behavior, as well as the melting flow, will be foreseeable.

(d) The analysis of experimental data permits us to affirm that by polymerization with constant rate there can be obtained a poly(vinyl chloride) similar to the polymer with the same average molecular weight synthesized at a constant temperature with slightly weaker volumetric properties but somewhat better thermal stability.

REFERENCES

[1] H. Hedden, Dissertation, Stuttgart, 13 (1969).
[2] M. Macoveanu and K. H. Reichart, Angew. Makromol. Chem. , 44, 676 (1975).
[3] M. Macoveanu, V. Nagacevschi, and D. Feldman, in press.
[4] A. Crosato-Arnaldi, P. Gasparini, and G. Talamini, Makromolek. Chem., 117, 140 (1968).
[5] H. Abdel-Alim and A. E. Hamielec, J. Appl. Polym. Sci., 16, 783 (1972).
[6] H. Abdel-Alim and A. E. Hamielec, J. Appl. Polym. Sci., 18, 1603 (1974).
[7] R. A. Jusipov, I. V. Moskviceva, T. M. Kartaşova, V. I. Tomaşcik, I. B. Kotlear, and A. O. Belopolski, Tr. Khim. Khim. Tekhnol., 3, 67 (1972).
[8] D. Feldman, M. Macoveanu, and G. Robilă, 2nd International Symposium on Poly(Vinyl Chloride), July 5–9, Lyon (France), 1976.
[9] V. Butucea, A. Negrei, L. Renția, and V. Băncilă, Materials Plastice, 10, 6, 309 (1973).
[10] M. Asahina and M. Onosuka, J. Polym. Sci., A2, 500 (1964).
[11] A. Guyot and M. Bert, J. Appl. Polym. Sci., 19, 10, 2923 (1975).
[12] M. Macoveanu, E. Bezdadea, and D. Feldman, Eur. Polym. J., in press.
[13] G. Talamini and G. Vidotto, Makromolek. Chem., 100, 48 (1967).
[14] E. C. Buruiană, G. Robilă, E. C. Bezdadea, and A. A. Caraculacu, The First Microsymposium on Macromoleculare Chemie, Jasay, Nov. 14–15 (1975).

THERMOPLASTIC URETHANE CHEMICAL CROSSLINKING EFFECTS*

C. S. SCHOLLENBERGER and K. DINBERGS

The B. F. Goodrich Company,
Research and Development Center,
Brecksville, Ohio 44141

SYNOPSIS

This paper reviews the chemical structures and attendant polymer morphology that have been advanced to explain the virtual crosslink (VC) phenomenon in segmented polyurethane elastomers. The different chemical methods employed to produce useful vulcanizates from the classical polyurethane elastomer use-forms (castable liquid, millable gum) through covalent crosslinks (CC) are discussed. And finally, the nature of the polymer property changes occurring when a CC network is superposed on the VC network of a representative thermoplastic poly(ester–urethane) elastomer composition through the use of a free radical curing agent (organic peroxide) is described. The CC–VC network polymer of the subject polyurethane composition proved to have greater resistance to: solvation and its effects, stress relaxation, heat distortion, and compression set, and showed much higher modulus values than the VC network polymer. But the CC–VC network polyurethane also showed less extensibility, tear strength, low temperature flexibility, and flex life. The degrees of these changes are discussed and molecular explanations proposed in some cases.

BACKGROUND

Thermoplastic urethane elastomers have become well-established commercial materials since their inception in The B. F. Goodrich research laboratories in the 1950s. Wide interest has been generated in this interesting polymer class as new applications steadily appear, supplementing the many important uses they have already found. The "virtual crosslink" phenomenon responsible for the unique physical state of these elastomers has now been extensively applied in other nonurethane thermoplastic elastomer systems.

As the name implies, thermoplastic urethane elastomers are at the same time both thermoplastic and highly elastic rubbers. They consist of essentially linear primary polymer chains. The structure of these primary chains comprises a preponderance of relatively long, flexible, "soft" chain segments which have been joined end-to-end by rigid "hard" chain segments through covalent chemical bonds. The "soft" segments are diisocyanate-coupled, low melting polyester or polyether chains. The "hard" segments include single diurethane bridges resulting when a diisocyanate molecule couples two polyester or polyether

* Previously published in *J. Elast. Plastics,* 7 (1975).

Journal of Polymer Science: Polymer Symposium 64, 351–368 (1978)
© 1978 John Wiley & Sons, Inc.   0360-8905/78/0064-0351$01.00

molecules, but more particularly they are the longer, high melting urethane chain segments formed by the reaction of diisocyanate with the small glycol chain extender component.

The polar nature of the recurring rigid, hard, urethane chain segments results in their strong mutual attraction, aggregation, and ordering into crystalline and paracrystalline domains in the mobile polymer matrix. The nature and morphology of such domains in urethanes has been the subject of much study. The abundance of urethane hydrogen atoms, as well as carbonyl and ether oxygen partners in urethane systems, permits extensive hydrogen bonding among the polymer chains. This hydrogen bonding apparently restricts the mobility of the urethane chain segments in the domains and thus their ability to organize extensively into crystalline lattices. As a consequence, semi-ordered regions result which have been described as "paracrystalline." Association of the π electrons of the polymer aromatic structures represents another binding force. The more weakly attractive van der Waals forces are also operative in all parts of the polymer chains. The polymer chains are long enough to get entangled in each other.

The lateral effect of all the foregoing states and forces, particularly paracrystallinity, and hydrogen bonding, is to tie together or "virtually crosslink" the linear primary urethane chains. That is, the primary urethane chains are crosslinked in effect, but not in fact. Concurrently, of course, the virtual linkages also lengthen the primary urethane chains. The overall consequence is a labile network of polymer chains which displays the superficial properties of a strong rubbery vulcanizate over a practical range of use temperatures.

Virtual crosslinking is a phenomenon that is reversible with heat and, depending upon polymer composition, with solvation, offering many attractive processing alternatives for thermoplastic urethanes. Thermal energy great enough to (reversibly) break virtual crosslinks, but too low to appreciably disrupt the stronger covalent chemical bonds that link the atoms in the primary polymer chains, can be applied to extrude or mold the polymers. And a solvent which solvates the polymer chains, reversibly insulating the chains and virtual crosslinks, carries the primary polymer chains into solution separate and intact as, for example, for coating applications.

Figure 1 depicts the type of molecular chain arrangement thought to exist in thermoplastic polyurethane elastomers.

The development of urethane elastomers began and grew around covalently crosslinked urethane systems, in keeping with the classical method of utilizing elastomeric materials. These castable liquid and millable gum systems involved covalent chemical crosslinking of the polymer chains through allophanate and/or biuret and urethane crosslinks [1-6] (Fig. 2). The crosslinking of urethane elastomers has been well reviewed by Meyer [7].

It was then shown that free radical-producing chemicals can also be used to crosslink urethane millable gum chains and to convert these gums to a useful state. Meyer points out that the millable polyurethane gums are prepared by coupling polyester or polyether macroglycols with nearly equal molar proportions of a diisocyanate, and that reinforcing pigments (e.g., carbon black, silica) are

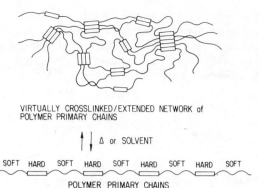

FIG. 1. Thermoplastic urethane elastomer chain organization.

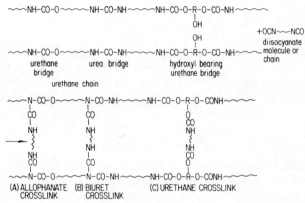

FIG. 2. Urethane chain crosslinking via (A) allophanate, (B) biuret linkage, and (C) urethane bonds.

contributors to the high (5000–7000 psi) tensile strength values possible in polyurethane millable gum vulcanizates [7].

Added organic peroxide may be used to cure poly(ester–urethane) millable gums [8–11]. The crosslinking mechanism in these polymers is considered to involve the generation of free radical positions in the polyurethane chains and the coupling of these positions to form carbon-to-carbon covalent chemical crosslinks (Fig. 3).

Alternatively, the sulfur-accelerator type of free radical curing system employed to vulcanize natural rubber and conventional unsaturated synthetic

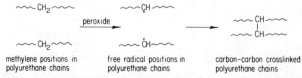

FIG. 3. Covalent chemical crosslinking of polyurethane millable gum chains by organic peroxide.

rubbers can be used to crosslink special-structured polyurethane millable gums [5, 12–15]. Such polymers are rendered responsive to sulfur-accelerator cure by building occasional, special cure sites into the polymer chains which are more reactive in the free radical cure mechanism. The preferred active structure is an ethylenically unsaturated appendage on the polyurethane chain [15] (which contains at least one methylene hydrogen atom on the carbon atom adjacent to the double bond),

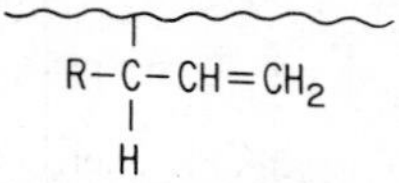

$$R-\underset{\underset{H}{|}}{\overset{|}{C}}-CH=CH_2$$

Such structures are readily incorporated in the polyurethane chains via ethylenically unsaturated glycols, e.g., glyceryl-α-allyl ether, trimethylolpropane monoallylether, as chain extender [5, 15–18] or as macroglycol polymerization initiator [17, 18].

The sulfur-accelerator cure of urethane millable gums can be rapid [19]. It likely proceeds according to Figure 4 if analogy to the same cure of conventional hydrocarbon rubbers holds. Sites suggested as likely points for free radical formation in polyurethane chains are the α-methylenic positions in the adipyl moieties of adipate-based polymers [19, 20] and the methylene group between the two phenyl rings in the diurethane bridge structure of MDI-based polymers [18, 21].

INTRODUCTION

Conventional elastomeric materials such as natural rubber and the various synthetic rubbers, including the classical urethane rubbers, are chemically crosslinked by vulcanization to develop useful mechanical properties. But thermoplastic urethane elastomers exhibit elasticity and high level mechanical properties without such crosslinking. However, they also can be covalently crosslinked by chemical agents. So it seemed of interest to determine the effects on polymer properties of superposing covalent chemical crosslinks (CC) on a thermoplastic urethane elastomer network already bonded by virtual crosslinks (VC). The consideration of these effects is the focal point of this paper. Weisfeld et al. earlier estimated the contributions of primary versus secondary valence crosslinks in polyurethanes by modulus–temperature relations [20].

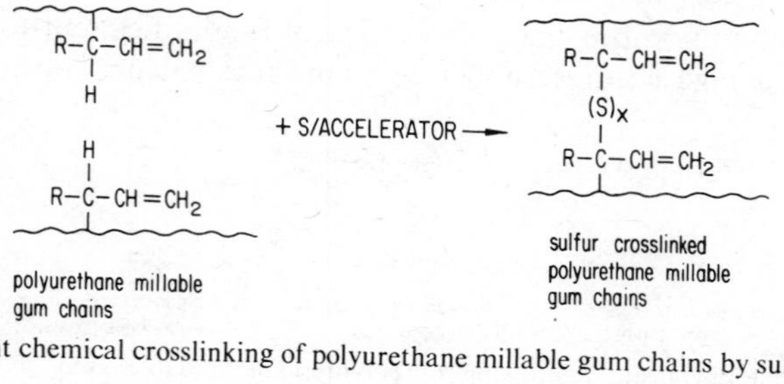

FIG. 4. Covalent chemical crosslinking of polyurethane millable gum chains by sulfur-accelerator systems.

EXPERIMENTAL SECTION

Materials

The thermoplastic urethane elastomer involved in this study was an essentially linear poly(ester–urethane) prepared by the reaction of 1.30 moles of diphenylmethane-p,p'-diisocyanate (MDI), 1.00 mole of poly(tetramethylene adipate) glycol (MW ca. 1000), and 0.30 mole of 1,4-butanediol under random melt polymerization conditions without benefit of added catalyst. The structures of these monomers and resulting polymer are shown in Figure 5.

Covalent crosslinking (CC) of such polymer was achieved by adding 2 phr (full strength basis) of a conventional, commerical, organic peroxide together with minor process aids but no pigment or filler to the thermoplastic urethane elastomer on the rubber mill, then curing samples of this "pure gum" compound by compression molding for the same time (<10 min) at 210–220°C in an electrically heated press. Uncured (VC polyurethane) test samples were prepared for property measurements by compression molding the identical thermoplastic urethane elastomer neat.

Test Methods

Specific test methods applied in comparing VC and CC–VC network polyurethane properties follow.

Solvation Resistance (ASTM-D741)

The 0.075 × 1 × 2″ samples were died from compression molded/cured sheets. They were immersed in the test liquid at 25°C for 7 days, then volume and weight changes were determined.

Stress Relaxation

This property was measured with the Instron Tester on samples extended 20% in 2.4 sec. The test was continuous and testing was done at 25 and 100°C. The 0.025 × 0.25 × 6″ test samples were died from compression molded/cured sheet and were bench-marked 4″ apart. The ends of these strip samples were clamped

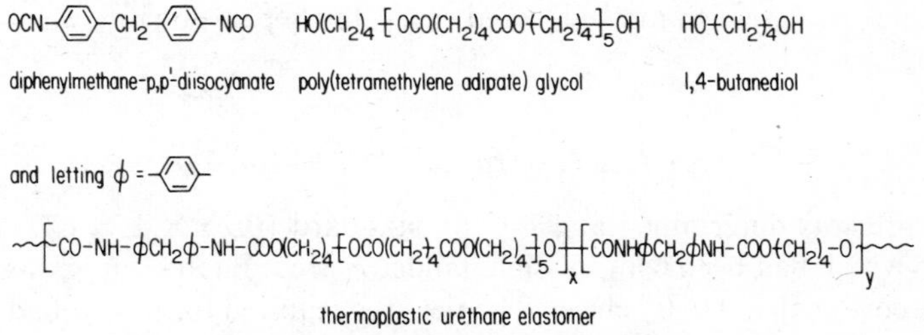

thermoplastic urethane elastomer

FIG. 5. Structure of monomers and polymer of this study.

in the Instron, jaws even with bench marks, and the slack example was conditioned at test temperature for 10 min prior to loading and test.

Bulk Flow at Elevated Temperature

In this test $0.025 \times 0.25 \times 6''$ strips died from compression molded/cured sheets were looped around a $1/8''$ diameter stainless steel rod and stapled in two places to flatten the loop. The rod with the samples was placed in a circulating air oven whose temperature was then increased at an average rate of about $2°/min$. The samples were observed regularly through the glass door of the oven and periodic measurements of the length of the exposed samples were made.

Compression Set (ASTM-D395-B)

The compression set of VC and CC–VC network polyurethane samples was determined at 25 and 70°C according to the indicated test procedure on seven plied discs cut from 75 mil thick compression molded/cured sheet.

Low Temperature Flexibility (Gehman Low Temperature Modulus, ASTM-D1053)

Low temperature flexibility was determined on samples died from 75 mil thick compression molded/cured sheet. Cooling was in gaseous nitrogen via liquid nitrogen.

Stress–Strain Properties (ASTM-D412)

Stress–strain properties were measured at 25 and 100°C on $1/8''$ dumbbells died from 25 ml thick compression molded/cured sheet. Testing was performed in the Instron Tester at a jaw separation rate of 20 inches per min. Samples were conditioned for at least 10 min at test temperature in the slack condition prior to testing.

Tear Strength (Graves Tear Strength, ASTM-D624)

Tear strength was determined at 25 and 100°C on Die C samples cut from 75 mil thick compression molded/cured sheet. Jaw separation rate was 20 inches per min.

Flex Life (DeMattia, ASTM-D813)

Flex life was determined at 27°C on standard ($0.25 \times 1 \times 6''$, grooved) samples which had been compression molded/cured. Each sample was pierced in the groove with a $0.080''$ chisel. The flexures required for this wound to grow to $0.8''$ comprises flexures to failure.

Abrasion Resistance

Taber Abrasion. The Taber Abrader was used with the H-18 wheel and a 1000 g load to measure the abrasion of test discs cut from 75 mil thick compression molded/cured sheet. Results are expressed as sample weight loss (milligrams) after 5000 revolutions at 25°C.

Pico Abrasion

The Pico abrader [22] was also used to measure the abrasion resistance of compression molded/cured samples after 80 and 240 revolutions at 60 rpm and 25°C. Results are expressed as an index relative to the value, 100, which is assigned to the control (a high quality) rubber tire tread vulcanizate. Indexes above 100 mean better abrasion resistance then the control and vice versa.

The samples used in the foregoing tests were exposed for several days in a 50% relative humidity, 25°C environment prior to testing which was either conducted in the same area or promptly elsewhere at similar humidity.

RESULTS AND DISCUSSION

In the following discussion of experimental results we will first compare the subject elastomeric urethane composition in the VC and the CC–VC network states using tests which characteristically distinguish between labile and stable polymer network systems. These will include solvation resistance, stress relaxation, permanent set, bulk flow at elevated temperature, and low temperature flexibility. Then comparisons will be made in tests whose dependency on crosslinking is perhaps somewhat less obvious, namely: stress–strain properties, tear strength, flex life, and finally abrasion resistance. These will be discussed in the foregoing order.

Solvation Resistance

Extensive solvation of polymer chains by small solvent molecules masks the attractive forces (hydrogen bonding, van der Waals, crystallinity) tending to hold bulk polymer chains together, and provides an insulating layer around the chains. In the case of linear and branched polymer chains this solvation increases the distance of chain separation, reducing chain interaction in concentrated solutions, and allowing solvated chains to float free of one another in dilute solutions.

In the case of CC polymer compositions, no solvent is capable of dissolving such polymer networks short of the irreversible chemical destruction of at least part of the CC network bonds. However, the chains in CC polymer networks may still be solvated to various degrees, depending on the solvator. A good solvator will solvate the CC network chains extensively, thereby insulating them from one another. This causes them to separate some, but not completely, since the stable CC act as firm tie points for the chains and will not allow their full separation. Thus, the CC network polymer swells (reversibly) but does not dissolve. This phenomenon is the basis of a classical method for determining the crosslink density of CC polymer networks.

In the case of those VC polymers which are linear or branched, a given solvent may or may not strongly solvate both the soft and hard segments of the polymer chains [23, 24]. If it is capable of solvating both adequately, then the VC polymer dissolves in the solvent. However, if, for example, it solvates only the soft segments in the polymer chains, then the VC hold firm preventing dissolution, but the solvated soft segments separate more from one another. So the VC sample only swells (reversibly) in the solvator, as was the case with the CC network polymer.

Table I compares the effects of several different liquids on the subject thermoplastic poly(ester–urethane) elastomer composition in both the VC and CC–VC network states.

The data of Table I show that the VC network state of the subject polyurethane composition was more solvated than the CC–VC network state. Immersion liquids (1) through (11) solvated the VC network to varying degrees, with special effects by the aromatics, benzene and chlorobenzene. Immersion liquids (12) through (17) completely dissolved the VC network polymer.

In contrast none of the immersion liquids dissolved the CC–VC network polymer although the high swellers for the VC polymer, benzene and chlorobenzene, and the VC polymer solvents, (12) through (17), have the most pronounced swelling action on the CC–VC network polymer.

TABLE I

Solvation Resistance of Subject Thermoplastic Urethane Elastomer in VC and in CC–VC States (7 Days Immersion, 25°C)

Immersion Liquid	VC Polymer Volume	(% Increase) Weight	CC–VC Polymer Volume	(% Increase) Weight
1. Water	1.31	1.00	1.27	1.16
2. Methanol	24.85	16.60	22.41	15.02
3. Ethylene glycol	1.96	1.82	0.78	0.82
4. ASTM Oil No. 1	0.87	0.77	0.29	0.25
5. ASTM Oil No. 3	7.49	5.16	1.21	0.96
6. JP4 Jet Fuel	5.08	3.49	2.85	1.84
7. Gasoline (Sunoco 200)	19.31	13.40	16.08	11.14
8. Benzene	169	126	101	74
9. Xylene	64	47	51	37
10. Chlorobenzene	273	255	135	125
11. Perchloroethylene	30.21	41.92	26.67	37.17
12. Methylene chloride	dissolves		259	295
13. Methyl ethyl ketone	dissolves		171	116
14. Cyclohexanone	dissolves		261	205
15. Tetrahydrofuran	dissolves		247	183
16. Ethyl acetate	dissolves		125	94
17. Nitromethane	dissolves		121	115

Stress Relaxation

The chains in an ideal, stable, elastomeric polymer network sample should resume their approximate, original, preferred arrangement when a distorting force is released. Thus, the force required to distort the network a given amount should not decay with time distorted. But in practice this ideal behavior is not observed and applied stress does decay with time. The decay of applied stress indicates that a relaxation of the polymer chains within the sample has occurred: that some of the polymer chains have irreversibly moved somewhat in response to the applied stress by slipping from their initial positions past one another to new positions where associations with new chain neighbors form. The inability of a displaced polymer chain to regain its original position in an elastomer network is a consequence of broken network linkages, including VC, due to the effect of applied mechanical or thermal energy, for example, or to chemical attack such as oxidation.

Singh and Weissbein [25, 26] studied the stress relaxation of polyurethane elastomers which they had highly covalently crosslinked exclusively via urethane bonds (see Fig. 2C). But their polymer synthesis—the catalyzed reaction of an isocyanate terminated prepolymer (free of diisocyanate monomer) with a small triol—was not designed to produce a segmented polyurethane structure with strong virtual crosslinks. And their tests were mostly run at higher temperatures (130–160°C) than those of the present study. They found the continuous stress relaxation of their polymers to be pronounced and attributed the phenomenon to reversible thermal dissociation of urethane bonds in the CC polymer network, with MDI affording urethane bonds of superior stability.

Figure 6 information was taken from polymer samples which were tested 2 weeks after forming (VC) or cure (CC–VC) by compression molding. Thus, we have considered them to be "unaged" samples. The plots of Figure 6 indicate

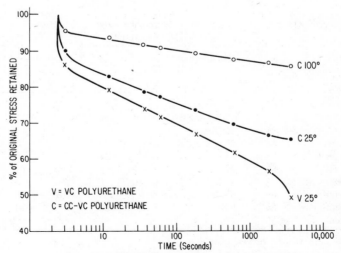

FIG. 6. Stress relaxation (continuous, 20% extension).

that at 25°C the subject VC polyurethane showed considerable stress relaxation, with a half-life (50% stress decay) of about 1 hr (Curve V). Presumably the continuous stress on the sample pulled some polymer chain hard segments out of the hard segment domains where they had aggregated, and allowed chain slippage to stable new positions in the VC polymer network to occur.

Figure 6 also shows that, in contrast, the CC–VC polyurethane stress decay at 25°C was only 35% in the same period, demonstrating the enhanced stability conferred on the VC network by the CC network (Curve C, 25°C).

The 100°C test results of Figure 6 do not include a curve for the VC polyurethane because thermal energy at 100°C overcame VC, melting the polymer sample.

It is seen that at 100°C the CC–VC polyurethane shows less stress relaxation than at 25°C (Curves C). This result may be a bit startling at first until it is recollected that the plots compare *percent retention of original stress* versus time. Apparently at 100°C we are primarily seeing the rather minor stress relaxation of the much more stable CC polyurethane network, the VC network having been largely dispelled by thermal energy during the 10 min sample conditioning at test temperature prior to sample extension in the test. (But some vestigial VC network may still remain after this conditioning so its relaxation during test may also be involved.)

Consideration of absolute stress level values at 20% elongation for unaged samples will perhaps help clarify the preceding apparent anomaly. These appear in Table II.

Table II data show that, in fact and as expected, the *absolute stress levels* of the CC–VC network at 20% elongation are actually *greater* at 25°C than those of the same material at 100°C for most of the test term. (So Curves C of Fig. 6 should not startle as they do not show absolute stress level, but rather *percent of initial stress retained.*)

TABLE II

Stress Levels for Unaged VC and CC–VC Polyurethane Network Samples at 20% Elongation vs Time at 25 and 100°C

Time at 20% Elong. (sec)	VC Polyurethane				CC–VC Polyurethane			
	A(psi) (at 25°C)	B(%)	A(psi) (at 100°C)	B(%)	A(psi) (at 25°C)	B(%)	A(psi) (at 100°C)	B(%)
0	112	100	Sample melted	—	150	100	116	100
3	96	86	"	—	135	90	110	95
12	89	79	"	—	124	83	108	93
36	83	74	"	—	118	79	106	92
60	80	71	"	—	116	77	105	91
180	75	67	"	—	110	74	103	89
600	69	62	"	—	105	70	102	88
1800	64	57	"	—	100	67	101	87
3600	55	49	"	—	98	65	100	86

(A) Absolute stress at 20% elongation.
(B) Percent of initial stress at 20% elongation retained.

In the case of the CC–VC sample, stress relaxation is actually considerably less at 100 than at 25°C, as Figure 6 shows, since at 100°C the more labile VC polymer network has been largely dispelled by thermal energy and we primarily measure only the stress relaxation of the more stable CC polymer network. But, note that Figure 8 absolute stress values for the CC–VC sample at 25 and 100°C gradually converge during the test term, finally coinciding at 1800 sec and beyond. This may comprise evidence for the presence and gradual relaxation of a vestigial VC network during 100°C testing already suggested.

The foregoing results suggest that at 20% elongation the CC contribution to network strength afforded a stress level of ca. 100 psi, so by difference the VC contribution was ca. 50 psi at 25°C and nil at 100°C. The initial 112 psi stress of the VC network polymer at 25°C then must be the sum of ca. 50 psi (the VC contribution) and a ca. 62 psi contribution from the attractive forces operative along the polyurethane chains which are not localized in the hard segment VC aggregates. These binding forces are not different in nature than those producing VC. They include hydrogen bonding, van der Waals forces, aromatic π electron association, and chain entanglements. They are merely more diffuse in the polymer system and operative along the entire length of the polymer chains.

Earlier study of thermoplastic polyurethane elastomer stress relaxation [27] revealed behavioral changes during long-term testing, suggestive of polymer chain reorganization and attendant morphology development. In the present study stress relaxation results suggest that a similar phenomenon occurred in the relaxed subject polyurethane during long-term (7 month) storage at 25°C. For, under these conditions the stress relaxation at 25°C of the VC network sample decreased substantially (Fig. 7, Curves V versus V′). And Table III shows that the absolute stress levels at 20% elongation of both the VC and CC–VC network polymers increased appreciably on aging. Current measurements show no increase and little change in the solution viscosity of the subject VC poly-

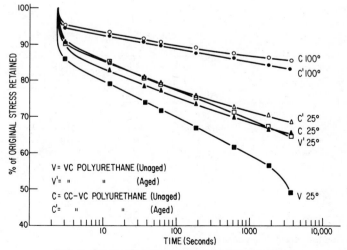

FIG. 7. Stress relaxation (continuous, 20% extension).

TABLE III

Stress Levels for Unaged and Aged VC and CC–VC Network Polyurethane Samples at 20%
Elongation vs Time, 25°C

Time at 20% Elong. (sec)	VC Polyurethane				CC–VC Polyurethane			
	Unaged		Aged		Unaged		Aged	
	A(psi)	B(%)	A(psi)	B(%)	A(psi)	B(%)	A(psi)	B(%)
0	112	100	164	100	150	100	197	100
3	96	86	147	90	135	90	178	91
12	89	79	139	85	124	83	167	85
36	83	74	133	81	118	79	159	81
60	80	71	129	79	116	77	156	79
180	75	67	123	75	110	74	149	76
600	69	62	116	71	105	70	143	73
1800	64	57	110	67	100	67	138	70
3600	55	49	105	64	98	65	134	68

(A) Absolute stress at 20% elongation.
(B) Percent of initial stress at 20% elongation retained.

urethane of this study, so we conclude that the above changes on aging were not due to polymer molecular weight increase.

Table IV data show that the enhancement of network strength on aging which we ascribe to VC development is also seen in the CC–VC network polymer in its increased 100°C absolute stress levels at 20% elongation. These increased stress levels apparently represent the contribution of a more fully developed, and thus enhanced VC network. But even so, Curves C and C′ at 100°C of Figure 7 almost coincide, showing that the enhanced network was dispelled by heat in a manner characteristic of the unaged VC network.

TABLE IV

Stress Levels for Unaged and Aged CC–VC Network Polyurethane Samples at 20% Elongation
vs Time, 100°C

Time at 20% Elong. (sec)	CC–VC Polyurethane			
	Unaged		Aged	
	A(psi)	B(%)	A(psi)	B(%)
0	116	100	154	100
3	110	95	146	95
12	108	93	142	92
36	106	92	140	91
60	105	91	138	90
180	103	89	135	88
600	102	88	133	86
1800	101	87	130	84
3600	100	86	128	83

(A) Absolute stress at 20% elongation.
(B) Percent of initial stress at 20% elongation retained.

TABLE V
Bulk Flow at Elevated Temperature of VC and CC–VC Network Polyurethanes

Test			
Time Elapsed (min)	Temperature Achieved (°C)	Increase in Sample Length (cm)	
		VC Polyurethane	CC–VC Polyurethane
0	45	—	—
23	85	perceptible	none
27	89	0.5	none
32	97	2	none
35	102	5	none
37	104	sple. melts & falls from support	none
139	290	—	none

Bulk Flow at Elevated Temperature

The subject VC and CC–VC network polyurethanes were compared in our heat distortion test as 25 mil strips looped around a thin, horizontal, metal supporting rod. The data of Table V show that the VC network polymer sample noticeably elongated under its own weight by 89°C (27 min) of exposure and melted off the support by 104°C (37 min). In contrast, the CC–VC network polymer sample remained intact on the support and retained its original dimensions to the temperature limit of the test oven, 290°C (139 min). Some sample darkening occurred during the longer exposures at high temperature. These results indicate that as expected, the CC–VC network polymer displayed appreciably greater integrity at high temperatures than the VC network polymer.

Compression Set

The permanent set of the VC and CC–VC network polyurethanes due to prolonged sample compression was compared at two temperatures. The results of Table VI indicate the much greater retractive tendency, particularly at elevated temperature, of the CC–VC network polyurethane sample.

Low Temperature Flexibility (Gehman Low Temperature Modulus)

Comparison at low temperatures shows the VC network polyurethane to be considerably more flexible than the CC–VC network polyurethane in the range

TABLE VI
Compression Set of VC and CC–VC Network Polyurethanes

	Compression Set After 22 Hours (%)	
	At 25°C	At 70°C
VC Network Polyurethane	39	98
CC–VC Network Polyurethane	15	11

TABLE VII
Gehman Low Temperature Stiffness Values for VC and CC–VC Network Polymers

°C	VC Polyurethane	CC–VC Polyurethane
T_2	−22	−10
T_5	−28	−19
T_{10}	−30	−21
T_{100}	−35	−28

−10 to −35°C (Fig. 8). Apparently the imposed CC network restricts the polymer chain segment mobility characteristic of the VC network, thus decreasing sample flexibility in this temperature range. Flexibility above −10°C and below −35°C is similar for the two network polymers, the latter possibly because the glass transition of the (common) major component, the polyester glycol, exerts a dominant influence on overall polymer chain flexibility at this temperature and below.

Gehman low temperature modulus values are often expressed in terms of relative sample stiffness at various temperatures (T_n values) where T is temperature (°C) and n is the multiple of sample stiffness at T relative to the stiffness at 25°C. Such data from Figure 8 are shown in Table VII.

Stress–Strain Properties

Figure 9 shows the stress–strain properties of the VC and CC–VC network polyurethanes at 25 and 100°C. At 25°C the VC network polyurethane is seen to be quite extensible. It develops modulus very gradually with extension, reflecting its relatively low concentration of urethane hard segments and thus VC. However, it displays high tensile strength (3700 psi) at break. The VC network polyurethane melted in the 100°C stress–strain test, so no curve appears for this test.

At 25°C the CC–VC network polyurethane showed a rapid increase in modulus with extension, reflecting the presence and effectiveness of the added

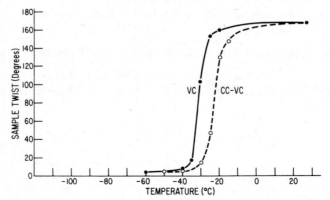

FIG. 8. Low temperature flexibility of VC and CC–VC network polyurethanes.

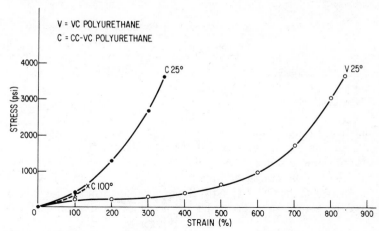

FIG. 9. Stress–strain values for VC and CC–VC polyurethanes at 25 and 100°C (Instron, 20″/min, 25 mil 1/8″ dumbbell).

CC contribution to the polymer network (Curve C, 25°C). The tighter CC–VC network is seen to be less extensible than the VC network but to develop similar breaking strength (3600 psi). At 100°C the CC–VC network polyurethane was fully intact and easily tested but showed much reduced extensibility and strength at break (Curve C, 100°C). Nonetheless, the tensile strength at break of this pure gum vulcanizate was a respectable 600 psi and indicative of substantial sample integrity at this elevated temperature.

Curve C (100°C) is seen to lie a bit below Curve C (25°C). This is in keeping with stress relaxation results which suggest that at 100°C the VC of the subject polyurethane composition are largely dispelled by thermal energy. So what we see in Curve C (100°C) is interpreted to be predominantly a CC contribution to modulus and the difference between the two curves the lost VC contribution.

Tear Strength

Table VIII compares the Graves tear strengths of the VC and CC–VC network polyurethanes at 25 and 100°C. The data show that at 25°C the subject (low urethane content, pure gum) polyurethane showed only fair but marginally better tear strength as a VC network polymer than as a CC–VC network polymer. This difference is consistent in kind with the known effects of CC on gum

TABLE VIII
Tear Strength of VC and CC–VC Network Polyurethanes at 25 and 100°C

	Tear Strength (pli)	
	25°C	100°C
VC Polyurethane	167	sample melted
CC–VC Polyurethane	144	42

TABLE IX
Flex Life of VC and CC–VC Network Polyurethanes at 27°C

	VC Polyurethane	CC–VC Polyurethane
Cycles to Failure	10,000[a]	<500[b]

[a] 0.08″ wound grew to 0.83″.
[b] Sample broke completely.

elastomer tear strength, including that of polyurethanes [2]. Apparently the addition of the more stress-stable CC to the more stress-labile VC of the subject polyurethane composition reducing chain mobility, allows greater stress concentrations within the sample and readier rupture under stress.

In tests at 100°C the CC–VC network polyurethane still retained about 30% of its 25°C tear strength whereas the VC network polyurethane again melted.

Flex Life

The De Mattia flex test was used to compare the resistance to crack growth of the subject polyurethane composition in the VC and the CC–VC network states at 27°C. The data of Table IX demonstrate that the VC network polymer showed only fair performance but still much better than that of the CC–VC network polyurethane sample. This difference is explained in terms of the greater ability of the more stress-labile VC network to relax (see Stress Relaxation) in response to applied stress, and thus to reduce stress concentrations (see Tear Strength) and attendant network rupture. A lower CC concentration presumably would improve CC–VC polymer performance in this test.

Abrasion Resistance

Comparison of the abrasion resistance of the subject polyurethane composition in the VC and the CC–VC network states using the Taber abrasion and Pico abrasion tests yielded opposing results. Contradiction in degree and even kind is not uncommon when different types of abrasion tests are used to compare elastomer abrasion resistance. Each test applies a characteristic abrasive action and the sample may resist one type of abrasive action much better than another.

The data of Table X show that the CC network polyurethane abraded 30%

TABLE X
Abrasion Resistance of VC and CC–VC Network Polyurethanes at 25°C

	Taber Abrasion Resistance (mg lost/5000 revolutions)	Pico Abrasion Relative Index (%)	
		80 rev.	240 rev.
VC Polyurethane	136	71	54
CC–VC Polyurethane	179	82	80

more than the VC network polyurethane in the Taber test but 48% less in the Pico abrasion test. Pico test results indicate that abrasion resistance worsened with increasing uninterrupted test time in the case of the VC network polyurethane. This suggests that the softening action of frictional heat on the test sample may be an especially important factor in the Pico abrasion resistance of the VC network polyurethane. The abrasion resistance of the subject polyurethane composition in the VC and CC–VC network states was not impressive in either the Pico or Taber abrasion tests.

REFERENCES

[1] E. Müller, O. Bayer, S. Petersen, H. Piepenbrink, W. Schmidt, and E. Windemuth, *Rubber Chem. Tech.*, *26*, 493–509 (1953).

[2] K. A. Pigott, B. F. Frye, K. R. Allen, S. Steingiser, W. C. Darr, J. H. Saunders, and E. E. Hardy, *Chem. Eng. Data*, *5*, 391–395 (1960).

[3] J. H. Saunders, *Rubber Chem. Tech.*, *33*, 1259–1292 (1960).

[4] N. V. Seeger, T. V. Mastin, E. E. Fauser, F. S. Farson, A. F. Finelli, and E. A. Sinclair, *Ind. Eng. Chem.*, *45*, 2538–2542 (1953).

[5] E. F. Cluff and E. K. Gladding, *J. Appl. Polym. Sci.*, *3*, 290–295 (1960).

[6] L. C. Kreider and R. F. Nichols, assignors to the B. F. Goodrich Co., U.S. Patent 2,785,150, March 12, 1957.

[7] D. A. Meyer, Chapter 10, "Vulcanization of Polyurethane Elastomers," in *Vulcanization of Elastomers*, G. Alliger, I. J. Sjothun, Eds., Reinhold, New York (1964).

[8] E. E.Gruber and O. C. Keplinger, *Ind. Eng. Chem.*, *51*, (*2*), 151–154 (1959).

[9] The General Tire and Rubber Co., Akron, Ohio, Technical Bulletins: "Genthane S" (GT-S3); "Genthane SR" (GT-SRI).

[10] Farbenfabriken Bayer, Leverkusen, Germany, Technical Bulletin, "Urepan E, an Ester Based Urethane Rubber for Crosslinking with Peroxides" (July 1, 1961).

[11] Naugatuck Chemical Co., Division of United States Rubber Co. (Uniroyal), Naugatuck, Connecticut, Technical Bulletins: "Vibrathane 5003"; "Vibrathane 5004."

[12] Thiokol Chemical Corporation, Trenton, New Jersey, Technical Bulletin, "Thiokol Facts," *5*, (2), (1963).

[13] M. M. Swaab, *Rubber Age*, *92*, (*4*), 567–570 (1963).

[14] E. I. duPont de Nemours & Co. Elastomer Chemicals Dept., Wilmington, Delaware, Technical Bulletins, (a) "Adiprene C, a Urethane Rubber," Development Products Report, No. 4 (July 15, 1957); (b) "Adiprene CM, a Sulfur Curable Urethane Rubber," by S. M. Hirsty, Report A-66693 (1969).

[15] D. B. Pattison, assignor to E. I. duPont de Nemours and Co., U.S. Patent 2,808,391, October 1, 1957.

[16] W. Kallert, *Kaut. Gummi Kunst.*, *19*, 363–371 (1966).

[17] H. S. Kincaid, G. P. Sage, and F. E. Critchfield, "Polycaprolactone Millable Urethane Elastomers," American Chemical Society (Rubber Division) Meeting, Montreal, May, 1967.

[18] P. Wright and A. P. C. Cumming, *Solid Polyurethane Elastomers*, Gordon and Breach, New York, 1969.

[19] S. V. Urs, *Ind. Eng. Product Res. and Devel.*, *1*, (*3*), 199–202 (Sept. 1962).

[20] L. B. Weisfeld, J. R. Little, and W. E. Wolstenholme, *J. Polym. Sci.*, *56*, 455–463 (1962).

[21] C. S. Schollenberger, D. Esarove, *Polyurethanes, The Science and Technology of Polymer Films*, Vol. 2, O. J. Sweeting, Ed., Wiley New York, 1971, Chap. 12.

[22] E. B. Newton, H. W. Grinter, and D. S. Sears, *Rubber Chem. Tech.*, *34*, (*1*), 1–15 (1961).

[23] S. L. Cooper and A. V. Tobolsky, *J. Appl. Polym. Sci.*, *10*, 1837–1844 (1966).

[24] S. L. Cooper, *J. Polym. Sci.* (*A-1*), *7*, 1765–1772 (1969).

[25] A. Singh and L. Weissbein, *J. Polym. Sci.*, *A-1*, *4*, 2551–2561 (1966).

[26] A. Singh, *Advances in Urethene Science and Technology,* Vol. I, Chap. 5, K. C. Frisch and S. L. Reegen, Eds., Technomic Publishing Co., Inc., Westport, 1971.

[27] C. S. Schollenberger and K. Dinbergs, *J. Elastoplastics, 5,* 222–251 (1973).

Author Index

Bailey, W. J., 17
Buruiana, E., 189
Butler, G. B., 71

Caraculacu, A., 189
Cincu, C., 245
Cobzaru, V., 27

Deleanu, Th., 125
Denes, F., 281
Dimonie, M., 245
Dimonie, V., 125
Dinbergs, K., 351
Dobrescu, V., 27
Donescu, D., 125
Dragan, Gh., 141
Dumitrescu, Sv., 209

Endo, T., 17

Feldman, D., 339

Gallagher, R., 329
Gavat, I., 125
Gosa, K., 125

Hagiopol, C., 125
Haiduc, I., 43
Hsu, T. D., 229
Hubca, Gh., 245
Hwa, J. C. H., 329

Israel, S. C., 229

Kennedy, J. P., 117

Lindenmayer, P. H., 181

Macoveanu, A., 339
Mahmud, M. U., 229
Memetea, T., 57
Munteanu, M., 125

Negulescu, I., 281
Negulianu, C., 149

Oprescu, C., 245

Percec, V., 209

Robila, G., 189

Salamone, J. C., 229
Schneider, I. A., 95
Schollenberger, C. S., 351
Shah, S. C., 267
Simionescu, C., 149, 209
Simionescu, C. I., 281, 305
Smid, J., 267
Stannett, V., 57

Tsai, C. C., 229

Varma, A. J., 267
Vasilu Oprea, C. V., 149
Vogl, O., 1

Watterson, A. C., 229
Wisniewski, A. W., 229
Wong, L., 267

Published Polymer Symposia

1963   No. 1. First Biannual American Chemical Society Polymer Symposium
Edited by H. W. Starkweather, Jr.

No. 2. Fourth Cellulose Conference
Edited by R. H. Marchessault

No. 3. Morphology of Polymers
Edited by T. G. Rochow

1964   No. 4. Macromolecular Chemistry, Paris 1963 (Published in 3 parts)
Edited by M. Magat

No. 5. Rheo-optics of Polymers
Edited by Richard S. Stein

No. 6. Thermal Analysis of High Polymers
Edited by Bacon Ke

No. 7. Vibrational Spectra of High Polymers
Edited by Giulio Natta and Giuseppe Zerbi

1965   No. 8. Analysis and Fractionation of Polymers
Edited by John Mitchell, Jr. and Fred W. Billmeyer, Jr.

No. 9. Structure and Properties of Polymers
Edited by Arthur V. Tobolsky

No. 10. Transport Phenomena in Polymeric Films
Edited by Charles A. Kumiss

No. 11. Fifth Cellulose Conference
Edited by T. E. Timell

1966   No. 12. Perspectives in Polymer Science
Edited by E. S. Proskauer, E. H. Immergut, and C. G. Overberger

No. 13. Small Angle Scattering from Fibrous and Partially Ordered Systems
Edited by R. H. Marchessault

No. 14. Transitions and Relaxations in Polymers
Edited by Raymond F. Boyer

No. 15. U.S.-Japan Seminar in Polymer Physics
Edited by Richard S. Stein and Shigeharu Onogi

1967   No. 16. Macromolecular Chemistry, Prague 1965 (Published in 8 parts: Parts 1–5, 1967; Parts 6 and 7, 1968; Part 8, 1969)
Chairmen: O. Wichterle and B. Sedláček

No. 17. Electrical Conduction Properties of Polymers
Edited by A. Rembaum and R. F. Landel

No. 18. The Meaning of Crystallinity in Polymers
Edited by Fraser P. Price

No. 19. High Temperature Resistant Fibers
Edited by A. H. Frazer

No. 20. Supramolecular Structure in Fibers
Edited by Paul H. Lindenmayer

1968   No. 21. Analytical Gel Permeation Chromatography
Edited by Julian F. Johnson and Roger S. Porter

No. 22. Macromolecular Chemistry, Brussels-Louvain 1967 (Published in 2 parts: Part 1, 1968; Part 2, 1969)
Chairman: G. Smets

No. 23. Macromolecular Chemistry, Tokyo–Kyoto 1966 (Published in 2 parts: Part 1, 1968; Part 2, 1969)
Chairmen: I. Sakurada and S. Okamura

No. 24. Polymer Reactions
Edited by E. M. Fettes

No. 25. The Computer in Polymer Science
Edited by Jack B. Kinsinger

1969 No. 26. Block Copolymers
Edited by J. Moacain, G. Holden, and N. W. Tschoegl

No. 27. New Concepts in Emulsion Polymerization
Edited by Jesse C. H. Hwa and John W. Vanderhoff

No. 28. Proceedings of the Sixth Cellulose Conference
Edited by R. H. Marchessault

1970 No. 29. Proceedings of the International Conference on Organic Superconductors
Edited by William A. Little

No. 30. Macromolecular Chemistry, Toronto 1968
Edited by L. E. St. Pierre

No. 31. Polymers and Polymerization
Edited by Charles G. Overberger and Thomas G. Fox

1971 No. 32. Molecular Order—Molecular Motion: Their Response to Macroscopic Stresses
Edited by H. H. Kausch

No. 33. Poly(vinyl Chloride): Formation and Properties
Edited by B. Sedláček

No. 34. Polymers at Interfaces
Edited by M. J. Schick

No. 35. Viscoelastic Relaxation in Polymers
Edited by M. Shen

No. 36. Proceedings of the Seventh Cellulose Conference
Edited by E. C. Jahn

1972 No. 37. Proceedings of the Symposium on Graft Polymerization onto Cellulose
Edited by J. C. Arthur, Jr.

No. 38. Morphology of Polymers
Edited by B. Sedláček and E. F. Casassa

No. 39. Thermodynamic Interactions in Polymer Solutions
Edited by E. F. Casassa and B. Sedláček

1973 No. 40. Mechanism of Inhibition Processes in Polymers: Oxidative and Photochemical Degradation
Edited by B. Sedláček

No. 41. Transport Phenomena Through Polymer Films
Edited by C. A. Kumins

No. 42. International Symposium on Macromolecules, Helsinki, 1972 (Published in 3 Parts)
Edited by O. Harva and C. G. Overberger

No. 43. 2nd International Symposium on Polymer Characterization
Edited by F. A. Sliemers and K. A. Boni

1974 No. 44. Organized Structures in Polymer Solutions and Gels
Edited by C. G. Overberger and B. Sedláček

No. 45. Ion-Containing Polymers
Edited by A. Eisenberg

No. 46. Recent Advances in Polymer Science
Edited by G. L. Wilkes and B. Maxwell

No. 47. Transformations of Functional Groups on Polymers
Edited by C. G. Overberger and B. Sedláček

No. 48. Rubber and Rubber Elasticity
Edited by A. S. Dunn

1975 No. 49. Proceedings of the Australian Polymer Symposium
Edited by J. H. Bradbury

No. 50. International Symposium on Macromolecules: Invited Lectures
Edited by J.-L. Millan, E. L. Madruga, C. G. Overberger, and H. F. Mark

No. 51. International Symposium on Macromolecules in Honor of Professor Herman F. Mark
Edited by E. F. Casassa, T. G Fox, H. Markovitz, C. G. Overberger, and E. M. Pearce

No. 52. Contributions from Students and Friends of Professor Champetier on the Occasion of his 70th Birthday
Edited by P. Sigwalt, T. G Fox, C. G. Overberger, and H. F. Mark

No. 53. Crosslinking and Networks
Edited by K. Dušek, B. Sedláček, C. G. Overberger, H. F. Mark, and T. G Fox

1976   No. 54. Polymer Science: Achievements and Prospects (In Honor of P. J. Flory)
Edited by H. Markovitz and E. F. Casassa

No. 55. Proceedings of the Eighth Australian Polymer Symposium
Edited by I. C. Watt

No. 56. Fourth International Symposium on Cationic Polymerization
Edited by J. P. Kennedy

No. 57. Degradation and Stabilization of Polyolefins
Edited by B. Sedláček, C. G. Overberger, H. F. Mark, and T. G Fox

1977   No. 58. Orientation Effects in Solid Polymers (to appear)
Edited by G. Bodor

No. 59. Recent Advances in the Field of Crystallization and Fusion of Polymers
Edited by J. P. Mercier and R. Legras

No. 60. Advances in Preparation and Characterization of Multiphase Polymer Systems
Edited by R. J. Ambrose and S. L. Aggarwal

No. 61. Advances in Scattering Methods
Edited by B. Sedláček, C. G. Overberger, H. F. Mark, and T. G Fox

1978   No. 62. International Symposium on Macromolecules
Edited by C. G. Overberger and H. Mark

No. 63. A Symposium in Memory of Fraser P. Price
Edited by R. S. Porter and R. S. Stein

No. 64. Unsolved Problems of Co- and Graft Polymerization
Edited by O. Vogl and C. I. Simionescu

All the above symposia can be individually purchased through the Subscription Department, John Wiley & Sons.